PREFACE

The effects of time on the strength and deformation characteristics of various materials assume importance in a variety of engineering problems, particularly in cases where very rapid or very long duration response is of concern. In the field of soil mechanics, the subject of time related behavior aroused interest early on, and was explicitly discussed by Terzaghi in 1931. From that point on, numerous investigations have shown that soil's time dependent mechanical behavior is highly complex and can impact a broad range of applications: *in situ* testing, seismic response, landslides, and embankment performance to name only a few. Donald Taylor is acknowledged as being the first to seriously address rate effects in soils during the early 1940's; this was followed by Arthur Casagrande and co-workers in the early 1950's. By 1959, the body of work on this topic was significant enough that ASTM held a special *Symposium on Time Rates of Loading in Soil Testing*. In the following decade, landmark work was done that is still widely referenced today. This included the experimental study of rate dependence under both undrained and one-dimensional consolidation conditions. Pioneering analytical modeling efforts were undertaken: Rate Process Theory was adapted by several investigators to model creep behavior. The subsequent formulation of viscoplasticity has facilitated three decades of advancement in constitutive modeling. The 1970's saw advances in both the practical and fundamental aspects of rate effects. Among other work, researchers observed a correspondence between undrained creep behavior and that under constant rates of loading, beginning a trend of unifying various aspects of time dependent soil behavior. For the past 20 years or more, constitutive modeling has perhaps dominated research on time dependence, followed closely by laboratory and field work concerning the nature of secondary compression and its effects on consolidation behavior and the modeling thereof. Vast strides in computational capacity have moved the geotechnical research community into more complex constitutive models and formulation of micromechanical models.

At present, two trends seem to guide research and application of time dependent soil behavior phenomena. First, there is an appreciation that high quality laboratory and field measurements are needed to study this behavior. These data can then be used to formulate more accurate constitutive models as well as micromechanical models. The second trend is that models being formulated increasingly try to link several time effects into one model, regardless of loading/deformation conditions. Other important variables such as temperature and degree of soil structure are also being acknowledged as important inputs to these models.

This volume provides a much needed update on the state of research and practice in the area of soil time effects. This is an international compilation, with contributions from 5 countries. A total of 17 papers were reviewed, and 14 were accepted for publication. Each of the papers included in these Proceedings has received two positive peer reviews; if only one out of two was positive, authors were required to make changes and the paper was re-reviewed by the editors to ensure required changes were made. All papers are eligible for discussion in the *Journal of Geotechnical Engineering*, and are eligible for ASCE awards. The papers include two state-of-the-art summaries, one on laboratory and field developments and one on modeling time dependent behavior. Papers describe rate dependence in a variety of soils, from tropical soils to subglacial till, and in new applications such as soils reinforced with geosynthetics and inclusion dowels. Modeling efforts are presented that include the influence of soil structure.

In addition to the significant efforts of the contributing authors, these Proceedings were made possible due to: the support of the Soil Properties Committee of the Geotechnical Engineering Division of ASCE, Dr. Richard J. Fragaszy, chair; the approval of the Executive Committee of the Geotechnical Engineering Division, Lawrence H. Roth, chair; and the staff at ASCE publications, particularly Shiela Menaker. The efforts of the reviewers are also acknowledged and their prompt attention to the paper reviews is appreciated.

GEOTECHNICAL SPECIAL PUBLICATION NO. 61

Measuring and Modeling Time Dependent Soil Behavior

Proceedings of sessions sponsored by
The Geo-Institute of the American Society of
Civil Engineers in conjunction with the
ASCE Convention in Washington, D.C.

November 10-14, 1996

Edited by
Thomas C. Sheahan and
Victor N. Kaliakin

Published by
ASCE
345 East 47th Street
New York, New York 10017-2398

Abstract:
This proceedings presents a series of papers on both laboratory and field investigations, as well as constitutive modeling efforts, that all deal with the time dependence of soils and its influence on their response to loading. Two state of the art summaries, one on laboratory and field investigations, and one on modeling, are presented to provide overviews of the current state of knowledge. Papers are included dealing with a variety of areas in which time effects play a role. These include soil reinforcement applications and cyclic behavior, as well as the range of identified time dependent behavior, i.e., creep, relaxation, constant strain rate and consolidation. A wide range of soil types are described, including sands, marine clays, weathered tropical clays, glacial till, high plasticity clays, and structured soils. The influence of other factors such as temperature, degree of soil structure and loading conditions (e.g., cyclic, plane strain and triaxial) are taken into account. The papers in the volume are intended to provide an update on the state of knowledge and new application areas in which time dependent behavior plays a critical role.

Library of Congress Cataloging-in-Publication Data

Measuring and modeling time dependent soil behavior : proceedings of sessions / sponsored by the ASCE Geotechnical Engineering Division in conjunction with the ASCE Convention in Washington, D.C., November 10-14, 1996 ; edited by Thomas C. Sheahan and Victor N. Kaliakin.
   p.   cm. -- (Geotechnical special publication ; no. 61)
   Includes indexes.
   ISBN 0-7844-0205-1
   1. Soils--Measurement--Congresses. 2. Soil mechanics--Congresses. 3. Soils--Mathematical models--Congresses. I. Sheahan, Thomas C. II. Kaliakin, Victor N. III. American Society of Civil Engineers. Geotechnical Engineering Division. IV. ASCE National Convention (1996 : Washington, D.C.) V. Series.
TA 710.5.M42   1996   96-36647
624.1'5136--dc20   CIP

The Society is not responsible for any statements made or opinions expressed in its publications.

Photocopies. Authorization to photocopy material for internal or personal use under circumstances not falling within the fair use provisions of the Copyright Act is granted by ASCE to libraries and other users registered with the Copyright Clearance Center (CCC) Transactional Reporting Service, provided that the base fee of $4.00 per article plus $.25 per page is paid directly to CCC, 222 Rosewood, Drive, Danvers, MA 01923. The identification for ASCE Books is 0-7844-0205-1/96/$4.00 + $.25 per copy. Requests for special permission or bulk copying should be addressed to Permissions & Copyright Dept., ASCE.

Copyright © 1996 by the American Society of Civil Engineers,
All Rights Reserved.
Library of Congress Catalog Card No: 96-36647
ISBN 0-7844-0205-1
Manufactured in the United States of America.

GEOTECHNICAL SPECIAL PUBLICATIONS

1) TERZAGHI LECTURES
2) GEOTECHNICAL ASPECTS OF STIFF AND HARD CLAYS
3) LANDSLIDE DAMS: PROCESSES, RISK, AND MITIGATION
4) TIEBACKS FOR BULKHEADS
5) SETTLEMENT OF SHALLOW FOUNDATION ON COHESIONLESS SOILS: DESIGN AND PERFORMANCE
6) USE OF IN SITU TESTS IN GEOTECHNICAL ENGINEERING
7) TIMBER BULKHEADS
8) FOUNDATIONS FOR TRANSMISSION LINE TOWERS
9) FOUNDATIONS AND EXCAVATIONS IN DECOMPOSED ROCK OF THE PIEDMONT PROVINCE
10) ENGINEERING ASPECTS OF SOIL EROSION, DISPERSIVE CLAYS AND LOESS
11) DYNAMIC RESPONSE OF PILE FOUNDATIONS— EXPERIMENT, ANALYSIS AND OBSERVATION
12) SOIL IMPROVEMENT - A TEN YEAR UPDATE
13) GEOTECHNICAL PRACTICE FOR SOLID WASTE DISPOSAL '87
14) GEOTECHNICAL ASPECTS OF KARST TERRAINS
15) MEASURED PERFORMANCE SHALLOW FOUNDATIONS
16) SPECIAL TOPICS IN FOUNDATIONS
17) SOIL PROPERTIES EVALUATION FROM CENTRIFUGAL MODELS
18) GEOSYNTHETICS FOR SOIL IMPROVEMENT
19) MINE INDUCED SUBSIDENCE: EFFECTS ON ENGINEERED STRUCTURES
20) EARTHQUAKE ENGINEERING & SOIL DYNAMICS (II)
21) HYDRAULIC FILL STRUCTURES
22) FOUNDATION ENGINEERING
23) PREDICTED AND OBSERVED AXIAL BEHAVIOR OF PILES
24) RESILIENT MODULI OF SOILS: LABORATORY CONDITIONS
25) DESIGN AND PERFORMANCE OF EARTH RETAINING STRUCTURES
26) WASTE CONTAINMENT SYSTEMS: CONSTRUCTION, REGULATION, AND PERFORMANCE
27) GEOTECHNICAL ENGINEERING CONGRESS
28) DETECTION OF AND CONSTRUCTION AT THE SOIL/ROCK INTERFACE
29) RECENT ADVANCES IN INSTRUMENTATION, DATA ACQUISITION AND TESTING IN SOIL DYNAMICS
30) GROUTING, SOIL IMPROVEMENT AND GEOSYNTHETICS
31) STABILITY AND PERFORMANCE OF SLOPES AND EMBANKMENTS II (A 25-YEAR PERSPECTIVE)
32) EMBANKMENT DAMS-JAMES L. SHERARD CONTRIBUTIONS
33) EXCAVATION AND SUPPORT FOR THE URBAN INFRASTRUCTURE
34) PILES UNDER DYNAMIC LOADS
35) GEOTECHNICAL PRACTICE IN DAM REHABILITATION
36) FLY ASH FOR SOIL IMPROVEMENT
37) ADVANCES IN SITE CHARACTERIZATION: DATA ACQUISITION, DATA MANAGEMENT AND DATA INTERPRETATION
38) DESIGN AND PERFORMANCE OF DEEP FOUNDATIONS: PILES AND PIERS IN SOIL AND SOFT ROCK
39) UNSATURATED SOILS
40) VERTICAL AND HORIZONTAL DEFORMATIONS OF FOUNDATIONS AND EMBANKMENTS

41) PREDICTED AND MEASURED BEHAVIOR OF FIVE SPREAD FOOTINGS ON SAND
42) SERVICEABILITY OF EARTH RETAINING STRUCTURES
43) FRACTURE MECHANICS APPLIED TO GEOTECHNICAL ENGINEERING
44) GROUND FAILURES UNDER SEISMIC CONDITIONS
45) IN-SITU DEEP SOIL IMPROVEMENT
46) GEOENVIRONMENT 2000
47) GEO-ENVIRONMENTAL ISSUES FACING THE AMERICAS
48) SOIL SUCTION APPLICATIONS IN GEOTECHNICAL ENGINEERING
49) SOIL IMPROVEMENT FOR EARTHQUAKE HAZARD MITIGATION
50) FOUNDATION UPGRADING AND REPAIR FOR INFRASTRUCTURE IMPROVEMENT
51) PERFORMANCE OF DEEP FOUNDATIONS UNDER SEISMIC LOADING
52) LANDSLIDES UNDER STATIC AND DYNAMIC CONDITIONS - ANALYSIS, MONITORING, AND MITIGATION
53) LANDFILL CLOSURES — ENVIRONMENTAL PROTECTION AND LAND RECOVERY
54) EARTHQUAKE DESIGN AND PERFORMANCE OF SOLID WASTE LANDFILLS
55) EARTHQUAKE-INDUCED MOVEMENTS AND SEISMIC REMEDIATION OF EXISTING FOUNDATIONS AND ABUTMENTS
56) STATIC AND DYNAMIC PROPERTIES OF GRAVELLY SOILS
57) VERIFICATION OF GEOTECHNICAL GROUTING
58) UNCERTAINTY IN THE GEOLOGIC ENVORONMENT
59) ENGINEERED CONTAMINATED SOILS AND INTERACTION OF SOIL GEOMEMBRANES
60) ANALYSIS AND DESIGN OF RETAINING STRUCTURES AGAINST EARTHQUAKES
61) MEASURING AND MODELING TIME DEPENDENT SOIL BEHAVIOR
62) CASE HISTORIES OF GEOPHYSICS APPLIED TO CIVIL ENGINEERING AND PUBLIC POLICY
63) DESIGN WITH RESIDUAL MATERIALS; GEOTECHNICAL AND CONSTRUCTION CONSIDERATIONS

CONTENTS

State of the Art: Importance of Strain Rate and Temperature Effects in Geotechnical Engineering
Serge Leroueil and Maria Esther Soares Marques 1

State of the Art: Modeling Aspects Associated with Time Dependent Behavior of Soils
Toshihisa Adachi, Fusao Oka, and Mamoru Mimura 61

Simulation of Pore Pressures in Triaxial Creep Tests
Horst G. Brandes and Armand J. Silva 96

Stabilization of a Creeping Slope Using Soil Nails
Peter R. Cali 109

Evaluation of Long-Term Time-Rate Parameters of Subglacial Till
C.L. Ho, J.C. Vela, P.U. Clark, and J.W. Jenson 122

Strain Rate and Structuring Effects on the Compressibility of a Young Clay
Serge Leroueil, Didier Perret, and Jacques Locat 137

Effects of Stress Ratio on Behavior of Quasi-Preconsolidated Compacted Clay under Plane Strain Compression
Hoe I. Ling and Fumio Tatsuoka 151

Rate and Creep Effect on the Stiffness of Soils
Diego C.F. Lo Presti, Michele Jamiolkowski, Oronzo Pallara, and Antonio Cavallaro 166

Prediction of Time-Dependent Behaviour of Remolded Soft Marine Clay in Axi-symmetric Undrained Conditions
Satoshi Murakami, Kazuya Yasuhara, and Kaoru Bessho 181

Soil Creep and Creep Testing of Highly Weathered Tropical Soils
Peter G. Nicholson, Philip W. Russell, and Clint F. Fujii 195

Strain Rate Effects on Stress-Strain Behavior of Clay as Observed in Monotonic and Cyclic Triaxial Tests
Satoru Shibuya, Toshiyuki Mitachi, Akihiko Hosomi, and Seong Chun Hwang 214

Drained Creep Behavior of Marine Clays
Armand J. Silva and Horst G. Brandes 228

Rate-Dependent Deformation of Structured Natural Clays
Kenichi Soga and James K. Mitchell 243

Creep Deformation and Stress Relaxation in Preloaded/Prestressed Geosynthetic-Reinforced Soil Retaining Walls
Fumio Tatsuoka, Taro Uchimura, Masaru Tateyama, and Katsumi Muramoto 258

Subject Index 273

Author Index 275

IMPORTANCE OF STRAIN RATE AND TEMPERATURE EFFECTS IN GEOTECHNICAL ENGINEERING

Serge Leroueil[1] and Maria Esther Soares Marques[2]

Abstract

This paper examines the importance of strain rate and temperature effects in geotechnical engineering. The first part considers the effects on the main characteristics of soil behavior. It emerges that compressibility of soils is controlled by an effective stress-strain-strain rate-temperature relationship; viscous effects on the limit state curve, the undrained shear strength and the preconsolidation pressure of clays are important, on the order of 10% per logarithm cycle of strain rate or a temperature change of about 12°C; viscous effects on the critical state and residual strength envelopes are extremely small or non existent; the shear modulus at small strains is strain rate dependent, but the viscous effects are smaller in cohesionless materials than in cohesive soils and have a tendency to decrease at very small strains. Creep and relaxation phenomena are also examined. The influence of viscosity on soil testing, both in laboratory and in situ, is then considered in the second part. Finally, implications of viscosity for several practical problems (embankments on soft clays; foundations; preloading of clay deposits by heating; energy storage; radioactive waste disposal; movements in slopes) are discussed.

1. Introduction

Viscous phenomena in soils are relatively well-known to the profession but generally ignored in geotechnical engineering practice. However, they are not negligible: for clays, the undrained shear strength and the preconsolidation pressure typically change by 10% per logarithm cycle of strain rate or per a 12°C change in temperature. In most geotechnical problems, there are differences between the strain rates existing in situ and those in the laboratory; there are also differences in temperature. For example, when considering the settlement of embankments on soft clays, the strain rates in situ are typically 2 to 4 orders of magnitude smaller than in laboratory tests, and the temperature is typically 10°C less. Possibly, the viscous phenomena should not be disregarded. Moreover, as strain rate and temperature effects

---

[1]Professor, Department of Civil Engineering, Laval University, Québec City, (Québec), Canada G1K 7P4

[2] Graduate student in a cooperative research program between Coordenação dos Programas de Pós-Graduação de Engenharia da Universidade Federal do Rio de Janeiro (COPPE/UFRJ), Brazil, and Laval University, Québec, Canada

are the two facets of the viscous behavior of soils, they must be considered simultaneously.

Thousands of papers and reports on the rheological behavior of materials considered in geotechnical engineering and its practical implications are scattered in the literature, and the review presented here obviously cannot be exhaustive. In particular, cyclic behavior, which is influenced by the number of cycles, and thus time, will be mentioned only when it is affected or influenced by rheology. Also, the present review focuses on the behavior of soils in saturated conditions, but many of the features described here also apply to other geotechnical materials such as unsaturated soils, soft and hard rocks and frozen soils. Finally, because each application and each soil would require special considerations, no recommendation will be made in relation to appropriate strain rate and temperature to be used in geotechnical tests.

The paper is divided into three main parts. The first part summarizes the effects of strain rate and temperature on the main characteristics of soil behavior: compressibility, limit state, shear modulus at very small strains, strength envelopes in the normally and overconsolidated domains and K_o. In the second part, implications of viscous phenomena for both laboratory and in situ tests are examined. The last part considers the implications for a variety of practical geotechnical problems from foundations for buildings to radioactive waste disposal. Reference is made to several studied materials, the basic properties of which are given in Appendix A.

Several review papers on the influence of time and strain rate on the compression of clays and on its practical implications have been recently presented (Mesri et al. 1994; Imai 1995; Leroueil 1994-96). The interested reader is encouraged to examine them for complementary and more detailed information.

2. Rheology of soils

2.1 *Generalities on soil behavior*

The concepts of yielding and critical state are extremely powerful for describing and understanding the behavior of most geotechnical materials. These concepts were first developed by researchers at Cambridge University, for reconstituted, isotropically consolidated clays (Roscoe et al. 1958; Roscoe and Burland 1968; Schofield and Wroth 1968). In the form proposed at that time, they do not reliably represent the behavior of natural geotechnical materials. Natural clays are generally anisotropic (Tavenas and Leroueil 1977; Ladd 1991; Diaz-Rodriguez et al. 1992); most natural soils are structured (Leroueil and Vaughan 1990) and viscous (Murayama and Shibata 1961; Bjerrum 1967; Singh and Mitchell 1968; Sällfors 1975; Mesri and Godlewski 1977; Vaid and Campanella 1977; Tavenas et al. 1978; Graham et al. 1983; Leroueil et al. 1985b; Tidfors and Sällfors 1989). Thus, the basic concepts first developed at Cambridge University need to be modified to incorporate these peculiarities.

In a stress[3] diagram (($\sigma'_1 - \sigma'_3$)/2 vs ($\sigma'_1 + \sigma'_3$)/2, or q = ($\sigma'_1 - \sigma'_3$) vs p' = ($\sigma'_1 + 2\sigma'_3$)/3), the limit state curve (often referred as yield locus in a 2-D stress

---

[3] In this paper, σ'_1 indifferently describes the major stress, the axial stress in triaxial and oedometer tests, or the vertical stress *in situ*. σ'_3 describes without distinction the minor stress, the radial stress in triaxial and oedometer tests, or the horizontal stress *in situ*.

diagram, or limit state surface in a 3-D stress diagram) delimits a zone in which soil behavior is mostly elastic from a zone in which it is mostly plastic or where there is failure. Limit state is defined here with the same meaning as the preconsolidation pressure in one-dimensional compression tests, and not in relation to the early development of plastic strains which occur at much smaller strains, as indicated later on. Figure 1 shows such a limit state curve (ABPK). Inside, the soil is said to be overconsolidated; AB is the strength envelope of the overconsolidated soil; outside the limit state curve and below the critical state line (CSL) or the strength envelope of the normally consolidated soil (BC on the figure), the soil is normally consolidated. For a soil specimen consolidated at a point such as D inside the limit state curve and subjected to a given effective stress path, the general behavior can be described as follows (Jardine et al. 1991): within zone I, soil behavior is linear elastic; within zone II, soil behavior is non-linear elastic; in zone III, the soil is elasto-plastic, with a plastic component increasing when the effective stress path approaches the limit state curve. The linear elastic region (I) is for strains smaller than a value usually between 0.001% and 0.03%, depending on the material considered and its history.

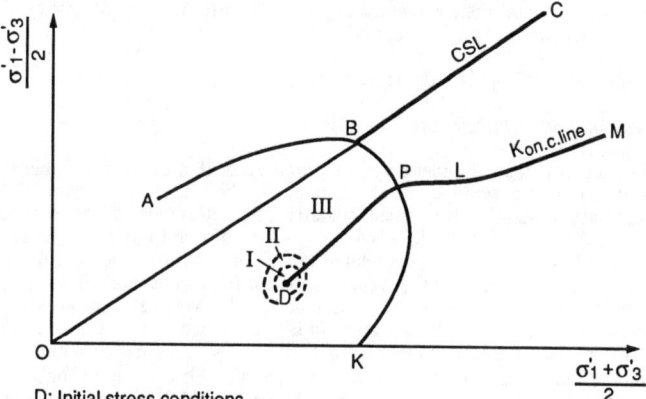

D: Initial stress conditions
ABPK: Limit state curve
  I: Zone of linear elastic behavior
  II: Zone of non-linear elastic behavior
  III: Zone of elasto-plastic behavior
  AB: Peak strength envelope of the overconsolidated soil
 OBC: Critical state line and strength envelope of the normally consolidated soil
DPLM: One-dimensional compression stress path
  LM: K_o line in the normally consolidated range

Fig. 1 - Basic elements of soil behavior

In one-dimensional compression, the stress path followed is as DPLM on Fig. 1. From D to P, the soil is inside the limit state curve, and the effective stress path is the "elastic response" of the overconsolidated soil to loading. According to the linear isotropic elastic theory, $\Delta\sigma'_3/\Delta\sigma'_1 = \nu'/(1 - \nu')$, with ν', Poisson's ratio of the soil. In the normally consolidated range, between L and M, the effective stress ratio $\sigma'_3/\sigma'_1 =$

$K_{0\ n.c.}$. This latter parameter is the coefficient of earth pressure at rest for normally consolidated soil. P to L is a transition.

In shear tests, at large strains, the soil comes to a state in which the deviatoric stress and the void ratio or the pore pressure no longer vary, the soil is said to be at its critical or steady state and is characterized by a friction angle $\phi'_{n.c.}$. In a stress diagram, the critical state line is also the shear strength envelope of the normally consolidated soil (CSL in Fig. 1). In a void ratio versus logarithm of mean effective stress p' diagram, the critical state line is generally linear.

When clayey materials are sheared along a well-defined surface and particles are oriented in the direction of shearing, the strength is the residual strength characterized by the residual friction angle ϕ'_r. The residual friction angle is smaller than $\phi'_{n.c.}$.

2.2   Influence of rheology on soil behavior

The effects of strain rate and temperature on the different characteristics of soil behavior previously defined are now considered.

2.2.1 - One-dimensional and isotropic compression

Effects of time and strain rate

When a soil is loaded under a constant total vertical stress in one-dimensional conditions, it continues to settle after full dissipation of the excess pore pressures. This is the secondary consolidation phase usually characterized by the secondary compression index $C_{\alpha e} = \Delta e / \Delta \log t$, with e = void ratio and t = time. Mesri and Godlewski (1977) showed that $C_{\alpha e}$ is related to the compression index C_c of the soil and more precisely, that the ratio $C_{\alpha e}/C_c$ is a constant for a given soil. This has been confirmed for a large variety of geotechnical materials by Mesri and co-workers, and many other researchers. Examples are given in Fig. 2 for two eastern Canadian clays. An interesting aspect is that, as indicated in Table 1, $C_{\alpha e}/C_c$ remains within a very narrow range for each soil type (Mesri 1987; Mesri et al. 1995). A direct consequence of secondary consolidation is that, when the soil is reloaded, it presents an apparent preconsolidation pressure associated with its void ratio and larger than the previously applied stress. An indirect consequence is that, as shown by Crawford (1964), Bjerrum (1967), Sällfors (1975) and Leroueil et al. (1983, 1985b), when the soil is reloaded slowly, a reduction of the apparent preconsolidation pressure of the soil is observed, i.e. the apparent preconsolidation pressure is dependent on the rate of strain during reloading.

Leroueil et al. (1985b) performed a variety of oedometer tests on different natural clays: Multiple Stage Loading tests with reloading at the end of primary consolidation (MSL_p) or after 24 h (MSL_{24}), Constant Rate of Strain (CRS) tests, Controlled Gradient tests and long-term creep tests. They found that the behavior is controlled by a unique vertical effective stress-vertical strain-vertical strain rate (σ'_1-ε_1-$\dot{\varepsilon}_1$) [or (σ'_1-e-$\dot{e}$)] relationship. This rheological model, originally proposed by Suklje (1957), was advocated, in particular, by Lowe (1974). It has also been recently confirmed experimentally by Imai and Tang (1992) for Yokohama Bay mud (see also Imai 1995). It can be described by compression curves deduced from CRS tests performed at different strain rates, as shown in Fig. 3a. Leroueil et al. (1985b) also showed that the (σ'_1-ε_1-$\dot{\varepsilon}_1$) relationship can be described by two curves, one giving the variation of the preconsolidation pressure σ'_p with strain rate[4]:

$$\sigma'_p = f(\dot{\varepsilon}_1) \quad (1)$$

and the other curve presenting the normalized effective stress-strain curve:

$$\sigma'_1/\sigma'_p(\dot{\varepsilon}_1) = g(\varepsilon_1). \quad (2)$$

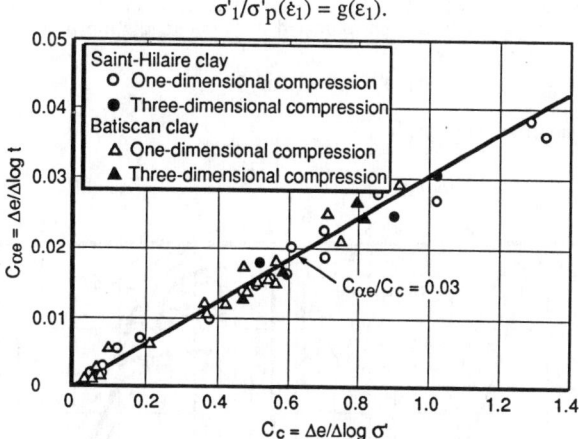

Fig. 2 - *Values of secondary compression index $C_{\alpha e}$ and compression index C_c [from Mesri et al. (1995)]*

Table 1 - *Viscous parameters for geotechnical materials*

Material	$C_{\alpha e}/C_c$	m' (see Eq. 3)	$\frac{\Delta \sigma'}{\sigma'} / \Delta \log \dot{\varepsilon}$
Granular soils, including rockfill	0.02 ± 0.01	100 - 33	2.3 - 7.2
Shale and mudstone	0.03 ± 0.01	50 - 25	4.7 - 9.6
Inorganic clays and silts	0.04 ± 0.01	33 - 20	7.2 - 12.2
Organic clays and silts	0.05 ± 0.01	25 - 17	9.6 - 14.8
Peat and muskeg	0.06 ± 0.01	20 - 14	12.2 - 17.5

Note: Types of material and $C_{\alpha e}/C_c$ values are from Mesri et al. (1995).

---

[4] In its geological meaning, the preconsolidation pressure is unique and constant. On the other hand, the "yield stress" separating small and large strains varies with strain rate and must be considered as a rheologic parameter. It should have a particular appellation but engineers usually refer to it as "preconsolidation pressure," and this term is also used in this paper.

This means that the ratio between the preconsolidation pressures or effective stresses measured at two different strain rates is a constant regardless of the strain or void ratio considered. This also means that the horizontal distance between 2 compression curves corresponding to different strain rates is constant in a ε_1 (or e)-$\log\sigma'_1$ diagram.

Figures 4a and b show the two curves represented by Eqs 1 and 2 for Berthierville clay. As seen in Fig. 4a for Berthierville clay, and as observed for other clays, Eq. 1 can generally be approximated by a linear relation in a $\log\sigma'_p$-$\log\dot\varepsilon_1$ diagram.

$$\log\sigma'_p = A + (1/m') \log\dot\varepsilon_1 \qquad (3)$$

in which A and m' are constants.

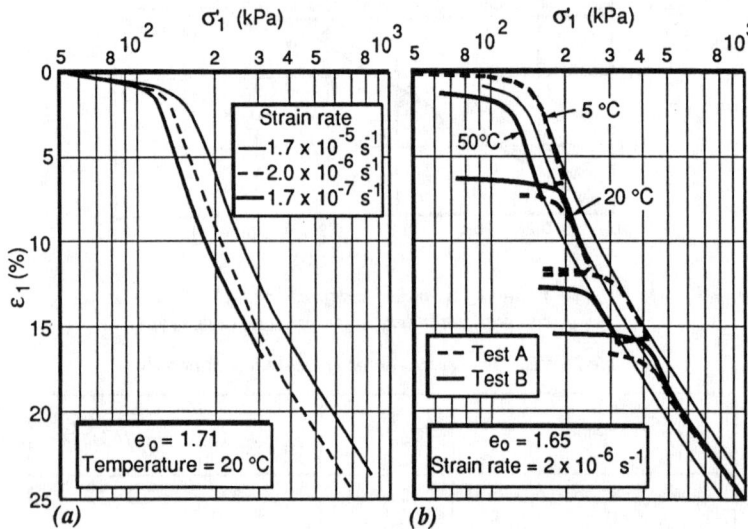

Fig. 3 - *Effects on one-dimensional compression of St-Polycarpe clay of a) strain rate and b) temperature [after Marques (1996)]*

Leroueil et al. (1983, 1985b) and Leroueil (1994-96) showed that the variation of the preconsolidation pressure with strain rate is very similar for numerous non-organic clays, i.e. 7 to 15% per logarithm cycle. The corresponding m' value is generally between 17 and 35. As can be easily demonstrated:

$$m' = 1 / (C_{\alpha e}/C_c) = C_c/C_{\alpha e} \qquad (4)$$

such m' values are equivalent to $C_{\alpha e}/C_c$ values between 0.059 and 0.029, which coincides with the results obtained by Mesri and co-workers (Table 1). Equation 4 also

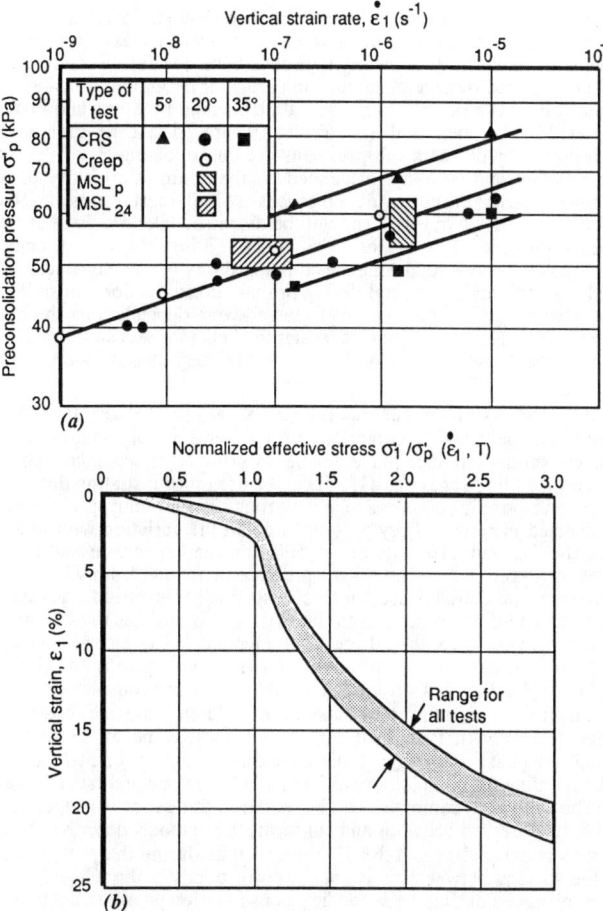

Fig. 4 - One-dimensional compression of Berthierville clay [from Boudali et al. (1994) and Kabbaj (1985)]: (a) preconsolidation pressure as function of strain rate and temperature; and (b) normalized effective stress-strain curve

implies that the $C_{\alpha e}/C_c$ and the strain rate (m') approaches used for describing the viscous behavior of soils during secondary consolidation are equivalent. Eqs 2 and 3, and the parameter m', which link effective stresses and strain rate, can be explained, at least qualitatively, by the rate-process theory and reflect a fundamental behavior of solids (see Section 2.2.9). The fact that $C_{\alpha e}/C_c$ is a constant for a given soil, or type of soil, is only an indirect consequence.

Although CRS tests performed at different strain rates showed different effective stress-strain curves, even when there were excess pore pressures in the specimens, the possibility of having a viscous behavior during primary consolidation has been questioned by several authors, in particular by Jamiolkowski et al. (1985) and Mesri and Choi (1985). However, the data presented by Mesri and Feng (1986) and reinterpreted by Leroueil et al. (1986) clearly showed that the effective stress-strain curves followed by different subspecimens in a consolidating clay layer were different from each other and strongly influenced by the strain rate history of the different subspecimens during the consolidation process. Similar test results presented by Mesri et al. (1995) are shown in Fig. 5 and will be discussed later on. Imai and Tang (1992) also observed similar behavior in a consolidating clay layer consisting of 7 subspecimens in series. Analysing the compressibility of sub-elements in a self-weight consolidating column of soft soil during primary consolidation, Sills (1995) observed that the effective stress-strain curves followed were dependent on the location of the considered sub-element and concluded that the behavior was strain-rate dependent. It is thus clear that there is a viscous effect during primary consolidation.

In a viscous model such as the $(\sigma'_1\text{-}\varepsilon_1\text{-}\dot{\varepsilon}_1)$ [or $(\sigma'_1\text{-}e\text{-}\dot{e})$] model previously described, soil behavior is controlled by the same rheological law, regardless of whether consolidation takes place during its primary or secondary phase. The test results presented by Mesri et al. (1995) can be used for illustrating this. These authors performed an isotropic compression test on four 125 mm-long specimens of St-Hilaire clay connected in series. They present in detail the variation with time of the axial strain of the four sub-elements of the 500 mm-long specimen and the excess pore pressures measured in between for the pressure increment from 97 to 138 kPa during the primary consolidation phase. In these tests, the compression was isotropic, but the drainage was one-dimensional. From these data, it is possible to estimate the effective stress-void ratio curve for the different sub-elements. These curves are drawn in Fig. 5a, from I (initial conditions) to P. Also drawn in the figure is a fictitious secondary consolidation phase (PF). Considering an end-of-primary consolidation reached in this test at a change-in-void-ratio rate of about 3×10^{-8} min^{-1}, the rates observed in the sub-elements during primary consolidation and a m' value of 33 in the secondary consolidation phase ($C_{\alpha e}/C_c = 0.03$, as shown in Fig. 2 for St-Hilaire clay), it is possible to define the complete set of isotaches (lines of equal rate of change-in-void-ratio). The following comments can be made: a) this set of isotaches coincides very well with the observed behavior and conforms to the model described by Eqs 1 and 2; b) for rates between 10^{-6} and 3×10^{-8} min^{-1}, thus during the primary consolidation phase, the spacing between the isotaches corresponds to the $C_{\alpha e}/C_c$ value of 0.03 (m' = 33) observed during the secondary consolidation phase, indicating that the clay does not make any distinction between primary and secondary consolidation; c) at rates larger than 10^{-6} min^{-1}, the variation of the effective stress per log cycle is larger than that associated with a m' value of 33, indicating that m' is possibly not a constant and could be higher at high void ratio or strain rates.

The views expressed by Mesri et al. (1994 and 1995) are quite different. They are represented in Fig. 5b in which the effective stress-void ratio curves obtained in the different sub-elements (I to P, as in Fig. 5a) as well as a fictitious secondary consolidation phase (PF) are drawn. The end-of-primary compression curve defined by Mesri et al. (1995) is also drawn on the figure. These views can be summarized as follows: a) during secondary consolidation, the $C_{\alpha e}/C_c$ and the $(\sigma'_1\text{-}\varepsilon_1\text{-}\dot{\varepsilon}_1)$ models are identical and can thus be described in the same manner in both approaches (Figs 5a and

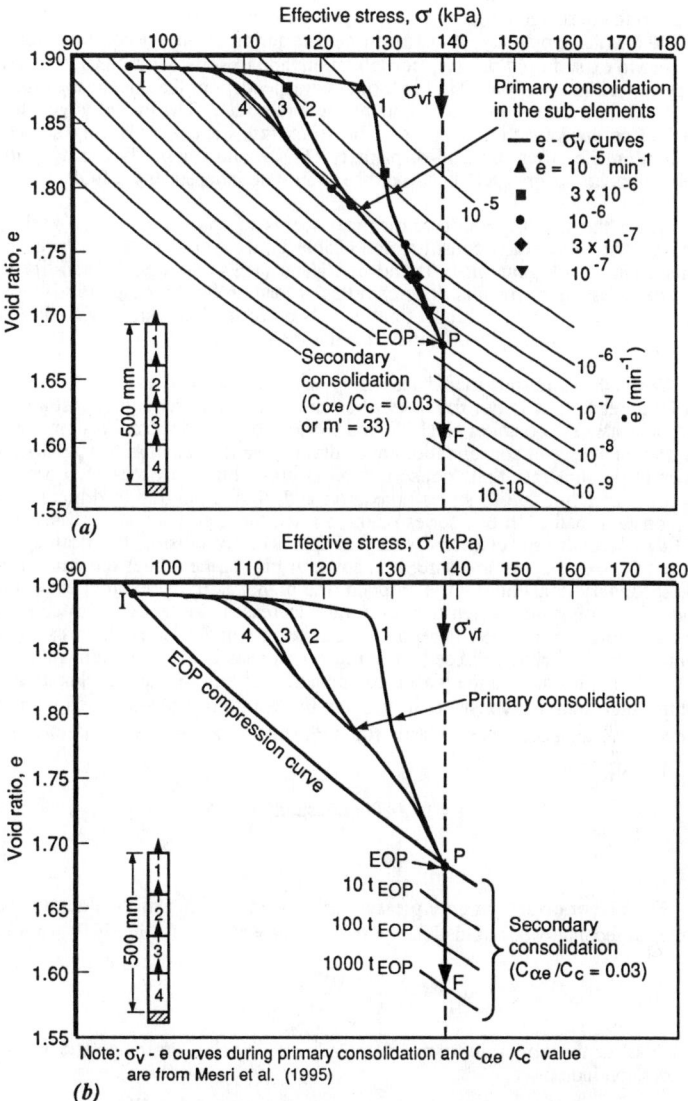

Fig. 5 - *Consolidation of the Saint-Hilaire clay for pressure increment from 97 kPa to 138 kPa (a) $\sigma' - e - \dot{e}$ approach (b) $C_{\alpha e}/C_c$ approach*

5b); b) Mesri et al. (1995) recognize that, during primary consolidation, the effective stress-strain behavior is not unique in a consolidating layer, but depends on the distance to the drainage boundary, as evidenced in the figure; c) Mesri et al. (1995) also show that the simple $C_{\alpha e}/C_c$ model that exists during secondary consolidation does not appear to exist during primary consolidation; and d) soil conditions at the end-of-primary consolidation (Point P in Fig. 5b) are independent of the thickness of the clay layer (Mesri and Choi 1985), and thus of the strain rate. The model advocated by Mesri and co-workers thus appears to be strain-rate dependent during secondary consolidation, i.e. after the End-of-primary (EOP), and implicitly during primary consolidation, i.e. before EOP, but would be strain-rate independent at EOP.

Since the $C_{\alpha e}/C_c$ = constant model seems to apply for a large variety of materials during secondary consolidation (Table 1), the (σ'_1-ε_1-$\dot{\varepsilon}_1$) model previously described should be general. Edil and den Haan (1994) showed its validity for a fibrous peat; tests performed at different rates of loading by De Waal (1986) indicate that it seems to be valid for sand; finally, tests performed on chalk by Ruddy et al. (1989) also indicate that the model may be valid for soft rocks.

While the strain-rate model accurately describes the behavior of clays when the strain is increasing, it is not the case when the axial strain remains constant, as in relaxation tests. Yoshikuni et al. (1994 and 1995a) performed special oedometer tests with different phases of consolidation (open drainage) and relaxation (closed drainage) in order to demonstrate that "consolidation is a combined process of (pore water pressure) generation due to stress relaxation and dissipation due to drainage, even under constant load". In one series of three tests, these authors stopped secondary consolidation under an effective stress of 314 kPa by closing the drainage and measured the excess pore pressure. As shown in Fig. 6, the higher the strain rate at which secondary consolidation is stopped, the higher is the generated excess pore pressure, indicating that, when the strain rate decreases, the tendency to creep also decreases. Holzer et al. (1973) report similar results under isotropic stresses. One explanation of the behavior observed in relaxation tests is that the strain-rate model (Eqs 1 to 3) applies only to the plastic component of strain, so that, in relaxation tests where the total strain ε is constant, the increase in plastic strain ε^p associated with creep is compensated by a decrease in elastic strain ε^e, and thus a decrease in effective stress (Fig. 7).

$$\varepsilon = \varepsilon^e + \varepsilon^p = \text{constant} \tag{5a}$$

and,

$$\dot{\varepsilon}^e = -\dot{\varepsilon}^p \tag{5b}$$

The rate of change in pore pressure $\dot{u}$ following closing of the drainage when the soil is in secondary consolidation at a strain rate $\dot{\varepsilon} = \dot{\varepsilon}^p$ (point A in Fig.7) would be:

$$\dot{u} = \frac{2.3\,(1+e_o)\,\sigma'}{C_s}\,\dot{\varepsilon}^p \tag{6}$$

in which σ' is the effective stress at the time of closing drainage and C_s is the recompression index.

When the applied load is decreased and the soil swells, viscous effects can also be observed. After primary swelling associated with pore water pressure equalization, secondary deformations in swelling or in compression develop. Such behavior has been observed and discussed by numerous authors (Mesri et al. 1978; Mesri and Feng 1991; Urciuoli 1992; Feijó and Martins 1993; Yoshikuni et al. 1993 and 1995b). It can be illustrated by the results reported by Feijó and Martins (1993). Special oedometer tests were performed on organic Sarapuí clay from Brazil, in which several specimens were first loaded in the normally consolidated range and then, at the end-of-primary consolidation, unloaded in one step to various overconsolidation ratios. The specimens were left under constant stress for more than 5 months, and the observations are shown in Fig. 8. For the specimens at OCR of 1.5 and 2, after some swelling, the soil starts to compress again in a manner similar to what is observed with normally consolidated clays. For OCR of 6, 8 and 12, the soil shows secondary swelling during the entire test. At an OCR of 4, secondary deformations are not significant. This kind of "neutral OCR" can, however, vary from one soil to another and certainly with the strain rate at the time of unloading.

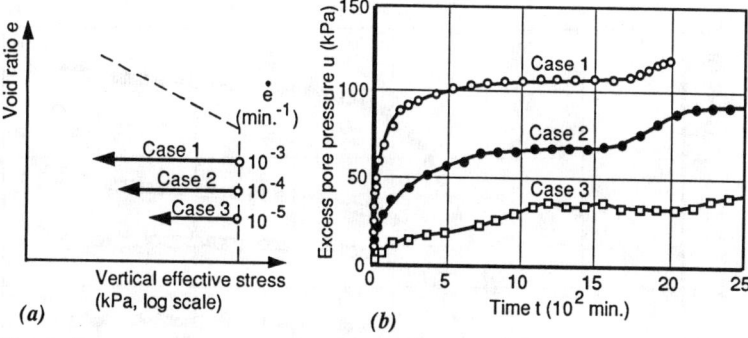

Fig. 6 - Excess pore pressure generation during relaxation tests under one-dimensional conditions [from Yoshikuni et al. (1994)]

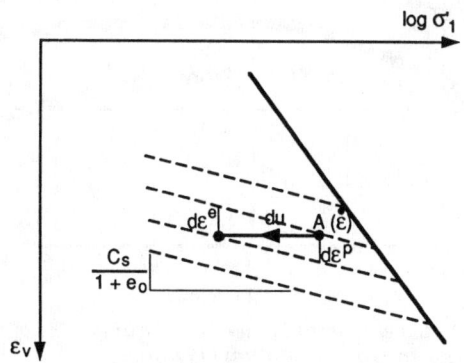

Fig. 7 - Strain components in relaxation test

Although the authors believe that the effect of strain rate (and probably the effect of temperature, which will be examined in the following section) is general, regardless of the phase of consolidation and the degree of structuration of the clay, there is evidence that suggests the (σ'_1-ε_1-$\dot{\varepsilon}_1$) model is not always valid. This seems to be due to structuring phenomena which can decrease the compressibility of some clays, in particular at very small strain rates and in young clays (Leroueil et al. 1996b). Also, even if Leroueil et al. (1985a) questioned the conclusions of Mesri and Choi (1985) concerning the small differences in void ratio observed between EOP e-log σ'_v curves obtained on specimens of different thickness, it is now thought that these small differences could also be explained, or partly explained, by structuring effects (Leroueil, 1994-96). The fact that some authors, such as Smith and Wahls (1969), did not observe a strain rate effect when comparing CRS tests performed at different strain rates could also be due to structuring phenomena.

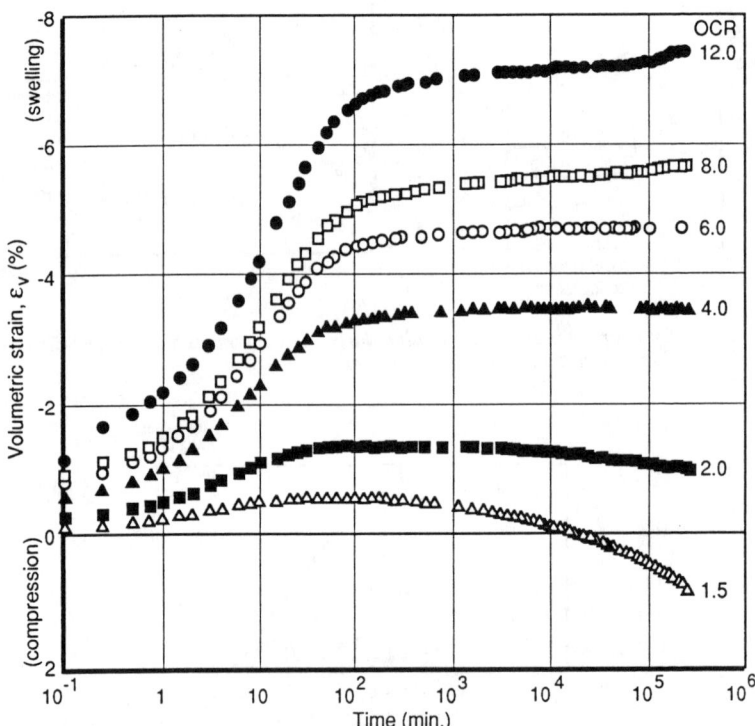

Fig. 8 - Volumetric strains observed after unloading of Sarapuí clay [after Feijó and Martins (1993)]

Effects of temperature

The influence of temperature on soil behavior was thoroughly studied in the 1960's, and a Specialty Conference was organized in Washington on that specific topic in 1969. Temperature has two major effects on soils: thermal expansion of solid particles and pore water, and thermally induced modification of the strength of contacts between particles or aggregates. The combined effect of these phenomena on the temperature-volume-effective stress behavior of soils in both drained and undrained conditions is explained by Mitchell (1993) and will not be described here in detail. Fig. 9 deduced from isotropic compression tests on samples initially consolidated under 200 kPa illustrates the behavior generally observed. In drained conditions, an increase in temperature is associated with an expulsion of pore water from the clay sample whereas a decrease in temperature is associated with an absorption. As explained by Mitchell (1993), "the weakening effect of the higher temperature is compensated by the strengthening effect of the lower void ratio". However, due to changes in the particle arrangement, the process is not reversible for soils normally or slightly overconsolidated. In undrained conditions, heating is associated with an increase in pore pressure, and consequently, with a decrease in effective stress. In these conditions where there is no significant particle rearrangement, the process is almost reversible.

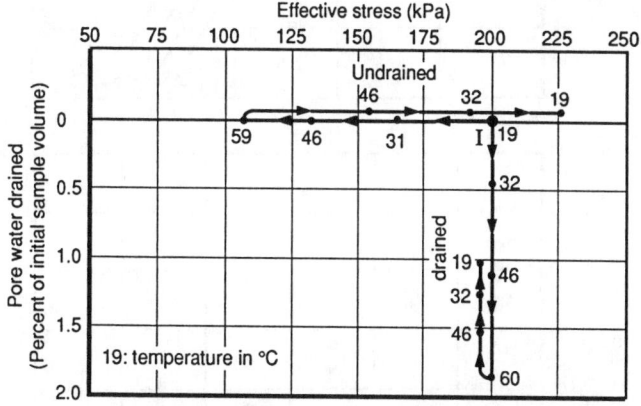

Fig. 9 - Effects of temperature changes on saturated illite under drained and undrained conditions (initial isotropic effective stress = 200 kPa) [after Mitchell (1993)]

In recent years, many studies have been conducted on the effects of temperature on the compressibility of natural clays (Eriksson 1989; Tidfors and Sällfors 1989; Boudali et al. 1994; Marques 1996). Typical results, obtained from sulphide clay from Luleå, Sweden, are shown on Fig. 10. It can be seen that there is a significant effect of temperature on the compressibility of the clay. With increasing temperature, the soil becomes more compressible in the overconsolidated range, the preconsolidation pressure decreases, and the entire compression curve moves towards smaller effective stresses. The two CRS tests A and B performed on the St-Polycarpe clay (Fig. 3b), with changes in temperature at different strains from 5°C to 20°C in Test A and from 50°C to 20°C in Test B, show the effective stress-strain curves jumping from one constant temperature curve to another, and thus confirm the influence of temperature.

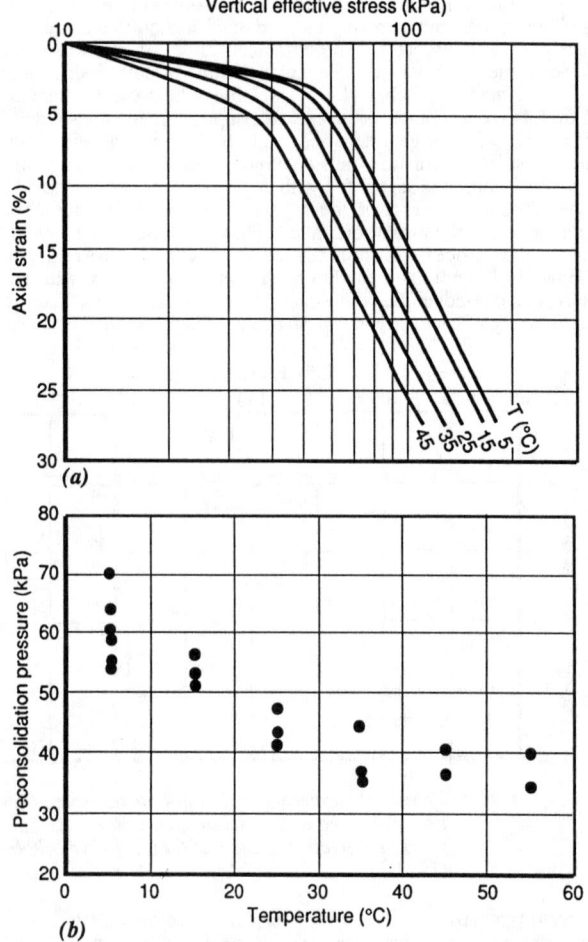

Fig. 10 - Oedometer tests performed on Luleå clay at various temperatures [from Eriksson (1989)]

Boudali et al. (1994) performed CRS tests at different strain rates and temperatures on 3 clays, among which was the Berthierville clay previously discussed (Fig. 4). Their results show the following: a) the preconsolidation pressure can be described as a function of strain rate and temperature (Fig. 4a), and b) the effective stress-strain curves obtained at various strain rates and temperatures reduce to a unique one when they are normalized with respect to the preconsolidation pressure corresponding to the strain rate and temperature used in each test (Fig. 4b). The model proposed by Leroueil et al. (1985b) for strain rate effects can thus be extended for including temperature effects. Equations 1 and 2 then become:

and
$$\sigma'_p = f(\dot{\varepsilon}_1, T) \qquad (7)$$

$$\sigma'_1/\sigma'_p(\dot{\varepsilon}_1, T) = g(\varepsilon_1). \qquad (8)$$

The test results shown in Figs 3 and 10 fit this model.

Figure 11 shows the preconsolidation pressure (or the effective stress at a given void ratio) normalized with respect to the preconsolidation pressure (or the effective stress at a given void ratio) measured at 20°C, as a function of temperature. The decrease in preconsolidation pressure with increasing temperature is significant, on average, almost 1% per °C between 5 and 40°C.

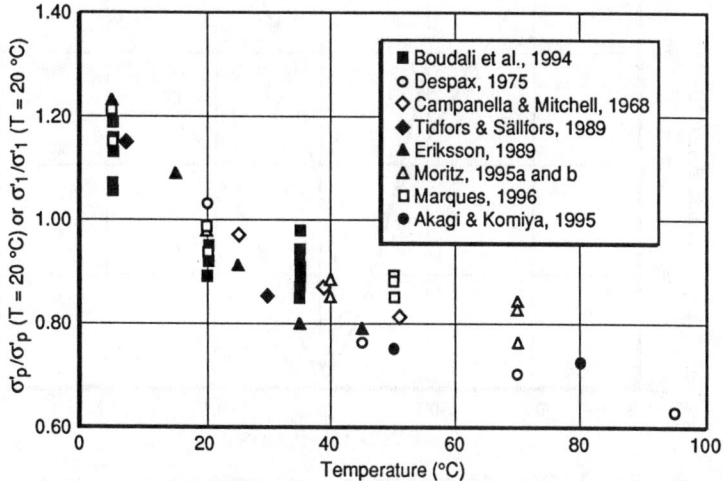

Fig. 11 - Variation of the normalized preconsolidation pressure, or vertical effective stress at a given void ratio, with temperature.

Equations 7 and 8 indicate that strain rate and temperature are two parameters which are intimately linked in the viscous deformation of soils and control the maximum effective stress a soil can support at given void ratio and structure. This results from fundamental physics and can be explained, at least qualitatively, by the rate-process theory (see Section 2.2.9).

For a soil consolidated in the normally consolidated range at a given temperature, cooling causes the soil to behave as if it were overconsolidated (Plum and

Esrig 1969; Burghignoli and Desideri 1988; Hueckel and Baldi 1990). This is also demonstrated in Fig. 3b and discussed further in Section 5.3 and Fig. 36. As a result and as for a change in effective stress, volumetric strains due to temperature changes in the overconsolidated range are relatively small and, to a large extent, reversible (Hueckel and Baldi 1990; Burghignoli et al. 1992; Boudali et al. 1994). Plum and Esrig (1969), Hueckel and Baldi (1990) and Burghignoli et al. (1992) also show that these volume changes depend on the overconsolidation ratio. For the natural Pasquasia clay (Fig. 12) and for temperatures increasing from 21°C to more than 90°C, volumetric strains are positive for OCRs smaller than approximately 6, and negative for higher OCRs. The similarity with the tendency for the soil to creep in compression when slightly overconsolidated and in swelling when strongly overconsolidated (see Fig. 8) is noteworthy. It has to be mentioned, however, that this phenomenon is not related to the strength of the material, as demonstrated by the increase in pore pressure with increasing temperature observed in a yellow tuff from Naples, by Aversa and Evangelista (1993).

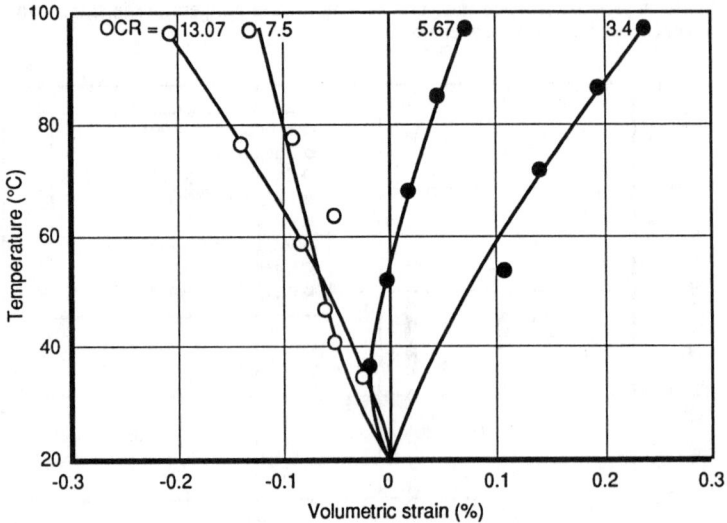

Fig. 12 - Drained heating tests at constant effective stress at various OCR (in terms of isotropic stress) - Natural Pasquasia clay [from Hueckel and Baldi (1990)]

2.2.2 - Limit state curve

Effects of strain rate

Sällfors (1975) and Tavenas and Leroueil (1977) indicate that the effects of time and strain rate on the preconsolidation pressure can be generalized to the entire limit state curve of the soil. Triaxial tests performed by Lo and Morin (1972) on St-Vallier clay (Fig. 13) demonstrate a strain rate effect on the strength envelope of the overconsolidated soil, i.e. the upper part of the limit state curve, similar to that on the preconsolidation pressure.

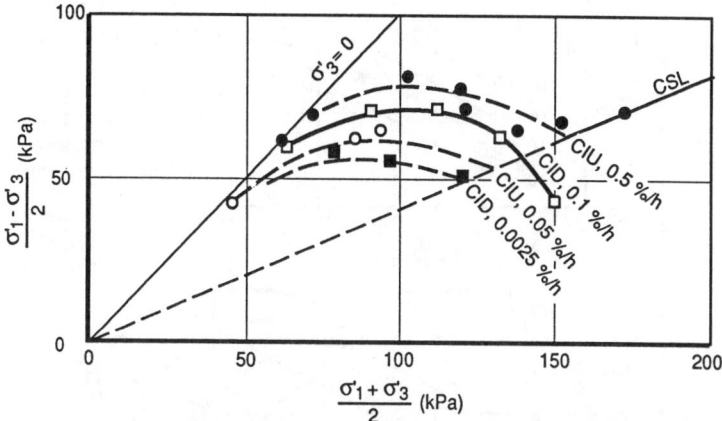

Fig. 13 - Effect of strain rate on the strength of Saint-Vallier clay [after Lo and Morin (1972)]

Marchand (1982) reported triaxial and oedometer tests performed on the Mascouche clay. It is a stiff plastic clay ($I_p = 29\%$) with a liquidity index of 1.14 and a preconsolidation pressure, as defined in conventional 24-hrs oedometer tests, equal to 300 kPa. The peaks obtained in the continuous CIU, CAU and CID tests, and the effective stress applied in the triaxial drained creep tests at failure are shown in the insert of Fig. 14 with their associated strain rate (minimum strain rate reached before failure in creep tests; see sections 2.2.4 and 2.2.5). All the data are relatively well-organized, confirming an effect of strain rate on the peak strength envelope of the overconsolidated clay. The one-dimensional CRS and creep test results, also presented in the figure, indicate an effect of strain rate on the preconsolidation pressure similar to that on the strength envelope and thus a rate effect on the entire limit state curve. Similar results were obtained for other eastern Canadian clays, the St-Jean-Vianney clay (Vaid et al. 1979; Leroueil and Tavenas 1979) and the St-Alban clay (Tavenas et al. 1978).

The viscous behavior of Berthierville clay (see Fig. 4 for its one-dimensional rheological law) below the strength envelope of the normally consolidated clay has been recently investigated at Laval University (Boudali 1995; Leroueil et al. 1996a). Continuous and step-loaded triaxial compression tests were performed along $K = \sigma'_3/\sigma'_1 =$ constant stress paths. From the test results and for the different stress paths followed, curves corresponding to different strain[5] rates were defined. The following conclusions were drawn:

---

[5] The deformations in these tests being two-dimensional, Boudali (1995) and Leroueil et al. (1996a) use the strain $\varepsilon_{vs} = \sqrt{\varepsilon_v^2 + \varepsilon_s^2}$, with ε_v, volumetric strain, equal to $(\varepsilon_1 + 2\varepsilon_3)$, and ε_s, shear strain, equal to $(2/3)(\varepsilon_1 - \varepsilon_3)$.

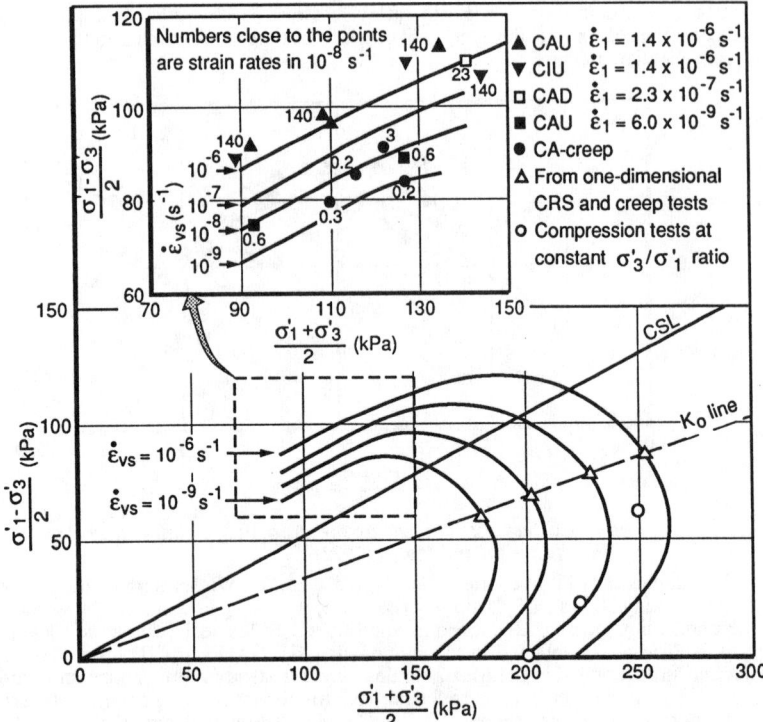

Fig. 14 - Influence of strain rate on the limit state curve of Mascouche clay [after Marchand (1982)]

- Along the different stress paths, yielding at a given strain rate is well defined, and the mean effective stress at yielding (p'_{y-K}) varies with the strain rate in the same manner as the preconsolidation pressure (Eq. 3):

$$\log (p'_{y-K}) = A' + (1/m') \log \dot{\varepsilon}_{vs} \qquad (9)$$

in which A' is a parameter which depends on the K value considered and m' is the inverse slope of the log (p'_{y-K}) vs log$\dot{\varepsilon}_{vs}$ relationship, which takes the same value for all K values and for the oedometer tests. This implies that the limit state curves obtained for different values of $\dot{\varepsilon}_{vs}$ have the same shape, as shown in Fig. 15a.

- The stress-strain curves obtained at different strain rates along a given K = constant stress path can be normalized with respect to the yield stress value corresponding to the considered strain rate, which can be described as follows:

$$p'_K/p'_{y-K} (\dot{\varepsilon}_{vs}) = g(K, \varepsilon_{vs}) \qquad (10)$$

Equations 9 and 10 are similar to those (Eqs. 3 and 2) relevant to one-dimensional compression and reflect the fact that the effective stress-strain-strain rate model proposed by Leroueil et al. (1985b) can be extended to the entire stress diagram, at least for stress paths at constant σ'_3/σ'_1 ratio and for sensitive eastern Canadian clays.

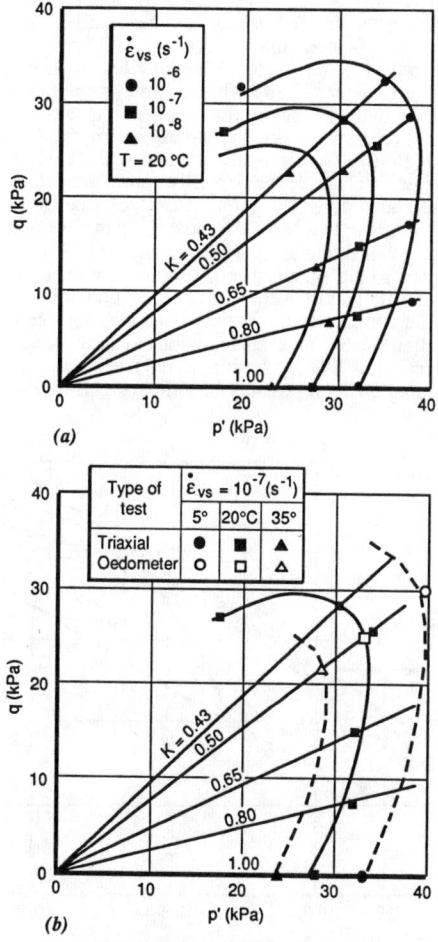

Fig. 15 - *Variation of the limit state curve of Berthierville clay with (a) strain rate and (b) temperature [after Boudali (1995)]*

The fact that m' takes the same value, independently of the K = constant stress path followed, is confirmed by the data presented by Mesri et al. (1995) showing that $C_{\alpha e}/C_c$ is the same in one-dimensional and isotropic conditions for two other clays (Fig. 2).

Hight (1983), Hight et al. (1987) and Sheahan et al. (1996) present anisotropically consolidated undrained triaxial tests performed on reconstituted Lower Cromer till (I_p = 13%) and Boston Blue clay (I_p = 23.7%) respectively. These tests were in compression only for the Boston Blue clay and both in compression and extension for the Lower Cromer till; in both cases, they were performed at different overconsolidation ratios and different strain rates. The results obtained in the two series of compression tests are very similar, and are summarized in Fig. 16 prepared after Sheahan et al. (1996). They are at variance with the behavior previously described. If a rate effect is clear for the normally consolidated clay and possibly for the slightly overconsolidated clay (OCR = 2), it is much smaller at larger OCRs, all the peak strengths being reached on about the same strength envelope. As this strength envelope, approximately characterized by a friction angle of 32° and zero cohesion for the Boston Blue clay, is also that obtained at large strains for the normally consolidated material, this could indicate that the behavior of reconstituted Lower Cromer till and Boston Blue clay at failure is essentially frictional, and thus not significantly influenced by strain rate (see Section 2.2.6). In terms of undrained shear strength, the change per logarithm cycle, typically of 8% for both materials when normally consolidated, decreases with increasing OCR to possibly become insignificant at large OCRs and small strain rates.

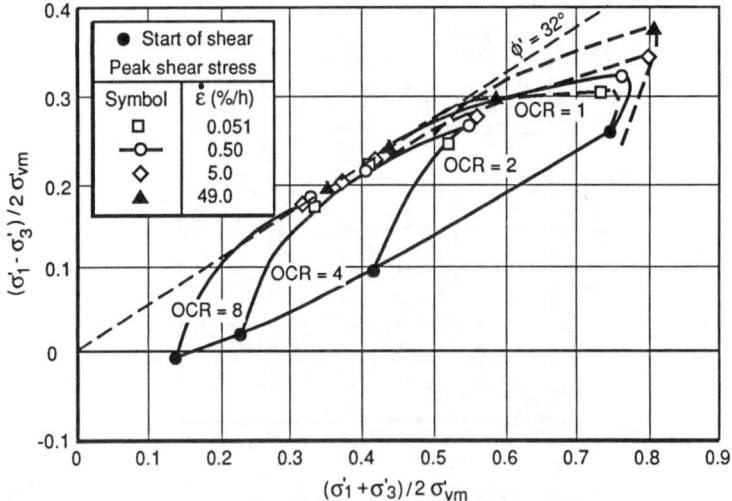

Fig. 16 - *Normalized effective stress paths and strengths as obtained on resedimented Boston Blue clay in undrained compression tests performed at different OCRs and strain rates [after Sheahan et al. (1996)]*

In extension, Hight (1983) found an increase in undrained shear strength with strain rate similar to that observed in compression tests on normally consolidated soil, and approximately independent of OCR for OCR<4. This later observation is attributed to the shape of the limit state curve of initially anisotropically consolidated soils.

Effects of temperature

To investigate the effect of temperature on the limit state curve, Boudali (1995) performed, in addition to the oedometer tests described in section 2.2.1 (Fig. 4), isotropic compression tests and undrained triaxial tests at temperatures of 5, 20 and 35°C on Berthierville clay. The isotropic compression tests were step-loaded, with a consolidation duration varying from 4 to 10 days in the normally consolidated range. The results, shown in Fig. 17 for a strain rate of 10^{-7} s^{-1}, clearly show the effect of temperature on the entire compression curve and, in particular, on the isotropic yield stress. The yield stresses obtained at the three temperatures are reported in Fig. 15b, together with the other yield stresses obtained at a strain rate of 10^{-7} s^{-1}. The pre-consolidation pressures obtained at the same temperatures and at the same strain rate (Fig. 4a) are also plotted on the figure, in an approximative manner, on the K = 0.5 line. Even if the data available are limited, it seems that Eqs. 9 and 10 could be extended to include the effects of temperature. Burghignoli and Desideri (1988) also observed a similar temperature effect on the isotropic compression curve of the reconstituted Todi clay.

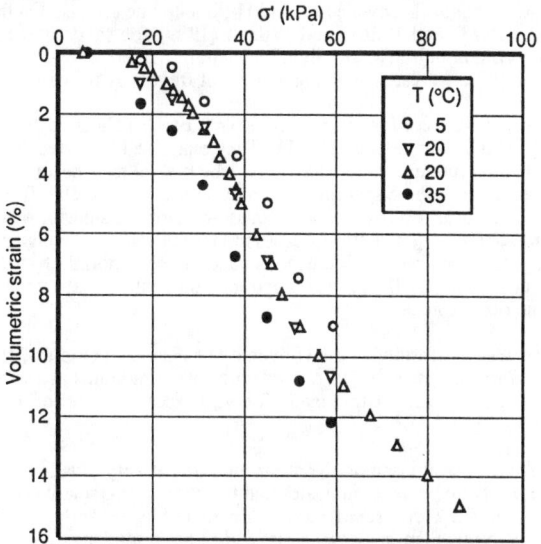

Fig. 17 - Effective stress-volumetric strain relations at a strain rate of 10^{-7} s^{-1} and at different temperatures, isotropic compression of Berthierville clay [from Boudali (1995)]

Boudali (1995) also performed undrained compression tests on Berthierville clay in which the samples were consolidated and sheared at temperatures of 5, 20 and 35 °C. The samples were consolidated in the overconsolidated range and compressed at a rate of 10^{-6} s^{-1}. The results show a general tendency for the peak shear strength to decrease when the temperature increases, indicating that the entire limit state curve is temperature-dependent. Hueckel and Baldi (1990) and Moritz (1995a) observed similar behavior for clays from Italy and Sweden, respectively. These results confirm an influence of temperature on the entire limit state curve, at least at an essentially constant void ratio. As noted by Hueckel and Baldi (1990), when temperature increase is associated with a significant void ratio decrease, the strength gain due to this decrease in void ratio could compensate for the strength loss due to a higher temperature.

2.2.3 - Very small strain domain

The strain-rate effect on the shear modulus G measured at small strains has been the object of numerous studies in recent years, in particular through resonant column tests and cyclic tests performed at different frequencies and amplitudes. Several authors (Isenhower and Stokoe 1981; Rampello and Silvestri 1993; Zavoral and Campanella 1994; Duffy et al. 1994; Stokoe et al. 1995) found a strain-rate effect in cohesive soils, but these effects were relatively small with a variation of the shear modulus on the order of 4% per logarithm cycle of frequency or strain rate. Dobry and Vucetic (1987) even considered that the roundness of the loop tips during a sinusoidally strain-controlled cyclic test could be due to rate effects. On the other hand, Iwasaki et al. (1978) and Bolton and Wilson (1989) for sands and Shibuya et al. (1995) for clays did not observe any significant strain-rate effect on the shear modulus. Stokoe et al. (1995) observed a very small effect of frequency for an undisturbed sand.

This apparently confused situation seemed to be clarified by Tatsuoka and Kohata (1995) and Tatsuoka et al. (1995). They examined the effect of strain rate on the very small strain modulus on the basis of cyclic triaxial tests performed on Toyoura sand, a mudstone from the Sagamihara site and a pleistocene clay from Osaka, and based on torsional shear tests as well as resonant-column tests performed on Pisa clay. These authors conclude that at strains of less than 0.001%, the behavior is strain-rate independent. At larger strains, it seems to depend on the material. It is still strain-rate independent at a strain of 0.01% for Toyoura sand, but not at strains of 0.005 or 0.008% for the other soils.

Recent tests performed at the University of Napoli (d'Onofrio, 1996; Santucci de Magistris, 1996) however show a rate effect on shear modulus at a very small strain of 0.0003% of 5.5% per logarithm cycle for Vallericca clay and of 3.4% for a silty clayey sand (Fig.18).

The effect of strain rate on the shear modulus at very small strains is thus not clear yet. The literature however indicates that the effect : a) is relatively small and less than 6% per tenfold increase in strain rate in cohesive soils; b) is smaller in cohesionless materials than in clayey soils; and c) has a tendency to decrease at very small strains (less than 0.001%).

The authors do not know of any research on the effects of temperature on the small strain shear modulus. However, there are data in the literature showing that at strains larger than approximately 0.1%, Young's modulus decreases when temperature increases (Murayama 1969; Hueckel and Baldi 1990).

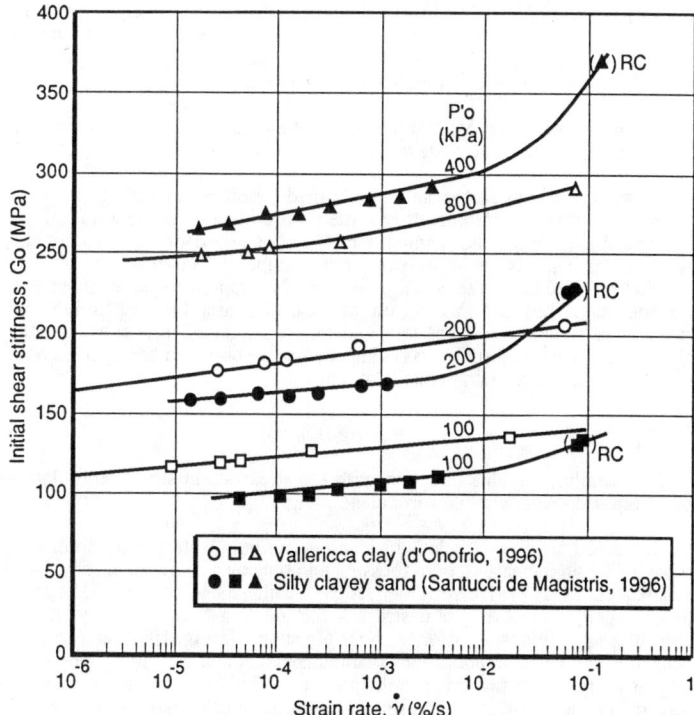

Fig. 18 - Effect of strain rate on shear modulus at a very small shear strain of 0.0003% in torsional shear tests [after d'Onofrio (1996) and Santucci de Magistris (1996)]

2.2.4 - In the normally consolidated domain

Creep

On the basis of creep tests performed in triaxial apparatus in drained or undrained conditions, Singh and Mitchell (1968) showed that the logarithm of strain rate linearly decreases with logarithm of time, that the slope of the log $\dot{\varepsilon}_1$ - log t relationship is essentially independent of the creep stress and that increases in applied deviatoric stress shift the relationship to higher strain rates. They also showed that, except at very low and very high stress levels, there is a linear relationship between the logarithm of the strain rate and the deviatoric stress at a given time. They thus proposed the following equation:

$$\dot{\varepsilon}_1 = A\, e^{\alpha \bar{q}} \left(\frac{t_1}{t}\right)^m \tag{11}$$

in which, $\dot{\varepsilon}_1$ is the axial strain rate at any time t, $\bar{q}$ is the stress level equal to the applied deviatoric stress divided by the deviatoric stress at failure, m is the slope of the log $\dot{\varepsilon}_1$ - log t curves which generally falls between 0.7 and 1.0, t_1 is a reference time generally taken equal to 1 min, and α and A are creep parameters.

It is noteworthy that Murayama et al. (1984) found a similar behavior in the Toyoura sand, with a m value close to 1.0.

For the creep tests performed in undrained conditions under high stress levels, the strain rate reaches a minimum after some time, and then increases up to failure of the soil specimen. Vaid and Campanella (1977) showed on Haney clay that the relationship between this minimum strain rate and the applied deviatoric stress is the same as that obtained in undrained shear between the strain rate used and the deviatoric stress at the peak. They also showed, on the basis of constant rate of strain undrained compression tests, undrained creep tests and special undrained tests in which either the rate of strain or the applied stress was changed, that the observed behavior is controlled by a deviatoric stress-strain-strain rate relation:

$$q = q(\varepsilon_1, \dot{\varepsilon}_1) \tag{12}$$

The similarity of this equation with the stress-strain-strain rate relationship deduced from oedometer tests is worth mentioning.

Undrained creep, under deviatoric stress or under isotropic conditions, is also accompanied by an increase in pore pressure and thus by a decrease in effective stress (Arulanandan et al. 1971; Holzer et al. 1973). Arulanandan et al. (1971) carried out undrained creep tests at different stress levels on San Francisco Bay mud. The effective stress conditions at different times are shown in Fig. 19. It can be seen that the locus of effective stress conditions maintains essentially the same shape with time but progressively moves towards smaller stresses. This behavior can be explained in the same manner as for relaxation tests in one-dimensional conditions (see Fig. 7 and Eqs. 5b and 6). The plastic component of the volumetric strain has a tendency to increase with time due to creep and, the total volumetric strain being constant, the elastic component and the mean effective stress have to decrease. This applies to the entire limit state curve.

As indicated, in particular by Sekiguchi (1977) and Sheahan (1995), there must be a lower limit to stress conditions which can be reached during creep tests. This is supported by observations showing that highly overconsolidated clays have a tendency to generate negative pore pressures in undrained creep tests, or swelling in drained creep tests (see Fig. 8). Sheahan (1995) calls the lower limit "static yield surface"; "inviscid yield surface" has also been used in the literature.

Undrained shear tests at different strain rates

The influence of strain rate on the behavior of normally consolidated clays during shearing has been studied by numerous researchers (Sekiguchi 1977; Vaid and Campanella 1977; Adachi and Oka 1982; Lefebvre and Leboeuf 1987; Sheahan et al. 1996; etc.). The observed behavior is generally as shown in Fig. 16 for an OCR equal to 1.0. The smaller the strain rate, the larger the pore pressure generated before failure and the smaller the peak shear strength. It results from the simultaneous increase in deviatoric stress and creep pore pressure with time. The friction angle associated with the peak shear strength either remains constant or slightly decreases when the strain rate decreases. As for the friction angle obtained at large strain, it appears to be strain rate independent (see Section 2.2.6)

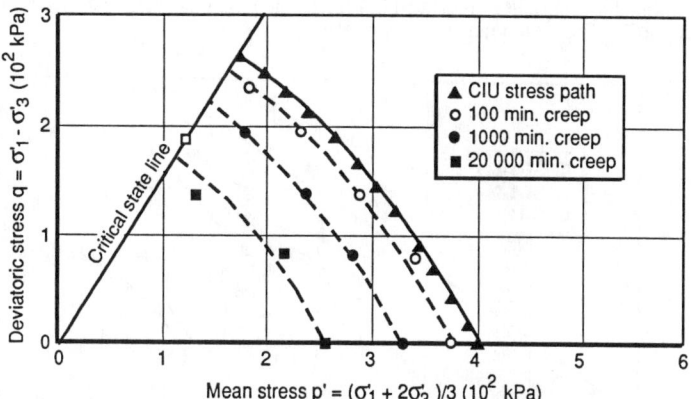

Fig. 19 - *Change in effective stress with time during undrained triaxial creep tests (($\sigma_1 - \sigma_3$) = cst) [after Arulanandan et al. (1971)]*

Undrained relaxation

These tests are characterized by the fact that the strain, which is the axial strain in triaxial tests, remains constant with time. Lacerda and Houston (1973) showed that, when a relaxation test is started after straining the soil under constant rate of strain $\dot{\varepsilon}_{10}$ up to a deviatoric stress D_0, a decrease of the deviatoric stress D is observed with time. This decrease varies linearly with the logarithm of time, after a certain value of time t_0.

$$D/D_0 = 1 - s \log\left(\frac{t}{t_0}\right), \text{ for } t > t_0 \quad (13)$$

The parameter s, the slope of the D/D_0 vs log t relationship, seems to be slightly dependent on the stress level at the beginning of the relaxation test. t_0 depends on the strain rate prevailing just before the relaxation test; it would be inversely proportional to $\dot{\varepsilon}_{10}$. This implies that the smaller the rate of shearing prior to the relaxation phase, the smaller the deviatoric stress decrease during the relaxation phase. This may be seen as similar to the results presented in Fig. 6 showing that the pore pressure generated during relaxation under one-dimensional conditions was decreasing with the strain rate prevailing just before the relaxation test. It indicates that all these test phases (drained creep, relaxation, shearing at different strain rates) contribute to the accumulation of visco-plastic strains and that the quantity accumulated during one specific phase decreases the ability of the soil to develop additional visco-plastic strains during the following phases.

The previously described behavior has been observed for different clays and Monterey sand by Lacerda and Houston (1973) and has been confirmed by Akai et al. (1975), Hicher (1988) and Sheahan et al. (1994).

Another characteristic of triaxial relaxation tests is that, as shown by Richardson and Whitman (1963), Murayama and Shibata (1964), Lacerda and Houston (1973), Akai et al. (1975), Hicher (1988) and Sheahan et al. (1994), pore pressure does not vary significantly during relaxation tests; the effective radial stress remains thus approximately constant as indicated in Fig. 20 showing relaxation tests performed by Lacerda (1976) on San Francisco Bay mud.

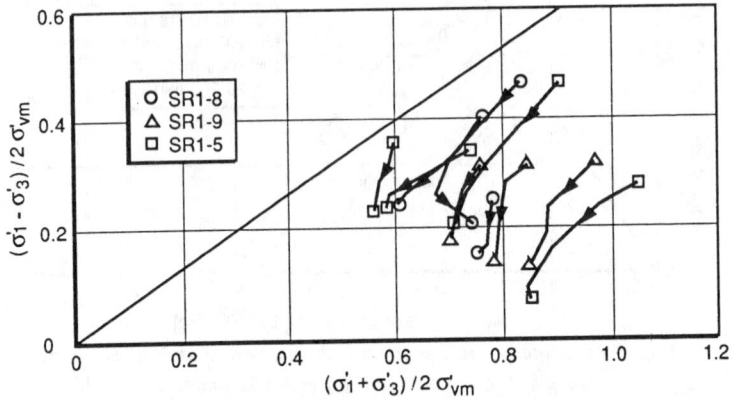

Fig. 20 - Effective stress paths followed during triaxial relaxation tests ($\varepsilon_1 = cst$) on normally consolidated San Francisco Bay Mud [data from Lacerda (1976), after Martins (1992)]

General influence of temperature

An increase in temperature weakens the soil with the consequence that, at least for normally or slightly overconsolidated clays, in creep tests strain rates increase, and in undrained tests pore pressure increases (Murayama and Shibata 1961; Mitchell 1964; Murayama 1969). In particular, as the pore pressure generated during undrained shear increases with temperature (Mitchell 1964), it implies that the critical state line (locus of the ultimate conditions reached during shearing) in a void ratio-mean effective stress diagram moves towards smaller mean effective stress as temperature increases (Lingnau et al. 1995). This aspect is illustrated in Section 3.2.1 and Fig. 24.

2.2.5 - Inside the limit state curve

Typical creep test results obtained by Tavenas et al. (1978) on St-Alban clay are shown in Fig. 21 for different effective stress conditions inside the limit state curve. They confirm the linear decrease of the logarithm of strain rate with logarithm of time. Tests performed by Bishop and Lovenbury (1969), Larsson (1977), and D'Elia (1991) on other clays, soft or stiff, also confirm this behavior. Tavenas et al. (1978) consider, however, that the stress function (exp ($\alpha \bar{q}$)) used in Eq. 11 should be generalized and written as $n(\sigma')$, with $n(\sigma')$ being a stress function reflecting the shape of the limit state curve of the soil. Eq. 11 then becomes:

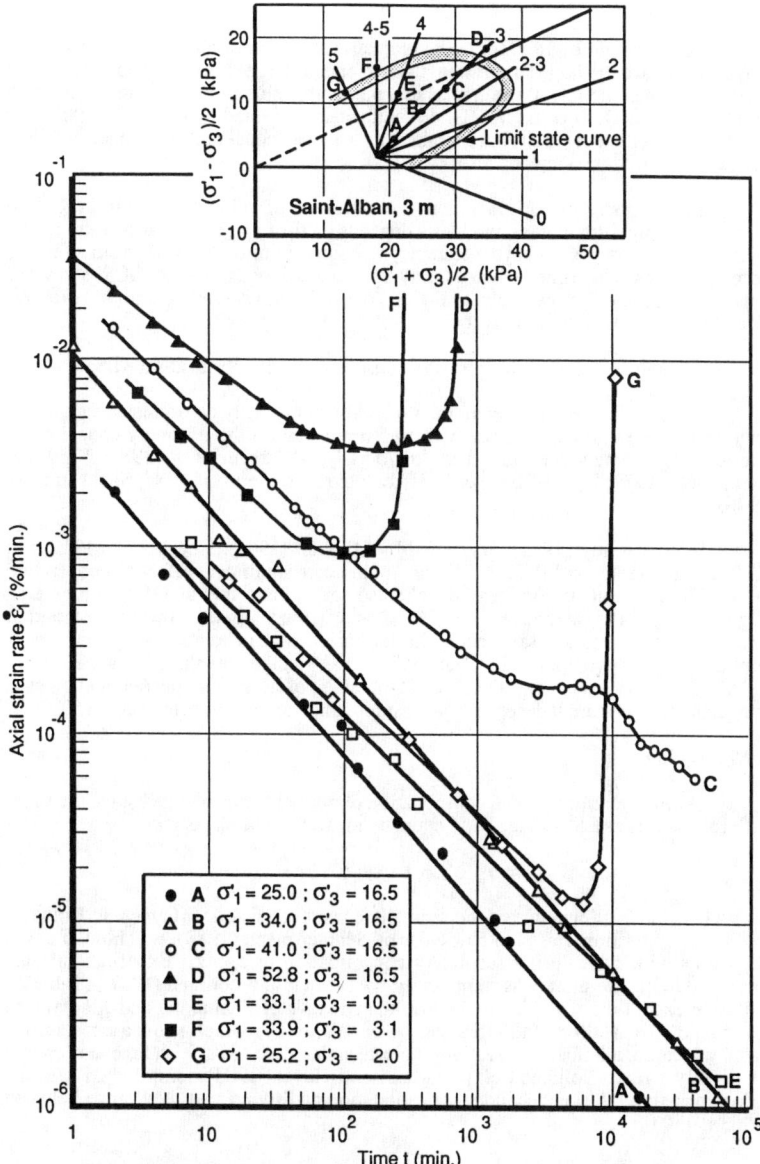

Fig. 21 - Axial strain rate-time relationship for creep tests on the Saint-Alban clay [after Tavenas et al. (1978)]

$$\dot{\varepsilon} = A\, n(\sigma') \left(\frac{t_1}{t}\right)^m \tag{14}$$

Eq. 14 would be valid for both volumetric and shear strain. For effective stress conditions above the large strain strength envelope of the soil, Eq. 14 is no longer valid after some time. The rate of strain reaches a minimum value and then increases to reach failure (see tests D, F, G in Fig. 21). From the data presented by Lefebvre (1981), the minimum strain rate is reached at a strain approximately equal to that obtained at failure in conventional triaxial drained tests.

The response of the soil in terms of variation of volumetric strain in drained tests or pore pressure in undrained tests depends on the overconsolidation ratio. Soils with relatively low OCRs have a tendency to compress or to generate positive excess pore pressures with time, whereas clay specimens having a high OCR show the opposite tendency. For example, in Fig. 21, the void ratio decreased with time for test C, and increased with time for test G.

2.2.6 - Strength envelope of the normally consolidated soil or critical state line

Tests performed at different rates of strain on normally consolidated samples of clay do not show any significant effect of strain rate on the normally consolidated friction angle when determined at $(\sigma'_1/\sigma'_3)_{max}$ (Bjerrum et al. 1958; Vaid and Campanella 1977; Badra-Blanchet 1981; Lefebvre and Leboeuf 1987; Sheahan et al. 1996).

Similarly, tests results presented by Mitchell (1964), Hueckel and Pellegrini (1989) and Hueckel and Baldi (1990) do not indicate significant effect of temperature on the friction angle of normally consolidated soil. Houston et al. (1985) observed a slight increase (1°) between 4 and 40°C and a much greater increase at higher temperatures. Testing a sand-bentonite mixture, Lingnau et al. (1995) found very similar strength envelopes at 26°C and 65°C, and a strength envelope lying well above at 100°C. It thus seems that the strength envelope of normally consolidated clays is essentially temperature independent, at least for temperatures lower than about 50°C.

2.2.7 - K_o line

During one-dimensional consolidation of normally consolidated soils, the stress path followed is as LM in Fig. 1 and corresponds well to the stress ratio

$$K_{o\,nc} = \sigma'_3/\sigma'_1 = 1 - \sin\phi' \tag{15}$$

where ϕ' is the friction angle of the normally consolidated soil (Mayne and Kulhawy 1982). The question, raised in particular by Schmertmann (1983), was how, if at all, K_o varies with time during secondary consolidation or aging. Experimental data presented in the literature was summarized by Tavenas and Leroueil (1987), (Table 2). With the exception of the heavily consolidated clays of Paninaglia and Montalto di Castro (σ'_1 = 1000 to 2400 kPa and I_L = 0.2 to 0.5) which show a constant K_o value with time, all other observations indicate an increase of K_o with time in normally or slightly overconsolidated clays. Mesri and Hayat (1993) compiled $\Delta K_o/\Delta \log t$ values measured during secondary consolidation and found that they range between 0.003 and 0.05.

Table 2 - Variation of K_o with time during secondary consolidation in normally consolidated clays [from Tavenas and Leroueil (1987)]

Clay tested	w_L (%)	I_p (%)	K_{op}	$\Delta K_o/\Delta$ lgt	Reference
Bangkok	150	90	0.5±	0.040	Bjerrum & Holmberg, 1971
San francisco	—	—	0.53	0.035	Kavazanjian & Mitchell, 1984
Panigaglia	65	40	0.59	−0.0046	Jamiolkowski et al., 1985
Soil A, undisturbed	138	78	—	0.009	Jamiolkowski et al., 1985
Soil A, remoulded	84	30	—	0.015	Jamiolkowski et al., 1985
Soil B, undisturbed	66	32	—	0.008	Jamiolkowski et al., 1985
Soil B, remoulded	50	17	—	0.005	Jamiolkowski et al., 1985
Montalto di Castro	75±6	55±5	0.56	0	Holtz et al., 1986
Saint-Alban	31-42	20-23	0.49	0.020	Mesri & Castro, 1987
Broadback	28-36	≈10	0.51	0.045	Mesri & Castro, 1987
Atchafalaya	82	49	0.66	0.038	Mesri & Castro, 1987
Batiscan	49	27	0.55	0.070	Mesri & Castro, 1987

2.2.8 - Residual strength envelope

Residual strength conditions apply only on surfaces where large displacements have occurred so that the particles are oriented in the direction of shearing. The residual strength envelope is characterized by a friction angle ϕ'_r and a cohesion equal to zero. Petley (1966), Kenney (1967), Lupini (1980) and Lemos (1986) studied the effect of the rate of shearing on the residual friction angle of clays. As indicated by the results shown in Fig. 22, the effect of rate of shearing is small, generally smaller that 3% per logarithm cycle of rate.

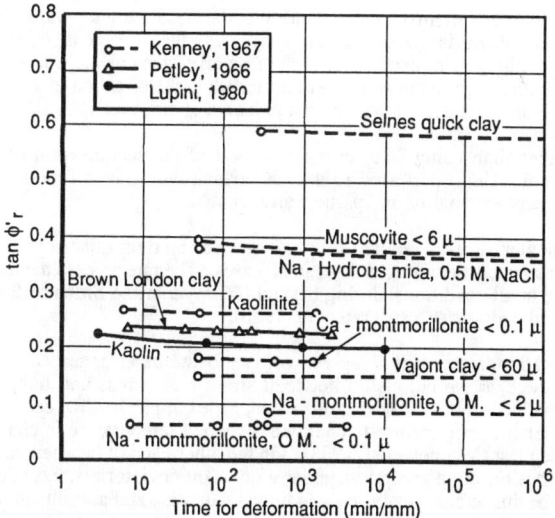

Fig. 22 - Influence of rate of shearing on residual strength

Because at large displacements, the arrangement of particles is stable, the rate of displacement under a given stress level is essentially constant and independent of time. It is a function of the soil and of the stress level only (Vulliet 1986).

For sands, Hungr and Morgenstern (1984) did not find any rate effect.

Considering the observations made for the strength envelope of the normally consolidated soil (Section 2.2.6) and the strain rate effect on ϕ'_r previously mentioned, it is thought that temperature should not have a significant effect on the residual strength envelope of soils.

2.2.9 - Soil deformation as a rate process

The rate process theory is based on the fact that molecules (or atoms or particles, and generally termed flow units) move from one equilibrium position to another when an energy barrier, termed activation energy, is exceeded. At a constant temperature and with no stress applied, energy barriers are crossed equally in all directions, and no movement is observed at a macroscopic level. On the other hand, if thermal energy is provided or if the height of barriers is reduced by the application of a stress, deformations can be observed. This theory was introduced in soil mechanics in the late fifties and sixties to explain the viscous behavior of soils (Murayama and Shibata 1961; Mitchell 1964; Wu et al. 1966; Mitchell et al. 1968). The rate process theory has also been synthetized by Mitchell (1993) and will not be repeated here. It is, however, important to remember its fundamental equation giving the rate of strain $\dot{\varepsilon}$ under an applied stress σl (Murayama and Shibata 1961; Mitchell 1964, 1993):

$$\dot{\varepsilon} = 2 n\lambda \frac{kT}{h} \exp\left(-\frac{\Delta F}{RT}\right) \sin h \left(\frac{\lambda}{2kT} \frac{\sigma l}{S}\right) \quad (16)$$

in which k is the Boltzmann's constant, T is the absolute temperature (°K), h is the Plank's constant, ΔF is the activation energy, λ is the average distance between two balanced positions of molecules, R is the universal gas constant, n is the number of molecules with activation in series per unit length in the direction of stress, and S is the number of molecules with activation in a unit area of cross section.

Rather than using "nλ" in the first part of the second term of the equation, Mitchell et al. (1968) preferred to use "X" which would be a function of "nλ", and could be dependent on time and particle arrangement.

Whether stresses applied to a soil are isotropic or anisotropic, shear stresses exist at the contacts between particles in all cases. The rate process theory thus remains appropriate in all conditions, during triaxial (Murayama and Shibata 1961; Mitchell et al. 1968) and oedometer tests (Wu et al. 1966).

Eq. 16 indicates that at a given void ratio, strain, arrangement of particles and structure, there is a relationship between strain rate, stress and temperature. This coincides with laboratory observations when, for example, in CRS oedometer tests, the strain rate or the temperature is changed, or when temperature is changed during a triaxial creep test (Mitchell et al. 1968). On the other hand, it is observed that the strain rate decreases during a creep test under a constant deviatoric stress (Eq. 11) ; this is thought to be due to changes in soil conditions with time and accumulation of strain.

Previous observations on behavior of soils call for some remarks regarding the amplitude of viscous effects: in "frictional processes" such as secondary consolidation in sands, drained shearing of sands, drained shearing of normally consolidated clays, shearing in residual conditions, strain rate effects are very small, in the order of 4% per logarithm cycle of strain rate or less; on the other hand, when considering the peak strength envelope of natural clays in their overconsolidated domain, or their preconsolidation pressure, parameters which certainly involve a cohesive component, strain rate effects are in the order of 10% per logarithm cycle of strain rate. They are more important in organic materials.

3 - Practical Implications for Laboratory Testing

3.1 - *Oedometer tests*

The implications of the effects of strain rate and temperature on oedometer test results have been described in detail elsewhere (Leroueil et al. 1985b; Leroueil 1988, 1994-96), and only the main aspects will be briefly presented here.

Assuming that primary consolidation is completed before 24 hours in conventional multiple stage loading oedometer test, the strain rate at the end of the loading periods, which can thus be associated with the 24-hrs compression curve, is given by the following equation (Leroueil 1988):

$$\dot{\varepsilon}_{24h}(s^{-1}) = 2 \times 10^{-7} \frac{C_c}{1 + e_o} \tag{17}$$

This strain rate is thus on the order of 5×10^{-8} s^{-1} for low compressibility clays and on the order of 10^{-7} s^{-1} for highly compressible clays. This is much smaller than the strain rates generally used in CRS tests which are usually between 1.0 and 4.0×10^{-6} s^{-1}. As a result, the preconsolidation pressure and the effective stress at any strain or void ratio measured in CRS tests are larger than those measured in conventional 24-hrs tests. A compilation made by Leroueil (1994-96) for a large variety of clays from six different countries indicates a ratio σ'_p CRS/σ'_p conv. typically equal to 1.25.

For multiple stage loading oedometer tests with reloading at the end-of-primary consolidation, the strain rate at the end of the loading periods varies depending on the characteristics of the tested clay and on the height of the tested specimen. As a consequence, the ratio σ'_p CRS/σ'_p EOP generally varies from clay to clay.

Another important implication is that, as shown in Fig. 5, the effective stress-strain curves followed by sub-elements in a consolidating clay layer are different from the compression curve generally defined by joining the points obtained at the end of the successive loading steps.

3.2 - *Shear tests*

3.2.1 - Consolidation and subsequent undrained compression

The influence of isothermal viscous effects on consolidation and subsequent shear testing is shown schematically in Fig. 23 for soil specimens consolidated at point A, in the normally consolidated domain (Fig. 23a), and subjected to undrained compression. If the specimen is tested immediately after reaching the end-of-primary consolidation, the stress path is such as AD in Fig. 23a and corresponds to A'D' in Fig. 23b, with D' on the critical state line of the soil. If, for a second specimen, the drainage valve is closed at the end of primary consolidation and the soil left under constant total stresses during some time, pore pressures develop (there is creep as shown in Fig. 19) and the effective stress moves from A to B. During shearing, the specimen first exhibits an overconsolidated behavior with an initial stress path at a constant mean effective stress p' (assuming a perfectly elastic behavior), and then moves towards D. Since the void ratio is the same in these two tests, the shear strength is the same. If, for a third specimen, drainage is allowed during some time after the end-of-primary consolidation, the void ratio decreases and the soil conditions progressively move from A' to C' in Fig. 23b. The new limit state curve of the soil is then RN (Fig. 23a) and its projection in the e-p' diagram is R'N'. At point C', the soil is overconsolidated and behaves as such during the initial part of the undrained shear test (p' ≈ constant); then, it progressively goes to failure, at points E (Fig. 23a) and E'

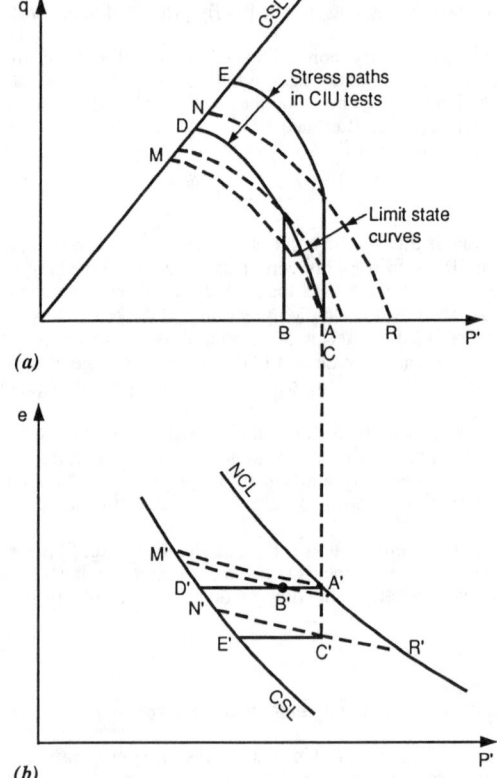

Fig. 23 - Effect of time on consolidation and subsequent undrained compression

(Fig. 23b) corresponding to its void ratio. This was demonstrated by Shen et al. (1973).

As the normally consolidated line is strain-rate dependent, the meaning of "normally consolidated soil" depends on the strain rate existing at the end-of-primary consolidation. However, for durations of consolidation from one to several days and rates of shear between 0.2 and 2% per hour, clay specimens subjected to consolidated undrained triaxial compression tests show a "normally consolidated behavior", with a stress path during shear similar in shape to AD in Fig. 23.

In order to avoid development of significant pore pressures after closing drainage, and in order to maintain tests of reasonable duration and limit the strength increase, Ladd (1991) recommends a time for consolidation prior to shear equal to 10 times the time necessary to reach the end-of-primary consolidation. Compared to the situation at the end-of-primary, the soil then has an overconsolidation ratio approximately equal to 1.1.

When discussing the influence of temperature on shear behavior of soils, numerous factors have to be considered: the preparation temperature when the soil is reconstituted; the consolidation temperature since it influences the void ratio at the beginning of the test; the temperature at which the soil specimen is sheared; the type of test performed, either drained or undrained; etc.

The influence of temperature on consolidation and subsequent shear testing is shown schematically in Fig. 24 for the simple case of heating a normally consolidated specimen at the end-of-primary consolidation. The specimens are first consolidated to point A, at temperature T_1. At the end-of-primary consolidation, they are thus at point A', on the normally consolidated line NCL_1 corresponding to T_1 (Fig. 24b). A specimen sheared in undrained conditions at this temperature reaches failure at point D (Fig. 24a), corresponding to point D' on the critical state line CSL_1 associated with temperature T_1 (Fig. 24b). If, for a second specimen, the valve is closed and the temperature increased to T_2, the pore pressure increases and, due to the difference in thermal expansion of water and soil particules, the void ratio increases slightly. The soil then goes to point B' on the normally consolidated line NCL_2 corresponding to the new temperature (Fig. 24b). During the shear test, the soil progressively goes to point E', on the critical state line CSL_2 corresponding to temperature T_2 (Fig. 24b). In the stress diagram (Fig. 24a), the corresponding point E is on the same strength envelope as the first specimen since temperature has no influence on the friction angle of the normally consolidated soil, but at a lower undrained shear strength. For a third specimen heated in drained conditions to T_2, there is a decrease in void ratio from A' to C' on the normally consolidated line NCL_2 corresponding to the new temperature (Fig. 24b). During undrained compression test performed at T_2 the soil goes to F' on the critical state line CSL_2 corresponding to temperature T_2.

Sherif and Burrous (1969) report test results in which specimens were consolidated at room temperature and then sheared in undrained, unconfined compression at the same or higher temperature. The situation was thus the same as that of the two first specimens previously considered (Fig. 24). The results presented in Fig. 25 demonstrate the influence of water content or void ratio; they also confirm that, at a given water content, the higher the shear temperature, the smaller is the undrained shear strength (strength at point E versus strength at point D in Fig. 24). Obviously, the excess pore pressure at failure has 2 components: one associated with the change in temperature before shearing and the other due to pore pressure generated during shearing. It is, however, worth noting again that, given an increase in temperature, the soil has a tendency to swell when its overconsolidation ratio is high (Section 2.2.1 and Fig. 12), so that a behavior opposite to that shown in Figs 24 and 25 could be expected in highly overconsolidated clays.

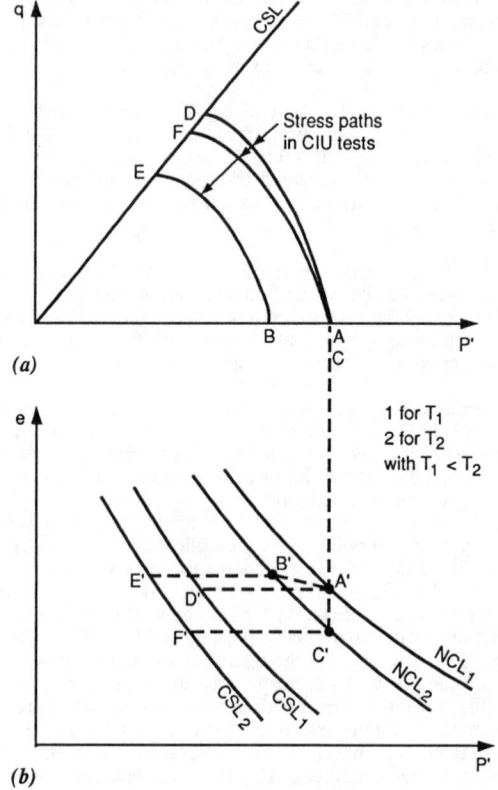

Fig. 24 - Effect of temperature change on consolidation and subsequent undrained compression

3.2.2 - Tests at different strain rates

Tests in the normally consolidated domain

Because the stress path is imposed in drained triaxial tests and, as indicated in section 2.2.6, the strength envelope in the normally consolidated range is almost uninfluenced by strain rate, the strength measured in such tests is essentially constant and independent of strain rate. On the other hand, due to creep, strains at failure can increase when strain rate decreases.

If undrained triaxial tests are performed in the normally consolidated domain, the behavior is as shown in Figs.16 and 26. With decreasing strain rate, the generated pore pressure increases, and the strength envelope is reached at a smaller mean effective stress value. As a consequence, the measured strength decreases with strain rate.

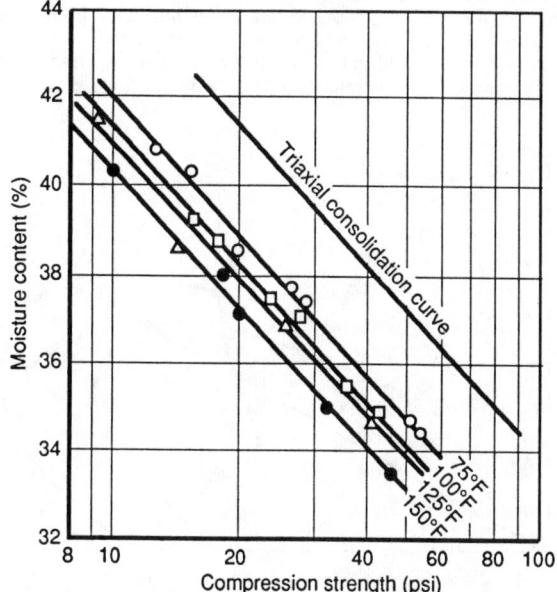

Fig. 25 - Effect of temperature on the undrained strength of kaolinite in unconfined compression [from Sherif and Burrous (1969)]

Tests in the overconsolidated domain

In the overconsolidated domain, the analysis becomes uncertain due to a rate effect on the strength envelope which apparently depends on the soil tested (see Figs 13 and 16).

When, in the overconsolidated domain, the strength envelope is strain-rate dependent, the drained shear strength is also strain-rate dependent. Lo and Morin (1972) provide a good example of such behavior (see Fig. 13). On the other hand, in soils like the resedimented Boston Blue clay, the drained shear strength would be essentially strain rate independent.

In undrained tests, the shear strength is influenced by the pore pressure generated, the intensity of which depends on strain rate, and by the possible lowering of the strength envelope as the strain rate decreases (Fig. 26). As shown, in particular by Berre and Bjerrum (1973), Vaid et al. (1979), Graham et al. (1983), Lefebvre and Leboeuf (1987) and Hanzawa et al. (1990), these effects result in a decrease in undrained shear strength of natural clays with decreasing strain rate. Graham et al. (1983) found a decrease in undrained shear strength in triaxial compression tests of 5 to 20%, with an average of about 12% per logarithm cycle of strain rate. Compiling data obtained from 26 different overconsolidated and normally consolidated clays, Kulhawy and Mayne (1990) found an average decrease typically equal to 10% (Fig. 27). This corresponds well with the values given in Table 1. Graham et al. (1983) found similar variations of the undrained shear strength in triaxial extension and direct simple shear tests.

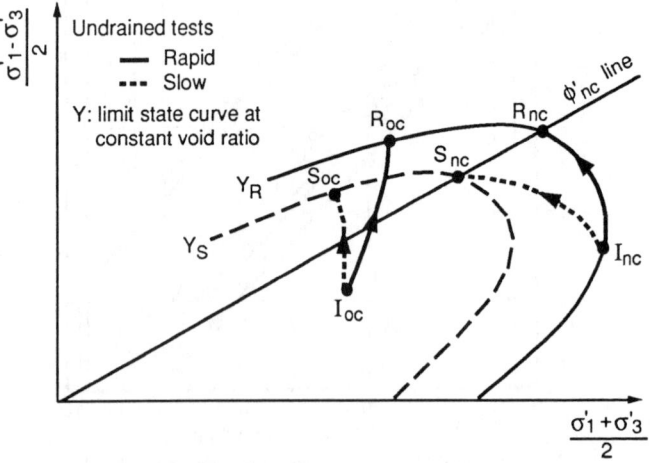

Fig. 26 - Undrained compression tests - Schematic behavior

From the studies performed by Hight (1983), Hight et al. (1987) and Sheahan et al. (1996) on reconstituted Lower Cromer till and Boston Blue clay, the strain rate effect on the compression undrained shear strength considerably decreases when the overconsolidation ratio increases. It is possible that Fig. 27 does not reflect the change in undrained shear strength with strain rate of overconsolidated and weakly structured clays.

Techniques for using a single specimen to study the effect of strain rate on the undrained shear strength have been proposed by Richardson and Whitman (1963) and Kenney (1966). In the first technique, the strain rate is step-changed during the test as shown in Fig. 28. In the second one, the test is stopped at different strains during the test but, due to the energy stored in the test equipment and particularly in the load ring, the specimen continues to strain at a decreasing rate; the relation between strain rate and deviatoric stress can then be evaluated. Graham et al. (1983) found that both techniques closely agree.

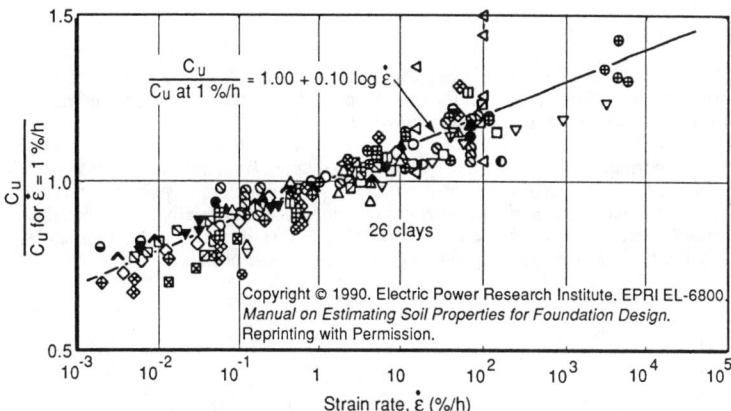

Fig. 27 - Influence of strain rate on the undrained shear strength measured in triaxial compression [from Kulhawy and Mayne (1990)]

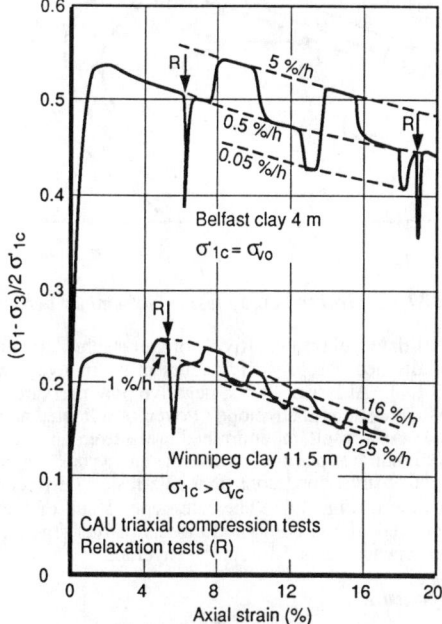

Fig. 28 - Stress-strain curves for triaxial compression tests with step-changed strain rates and relaxation procedures [after Graham et al. (1983)]

3.2.3 - Creep tests

Undrained creep

The behavior described in Section 2 can also help clarify the difference between undrained creep processes in the overconsolidated range and in the normally consolidated range (Leroueil and Tavenas 1979).

In normally consolidated clays, undrained creep from an initial stress condition I_{nc} (Fig. 29) is associated with the simultaneous development of pore pressure and of strain at a decreasing rate. As the strength envelope of the normally consolidated soil is essentially strain rate-independent (Section 2.2.6), failure at a point such as F_{nc} is simply the result of the time-dependent generation of pore pressure.

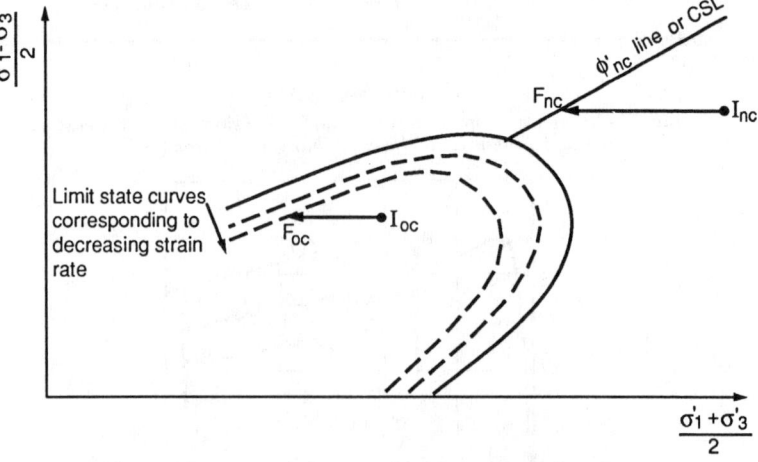

Fig. 29 - Undrained creep tests - Schematic behavior

In overconsolidated clays, positive pore pressures are not systematically generated during undrained creep. This occurs only for overconsolidation ratios (OCR) lower than 2 to 4. At larger OCRs, negative pore pressures can be generated (Larsson 1977). Also, the strength envelope of overconsolidated natural clays is often strain-rate dependent. As a result, an undrained creep from an initial stress condition I_{oc} produces a delayed failure at point F_{oc} where the creep pore pressure accumulated at time t_f results in effective stress conditions on the limit state curve corresponding to the creep strain rate at time t_f (Fig. 29). The rate-dependence of the undrained shear strength is then produced by the combined effects of the generation of creep pore pressure and the rate-dependence of the strength envelope.

Effect of temperature

The effect of a change in temperature during triaxial creep tests can be illustrated by the tests performed on saturated specimens of remolded illite and reported by

Mitchell et al. (1968). Partial results obtained on two specimens are presented in Fig. 30. After some adjustments during approximately one hundred minutes, the two specimens crept around 1000 min under essentially the same effective stresses (σ'_3 = 150 kPa, σ'_1 = 248 kPa) and temperature (19°C). During that period, specimen (1) was in undrained conditions (valve closed) while specimen (2)'s valve remained open. The temperature was then increased to 43°C. Fig. 30b shows that the increase in temperature is associated, for both drained and undrained tests, not only with an increase in axial strain but also in strain rate. According to Mitchell et al. (1968), this conforms with the rate-process theory. The difference in behavior between the two specimens stems from the fact that the drained specimen slightly decreased in volume and that the undrained specimen generated excess pore pressure (Fig. 30a).

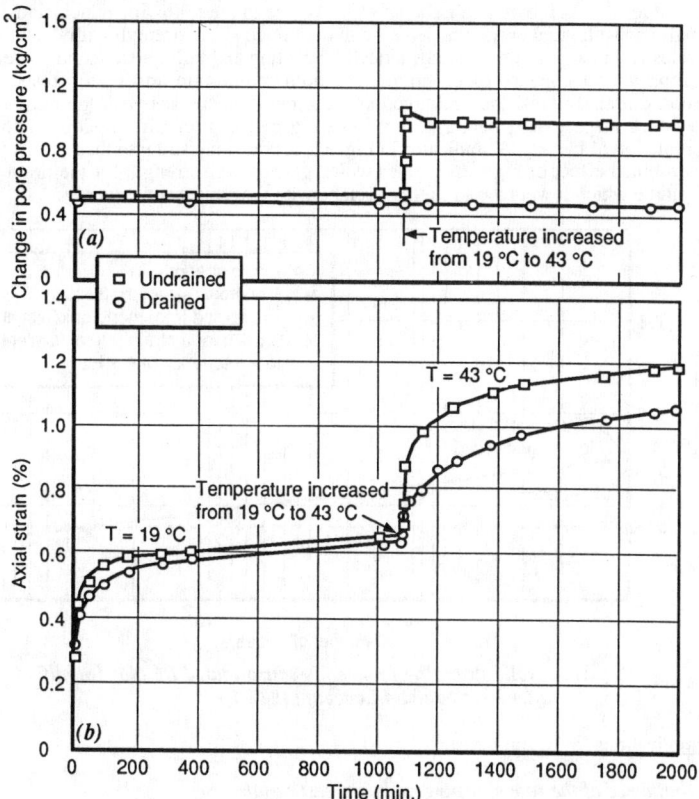

Fig. 30 - The effect of temperature change on the creep behavior of saturated remolded specimens of illite tested under drained and undrained conditions [after Mitchell et al. (1968)]

3.2.4 - Behavior of clays during cyclic loading

From monotonic and cyclic tests performed on two Japanese clays, Ishihara (1985) concluded that the dynamic failure envelope is above the monotonic one and attributed this behavior to the viscous nature of cohesive soils. Konrad and Wagg (1993) observed a similar behavior on a clayey silt, and showed that, when at the same strain rate, both monotonic and cyclic compression tests give the same strength envelope. Lefebvre and Leboeuf (1987) pointed out that at frequencies representative of seismic loading, the strain rates are much higher than those in usual monotonic tests. As a consequence, the strength mobilized in the first cycle is significantly higher than the one obtained in a monotonic test. However, it appears that the undrained shear strength of clays decreases with the number of cycles due to fatigue phenomena. From their test results, Lefebvre and Leboeuf (1987) came to the following conclusions: for normally consolidated clays, the decrease in undrained shear strength with the number of cycles is due to pore pressures generated by cycling and failure occurs on a strength envelope which appears to be unique for both monotonic and cyclic tests; for overconsolidated clays, the undrained shear strength decreases with the number of cycles to a large extent because of a lowering of the strength envelope with the accumulation of cycles. As indicated in Fig. 31, the dynamic strength thus results from the combined effect of high strain rates which gives a high strength for the first cycle and fatigue which lowers the strength envelope with the number of cycles.

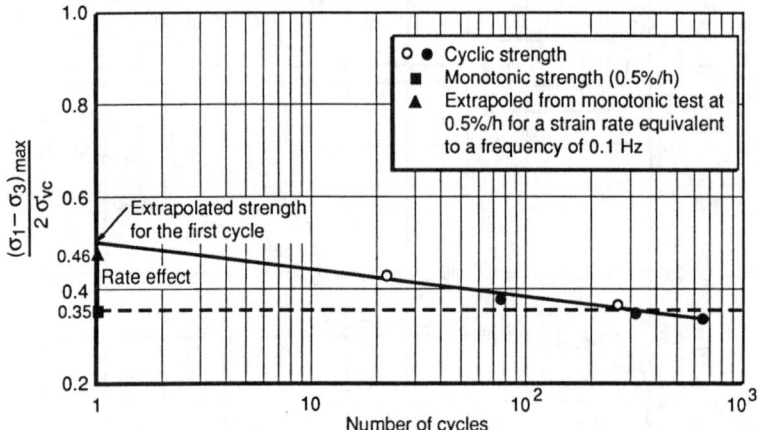

Fig. 31 - Cyclic strength of normally consolidated B6 clay [modified after Lefebvre and Leboeuf (1987)]

4 - Implications for in situ testing

4.1 - *Influence of the rate of testing on in situ test results*

4.1.1 - Generalities

According to what has been previously shown, the strength measured with the vane test or derived from the pressuremeter test, or the "strength parameters" such as the tip resistance measured with the piezocone should be influenced by the rate of testing. Indeed, these strengths or "strength parameters" generally increase as the rate of testing increases, as indicated by the rheological curve shown in Fig. 32. However, when the coefficient of consolidation of the soil in which the apparatus is inserted, the geometry of this apparatus and the rate of testing are such that partial consolidation occurs, the strength or "strength parameter" versus rate of testing relationship may diverge from the rheological curve, as indicated in Fig. 32. Chandler (1987) considers that, for practical purposes, undrained conditions can be assumed to apply if the degree of consolidation is lower than 10%. Although Chandler's conclusion was for the vane test, it could reasonably be extended to other in situ tests.

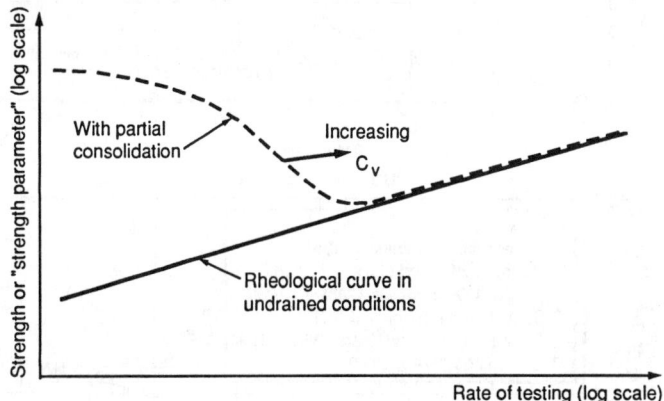

Fig. 32 - Combined effect of rheology and partial consolidation on the strength or "strength parameter" versus rate of testing relationship

The influence of the rate of testing will be examined for the vane test, the penetrometer test and the pressuremeter test in the following paragraphs. The rate of testing is not a strain rate since the latter varies with the radius of the blades and the thickness of the shear band for the vane, and with the distance of the considered soil element to the probe axis, normalized with respect to the probe radius, for the penetrometer and the pressuremeter. However, both rate of testing and strain rates are simply related, as shown in particular for the penetrometer by Ladanyi (1982). Experimental results from the literature are plotted in Figs 33a, b and c, respectively. The lines showing a decrease of 10% of strength or strength parameter considered per logarithm cycle of the rate of testing are also shown on the figures as a reference.

4.1.2 - Field vane test

Wiesel (1973), Torstensson (1977) and Roy and Leblanc (1987) studied the effect of rate of vane rotation in plastic Swedish clays and low plasticity eastern Canadian clays. The vane shear strengths obtained, after normalization at a rate of rotation of 0.2°/s, are presented in Fig. 33a. The shear strength decreases by 10% per log cycle of the rate of rotation for Swedish clays. On the other hand, the tests on St-Louis de Bonsecours and St-Alban clays show an increase in strength for rates of rotation slower than 0.1 °/s.

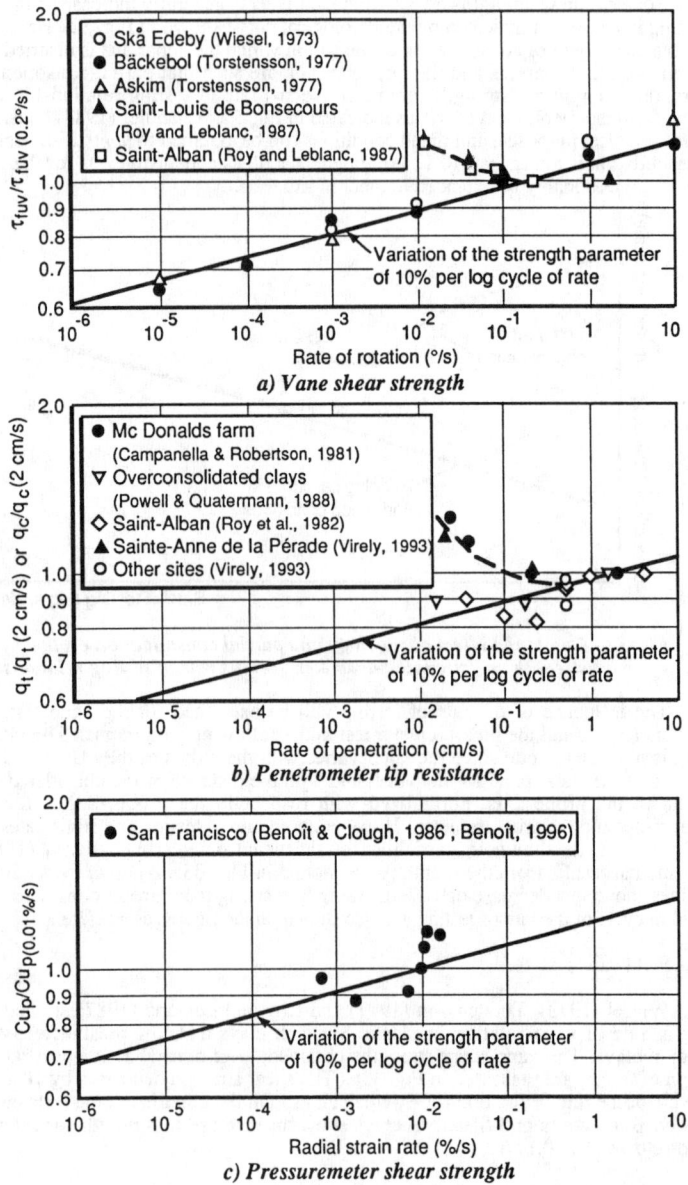

Fig. 33 - Effect of the rate of testing on "strength parameters"

Blight (1968) developed an approximate theory for evaluating partial consolidation and its effects on the measured strength during vane testing. According to this theory, and considering that undrained conditions can be assumed when the degree of consolidation is lower than 10%, the time to failure t_f should be:

$$t_f < \frac{0.05 \, D^2}{c_v} \qquad (18)$$

in which D is the vane diameter and c_v is the coefficient of consolidation of the clay.

According to Eq. 18, a rate of 0.1°/s for a deviation from undrained conditions would correspond to a coefficient of consolidation in the order of 5×10^{-6} m^2/s, which is reasonable for the eastern Canadian clays considered in Fig. 33a.

4.1.3 - Cone penetrometer or piezocone

Figure 33b shows the variation of the tip resistance measured during penetrometer or piezocone testing with the rate of penetration. In most cases, it follows the average rheological law for clays. As for the clayey silt from McDonalds farm (Campanella and Robertson 1981) and the stratified clay deposit of Ste-Anne de la Pérade (Virely 1993), however, there is an increase in tip resistance with rates smaller than about 0.5 cm/s which indicates partial drainage of the soil around the probe in these conditions.

It is worth noting that sleeve friction, resulting from the difference between total radial stress and pore pressure, appears to be much more sensitive to the rate of penetration than tip resistance or pore pressure (Campanella and Robertson 1981; Virely 1993).

It is also worth noting that Ladanyi (1982; 1985) uses the cone penetrometer for the determination of creep parameters of frozen soils.

4.1.4 - Self-boring pressuremeter

Benoît and Clough (1986) performed self-boring pressuremeter tests at different expansion rates in San Francisco Bay mud. The test results, reported as a function of average radial strain rate (Benoît, 1996) in Fig. 33c, show a significant scatter, but indicate a general increase in pressuremeter shear strength with expansion rate in clay.

Nutt and Houlsby (1995) performed pressuremeter tests in carbonate sand from the west coast of Ireland. Some were controlled rate-of-strain tests, with the rate being changed at different cavity strains. The results, shown in Fig. 34, clearly demonstrate the effect of strain rate. The increase in pressure for a tenfold increase in rate is of 3 to 4%, which is an expected value for this type of material (Table 1). Finally, it can be of interest to mention that Leidwanger et al. (1994) proposed the evaluation of creep characteristics of soils with a special pressuremeter in which a constant pressure can be applied during 2 months or more, and the volume change of the cell is measured.

4.2 - *Interpretation of self-boring pressuremeter tests in clays*

As shown by numerous authors, especially Lacasse et al. (1981), the undrained shear strength deduced from self-boring pressuremeter tests is higher than those obtained through more conventional tests such as the field vane test or the triaxial test. To examine that point, Hamouche (1995) performed self-boring pressuremeter tests in three clay deposits with overconsolidation ratios of 1.3 (Berthierville), 3 (Louiseville) and 5 (Mascouche). On Louiseville clay, a detailed study was also performed with a "true triaxial" cell, and a 3-D limit state surface was defined for the overconsolidated material (Boudali 1995). With the assumption that the soil is linear-elastic and undrained during the early stages of the test and that strains are constrained to a horizontal plane, it can be shown that the stress path then followed is at σ'_v = constant = σ'_{vo} in a p' = constant plane, with the mean effective stress p' equal to the mean effective stress in the soil before testing, p'_o. Based on these assumptions, the stress path followed during the pressuremeter test performed at Louiseville at a depth of 5.8 m is shown in Fig. 35. The stress path goes from the initial stress conditions I to the yield conditions P at which point the undrained shear strength derived from the pressuremeter test results was reached. Also shown in the figure is the limit state curve C_1 deduced from the "true triaxial" tests, for a mean effective stress equal to p'_o. It can be seen that P is quite different from the stress state R_1 corresponding to the intersection of the stress path followed during the in situ test and the laboratory limit state curve. However, during the self-boring pressuremeter tests, the strain rate up to failure was typically 10^{-4} s^{-1} and the temperature approximately equal to 10°C, whereas the laboratory tests were performed at a strain rate of 3×10^{-7} s^{-1} and a temperature of 20°C. When, according to the viscous behavior presented in Section 2.2.2, the limit state curve C_1 is corrected for a strain rate of 10^{-4} s^{-1} and then for a temperature of 10°C, it moves to the limit state curves C_2 and then C_3 in Fig. 35. The intersection of the stress path followed during the pressuremeter test with the yield curve C_3 corresponding to the same strain rate and temperature, i.e. R_3, then becomes close to the yield conditions P deduced from the in situ test. Viscosity of clay thus seems to be an important reason why the pressuremeter gives high undrained shear strength values.

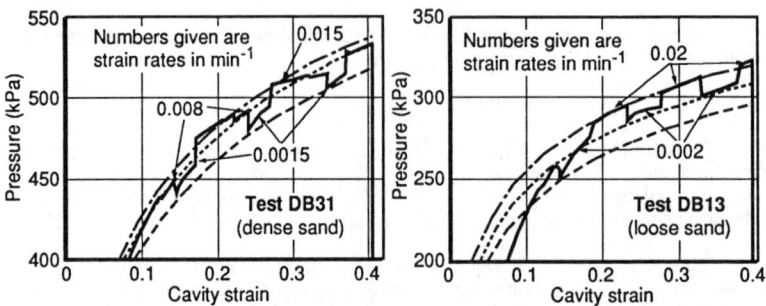

Fig. 34 - Controlled rate of strain pressuremeter tests performed on loose and dense Dogs Bay carbonate sand [after Nutt and Houlsby (1995)]

5 - Practical Problems

The viscous nature of soils has implications in numerous practical problems. Six particular cases are briefly described in the following paragraphs, but others such as piles subjected to sustained load, soil masses behind retaining walls, post-construction deformations of rockfill dams (Clements 1984; Dascal 1987) etc., could also have been considered.

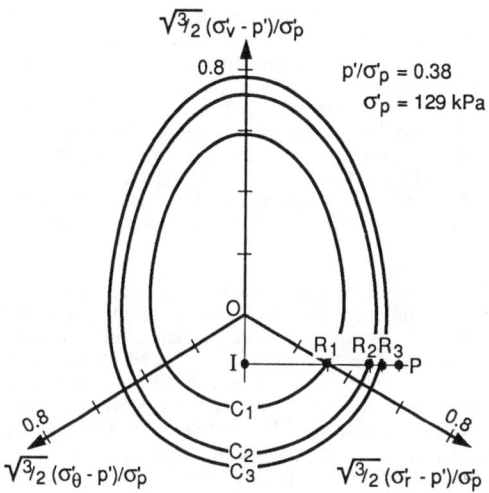

C_1: Laboratory limit state ($\dot{\varepsilon} = 3 \times 10^{-7}$ s^{-1} ; T = 20 °C)
C_2: Limit state curve corrected for strain rate ($\dot{\varepsilon} = 10^{-4}$ s^{-1} ; T = 20 °C)
C_3: Limit state curve at strain rate and temperature ($\dot{\varepsilon} = 10^{-4}$ s^{-1} ; T = 10 °C) corresponding to in situ test conditions
I : Initial stress conditions
P : Yielding during the pressuremeter test

Fig. 35 - Yielding of Louiseville clay (5.8 m), as deduced from in situ pressuremeter test and from laboratory tests performed in a "true triaxial" apparatus [from Hamouche (1995)]

5.1 - Embankments on soft clays

Evidence of the viscous nature of clays is given by the evolution of pore pressures under an embankment just at the end of construction. Crooks et al. (1984) and Kabbaj et al. (1988) reported at least 13 cases of embankments on soft clays (not all sensitive) in which an increase in pore pressure, and thus a decrease in vertical effective stress, was observed after the end of construction. This is attributed to the fact that, in these cases, the soil's tendency to creep was generating more pore pressures (as in undrained creep tests) than the clay mass was able to dissipate with its geometry, permeability and compressibility. The case of the Olga-C test embankment reported by St-Arnaud et al. (1992) and Leroueil (1994-96) is also of interest. This embankment presents four different sections: three with vertical wick drains and one without drains. In the sections with drains where the drainage length is thus relatively small, the pore pressures started decreasing just after the end of construction. In the section without drains where the drainage length is longer, the clay mass was not able to expell the excess pore pressures generated by creep, and the pore pressures continue to increase during 15 to 50 days after construction before starting to decrease. The behavior of the clay foundation of the Accomodation Building at Gloucester, Canada, is also of interest (Crawford and Bozozuk 1990). The building was built with its floor at an elevation 1.4 m higher than the original ground surface, which exceeded the overconsolidation of the clay and has generated long term settlements in the order of 30 cm. The vertical effective stress in the clay foundation reached the preconsolidation pressure during or shortly after construction, and stayed at this value, which is smaller than the expected final effective stress, during the thirty following years or so. During that time, the compression of the clay layer increased to 7.8%. Obviously, there was secondary compression under constant effective stress while excess pore pressures were still existing.

Generally, in the case of embankments on soft clay deposits, the strain rate in situ is smaller than that used for laboratory tests. The temperature is also smaller. The two effects thus compensate to some extent, which is different from the case of the self-boring pressuremeter (Section 4.2, Fig. 35) in which the two effects are added. The analysis made by Leroueil (1994-96) shows, however, that the strain-rate effect remains dominant where settlement is concerned. Considering a temperature difference between the laboratory and in situ of 12°C, Leroueil (1994-96) evaluates that the strain $\Delta\varepsilon_s^*$, which should be added to the strain estimated on the basis of conventional 24-hrs oedometer tests for evaluating the in situ settlement at the end of primary consolidation is:

$$\Delta\varepsilon_s^* = \frac{C_c}{m'(1+e_o)} \left[\log 6 \times 10^{-9} - \log(\dot{\varepsilon}_{EOP})\right] \qquad (19)$$

in which $\dot{\varepsilon}_{EOP}$ equals $(0.16 \text{ k } u'_o/\gamma_w H^2)$; C_c is the compression index under the final effective stress; e_o is the initial void ratio; k is the hydraulic conductivity of the clay; m' is a strain rate factor[6] taken equal to 32 by Leroueil (1988 and 1994-96); u'_o is the initial excess pore pressure; and H is the maximum drainage length.

Evidence of viscous effects on in situ settlements have been given by Larsson (1987), Kabbaj et al. (1988), Leroueil (1988) and Magnan (1992). Practical consequences have been recently described by Leroueil (1994-96) and will thus not be repeated here.

5.2 - *Foundations*

As in the case of embankments, foundations on clay deposits experience secondary settlements, even if the applied effective stress is smaller than the preconsolidation pressure determined in conventional 24-hrs oedometer tests (Bjerrum 1967).

Secondary settlements of foundations have also been observed on sand (Bolenski 1973 [ref. by Burland and Burbidge 1985]; Burland and Burbidge 1985; Amar et al. 1994). However, this occurs at a slower rate than in clays since $C_{\alpha e}/C_c$ is smaller (or m' larger, Table 1). According to Burland and Burbidge (1985), the long term settlement S_t of a foundation can be estimated as follows:

---

[6] Leroueil (1988 and 1994-96) took a m' value of 32, but it would be preferable to choose it as a function of soil type and according to Table 1.

$$S_t = S_o (1.3 + m_s \log(t/t_o)) \tag{20}$$

where S_o is the settlement at completion of construction and $m_s = 0.2$ for buildings. Amar et al. (1994) found larger values of m_s, but for loads close to 50% of the load at failure; Burland and Burbidge (1985) report m_s values as high as 0.8 for chimneys subjected to the fluctuating action of wind.

5.3 - *Preloading of clay deposits by heating*

As early as 1969, Plum and Esrig showed that heating followed by cooling gives the soil an overconsolidation. As indicated in Section 2.2.1 and in Fig. 3, this has been confirmed for different clays since that time. As shown schematically in Fig. 36, heating could thus be used for preloading soft soil deposits. In order to be more specific, it has been assumed that natural soil temperature is situated at 10°C and that the soil would be heated to 60°C. The corresponding compression curves are shown on the figure. Under a preloading stress σ'_{pL}, the void ratio would be e_{pL} at the end-of-primary consolidation under the natural temperature of 10°C but, by heating to 60°C, it becomes e_{pH}. If heating is stopped after consolidation, the temperature comes back to its natural value of 10°C, and the new preconsolidation pressure of the soil becomes σ'_{pH}. According to Fig. 11, for a temperature decrease from 60°C to 10°C, σ'_{pH} would be approximately 40% higher than the preloading stress σ'_{pL}. The construction load can then be increased to values approaching σ'_{pH} without generating important settlements. Another advantage of preloading by heating is an increase in hydraulic conductivity of the soil and a corresponding decrease in duration of consolidation. For a temperature increase from 10°C to 60°C, the duration of consolidation would be reduced by a factor of about 2.8.

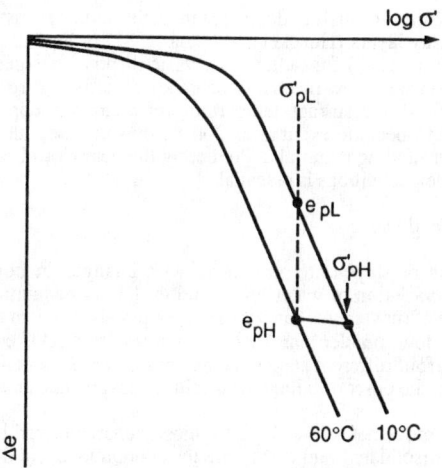

Fig. 36 - Preloading by heating

A site experiment on the effects of temperature changes has been carried out at Fiumicino, Italy (Miliziano 1992). Edil and Fox (1994) have also shown the effectiveness of precompression of peat by heating in both laboratory and in situ conditions. The Ministry of Transports of the Province of Québec is examining the possibility of heating as a means of preloading road embankments in combination with more conventional preloading techniques, in particular, vacuum. A specific site investigation is presently under way at Laval University, and the construction of a test embankment with heating and vacuum is planned for 1997.

5.4 - Energy storage

The possibility of storing energy in soft soil deposits has been considered in different countries, especially in Sweden (Adolfsson and Sällfors 1987; Tidfors and Sällfors 1989; Moritz 1995a and 1995b). The soil would be heated for a given period of time, for example, during the summer with solar energy, and then cooled at another period of time, say, during the winter for heating buildings. It can be imagined from what has been previously said that cyclic changes in temperature of large volumes of soils in essentially laterally-confined conditions will generate modifications in total stresses, pore pressures and void ratios which have to be predicted. A test site experiment is under way at Linköping, Sweden, to examine the behavior of soft clay masses (10 m x 10 m x 10 m) subjected to heating and heating-cooling cycles. Surface settlements observed for 2 test stores are shown in Fig. 37 (Moritz 1995a and 1995b). For store 2, large settlements were induced by heating to an approximatively constant temperature of 70°C. For store 1, subjected to heating-cooling cycles between 35 and 70°C, the general settling tendency is the same. However, it is interesting to note that the clay swells due to thermal expansion when the soil is heated.

5.5 - Radioactive waste disposal

The possibility of using deep ocean sediments (Houston et al., 1985), continental deep clay layers (Hueckel and Peano 1987) or hard rock with a bentonite buffer (Lingnau et al. 1995) for radioactive waste disposal has generated significant research on the thermo-hydro-mechanic behavior of soils and rocks. According to Houston et al. (1985), the surface temperature of a canister containing radioactive waste is expected to become as large as 200°C, thus creating an important thermal gradient in the surrounding materials. Predicting the behavior of the material in such unusual and complex conditions is essential.

5.6 - Movements in slopes

Movements of slopes are associated with changes in boundary conditions (loading, erosion, variation of water level) and to the viscous nature of geotechnical materials. The rate of movements can vary from extremely rapid, in the order of several m/s, to extremely slow, smaller than 10^{-9} m/s (Varnes 1978). Also, when con-sidering slope movements, four different stages can be considered (Leroueil et al. 1996c): pre-failure movements, the onset of failure, post-failure movements, and reactivation.

Pre-failure movements in a slope are mostly characterized by creep of the soil mass in the overconsolidated range; they are thus continuous and do not develop only in a shear band. Such pre-failure movements have been observed in several natural slopes in soft clays from Sweden and Canada (Mitchell and Eden 1972; Tavenas 1984; Moller et al. 1989; Ottosson and Johansson 1995).

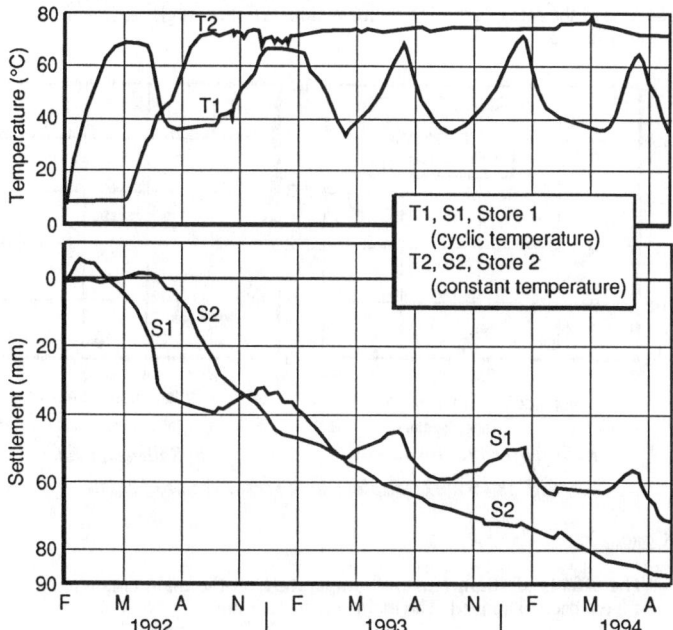

Fig. 37 - Temperature changes and settlement of the ground surface of stores 1 and 2 at Linköping [after Moritz (1995a and 1995b)]

The observations made by D'Elia, Esu and co-workers (see D'Elia 1991, and D'Elia et al. 1996) of the long-term behavior of very deep excavations in the area of the open pit mine of Santa Barbara, Italy, indicate that the general form of Eq. 11 could be valid in such a case. For natural slopes, however, it is difficult to define an origin for time and in addition, there are generally changes in effective stress with the seasons. What is certain is that strain rate increases when effective stress conditions approach the failure envelope because there is an increase in either shear stress or pore pressure (Mitchell and Eden 1972; Tavenas and Leroueil 1981). This is in agreement with the creep model presented in Section 2.2.2.

During the reactivation phase, movements are of rigid body over rigid base type, and the relevant strength parameter is the residual friction angle. Due to the viscous nature of soil, the rate of movement progressively increases when the shear stress approaches the shear strength of the soil. This is evidenced by observations of landslides in Italy and in France made by Bertini et al. (1986) and Cartier and Pouget (1988) respectively (Fig. 38a and b). However, it can be seen on these figures that, for a small change in the factor of safety of about 0.05 for the San Martino Basin and about 0.15 in the case of Sallèdes, the rate of displacement varies from very high values to essentially zero. Such a change in rate of displacement of several orders of magnitude for a small change in stress level of about 10% is in agreement with the small effect of the rate of shearing on the residual friction angle, as indicated in Section 2.2.8 and Fig. 22.

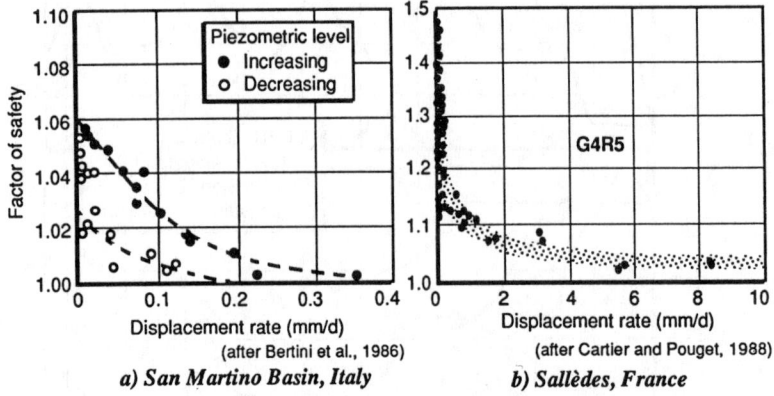

Fig. 38 - Rate of displacement versus factor of safety

6 - Conclusion

The effects of strain rate and temperature on the main characteristics of soil behavior have been examined. The main conclusions are:

- Compressibility of soils can be described by effective stress-strain-strain rate-temperature relationships. In fact, viscous phenomena probably apply only to the plastic component of strain, which explains the change in effective stress observed in undrained creep or relaxation tests.

- Effects of strain rate and temperature on the critical state and residual strength envelopes are small, if any.

- Viscous effects on the limit state curve, the undrained shear strength and the preconsolidation pressure of clays are important, in the order of 10% per logarithm cycle of strain rate or a change in temperature of about 12°C. Studies on reconstituted clayey soils indicate, however, that the strain rate effect on the undrained shear strength decreases when the overconsolidation ratio increases.

- Shear modulus G at small strains is strain-rate dependent, but the viscous effects remain small in cohesive soils (less than 6% per tenfold increase in strain rate), and have a tendency to be smaller in cohesionless soils and to decrease at very small strains (less than 0.001%).

These effects have important implications for the testing of soils, both in laboratory and in situ, and for a variety of practical problems. They thus cannot be ignored in geotechnical engineering practice.

The effects of strain rate and temperature on soil behavior have generally been studied, and also modelled separately. They are, however, and as shown by the rate process theory, intimately linked together and should not be considered separately. The development of numerical models integrating these two factors and other features of natural geomaterials such as anisotropy and structure should be encouraged.

7 - Acknowledgements

The authors highly appreciated the opportunity to put together the too often scattered information on the viscous behavior of soils and thank the organizers of this session for that. They are also indebted to J. Benoît, M. Boudali, K. Hamouche, W. Lacerda, I. Martins, G. Marchand, A. d'Onofrio, D. Perret, P. Santucci de Magistris, and F. Tavenas, with whom they discussed several aspects of the work presented here or who gave permission to present results from their theses. They also acknowledge D. Hight, D. DeGroot and L. Picarelli who kindly reviewed and commented a preliminary version of the paper.

8 - References

Adachi, T., and Oka, F. (1982). "Constitutive equations for normally consolidated clay based on elasto-viscoplasticity." *Soils and Foundations*, 22(4), 57-70.

Adolfsson, K., and Sällfors, G. (1987). "Energy storage in clay." *Swedish Board on Building Research*, Report R23:1987, Stockholm.

Akagi, H., and Komiya, K. (1995). "Constant rate of strain consolidation properties of clayey soil at high temperature." *Int. Symp. on Compression and Consolidation of Clayey Soils - IS-Hiroshima's 95*, Hiroshima, 1, 3-8.

Akai, K., Adachi, T., and Ando, N. (1975). "Existence of a unique stress-strain-time relation of clays." *Soils and Foundations*, 15(1), 1-16.

Amar, S., Baguelin, F., Canépa, Y., and Frank, R. (1994). "Experimental study of the settlement of shallow foundations." *Specialty Conf. on Vertical and Horizontal Deformations of Foundations and Embankments*, ASCE, Settlement '94, College Station, 2, 1602-1610.

Arulanandan, K., Shen, C.K., and Young, R.B. (1971). "Undrained creep behaviour of a coastal organic silty clay." *Géotechnique*, 21(4), 359-375.

Aversa, S., and Evangelista, A. (1993). "Thermal expansion of Neapolitan yellow tuff." *Rock Mechanics and Rock Engineering*, 26(4), 281-306.

Badra-Blanchet, P. (1981). "Effet de la vitesse de déformation sur le comportement de l'argile normalement consolidée." M.Sc. Thesis, Université Laval, Ste-Foy, Canada.

Benoît, J. (1996). Personal communication.

Benoît, J., and Clough, G. W. (1986). "Self-boring pressuremeter tests in soft clay." *J. Geotech. Engrg. Div.*, ASCE, 112(1), 60-78.

Berre, T., and Bjerrum, L. (1973). "Shear strength of normally consolidated clays." *Proc. 8th ICSMFE*, Moscow, 1, 39-49.

Bertini, T., Cugusi, F., D'Elia, B., and Rossi-Doria, M. (1986). "Lenti movimenti di versante nell ' Abruzzo Adriatico: caratteri e criteri di stabilizzazione." *16th Convegno Nazionale di Geotecnica*, Bologna, 1, 91-100.

Bishop, A.W., and Lovenbury, H.T. (1969). "Creep characteristics of two undisturbed clays". *Proc. 7th ICSMFE*, Mexico, 1, 29-37.

Bjerrum, L. (1967). "Engineering geology of normally consolidated marine clays as related to the settlement of buildings." *Géotechnique*, 17(2), 83-119.

Bjerrum, L., Simons, N., and Torblaa, I. (1958). "The effect of time on the shear strength of a soft marine clay." *Conf. on Earth Pressure Problems*, Brussels, 1, 148-158.
Bjerrum, L., and Holmberg, S. (1971). "Preliminary report on long term consolidation tests with measurements of horizontal stresses on the soft Bangkok clay." NGI internal report.
Blight, G.E. (1968). "A note on field vane testing of silty soils." *Canadian Geotechnical J.*, Vol.5(3), 142-149.
Bolton, M.D., and Wilson, M.R. (1989). "An experimental and theoretical comparison between static and dynamic torsional soil tests." *Géotechnique*, 39(4), 585-599.
Boudali, M. (1995). "Comportement tridimensionnel et visqueux des argiles naturelles." Ph.D. Thesis, Université Laval, Ste-Foy, Canada.
Boudali, M., Leroueil, S., and Murthy, B.R.S. (1994). "Viscous behaviour of natural soft clays." *Proc. 13th ICSMFE*, New Delhi, 1, 411-416.
Burghignoli, A., and Desideri, A. (1988). "Influenza della temperatura sulla compressibilità delle argille." *Gruppo Nazionale di Coodinamento per gli Studi di Ingegneria Geotecnica*, Convegno di Monselice, 193-206 (Ref. by Burghignoli et al. 1992).
Burghignoli, A., Desideri, A., and Miliziano, S. (1992). "Deformability of clays under non isothermal conditions." *Rivista Italiana di Geotecnica*, Anno XXVI, N.4, 227-236.
Burland, J.B., and Burbidge, M.C. (1985). "Settlement of foundations on sand and gravel." *Proc. Instn. Civ. Engrs, 78, Part 1, 1325-1381*.
Campanella, R.G., and Mitchell, J.K. (1968). "Influence of temperature variations on soil behaviour." *J. Soil Mech. and Found. Engrg. Div.*, ASCE, 94(3), 709-734.
Campanella, R.G., and Robertson, P.K. (1981). "Applied cone research." *ASCE National Convention. Cone Penetration Testing and Experience*, St.Louis, 342-362.
Cartier, G., and Pouget, P. (1988). "Etude du comportement d'un remblai construit sur un versant instable. Le remblai de Sallèdes (Puy de Dôme)." *Laboratoire Central des Ponts et Chaussées*, Rapport N° 153, Paris.
Chandler, R.J. (1987). "The in situ measurement of the undrained shear strength of clays using the field vane." *Int. Symp. on Laboratory and Field Vane Shear Strength Testing*, ASTM, Tampa, STP 1014, 13-44.
Clements, R.P. (1984). "Post-construction deformations of rockfill dams." *J. Geotech. Engrg. Div.*, ASCE, 110(7), 821-840.
Crawford, C.B. (1964). "Interpretation of the consolidation test." *J. Soil Mech. and Found. Engrg. Div.*, ASCE, 90(5), 87-102.
Crawford, C.B., and Bozozuk, M. (1990). "Thirty years of secondary consolidation in sensitive marine clay."*Canadian Geotechnical J.*, 27(3), 315-319.
Crooks, J.H.A., Becker, D.E., Jefferies, M.G., and McKenzie, K. (1984). "Yield behaviour and consolidation - 1: pore pressure response." *Symp. on Sedimentation Consolidation Models: Predictions and Validation*, ASCE, San Francisco, 356-381.
Dascal, O. (1987). "Postconstruction deformations of rockfill dams." *J. Geotech. Engrg. Div.*, ASCE, 113(1), 46-59.
D'Elia, B. (1991). "Deformation problems in the Italian structurally complex clay soils." *10th European Conf. on Soil Mechanics and Found. Engrg.*, Florence, 4, 1159-1170.
D'Elia, B., Picarelli, L., Leroueil, S., and Vaunat, J. (1996). "Geotechnical characterization of slope movements in structurally complex clay soils and stiff jointed clays." Submitted to *Géotechnique* for publication.

d'Onofrio, A. (1996). "Comportamento meccanico dell'argilla di Vallericca in condizioni lontane dalla rottura." Ph.D. Thesis, Università di Napoli "Federico II", Italy.
De Waal, J. A. (1986). "On the rate type compactation behavior of sandstone reservoir rock." Ph.D. Thesis, Delft Technical University, Delft, The Netherlands.
Despax, D. (1975). "Etude de l'influence de la température sur les propriétés mécaniques des argiles saturées." *Internal Report,* Université Laval, Ste-Foy, Canada.
Diaz-Rodriguez, J.A., Leroueil, S., and Aleman, J.D. (1992). "Yielding of Mexico City clay and other natural clays." *J. Geotech. Engrg. Div.*, ASCE, 118(GT7), 981-995.
Dobry, R., and Vucetic, M. (1987). "Dynamic properties and seismic response of soft clay deposits." *Int. Symp. on Geotechnical Engineering of Soft Soils,* Mexico City, 2, 49-85.
Duffy, S.M., Wheeler, S.J., and Bennell, J.D. (1994). "Shear modulus of kaolin containing methane bubbles." *J. Geotech. Engrg. Div.*, ASCE, 120(GT5), 781-796.
Edil, T.B., and den Haan, E.J. (1994). "Settlement of Peats and Organic Soils." *Specialty Conf. on Vertical and Horizontal Deformations of Foundations and Embankments,* ASCE, Settlement '94, College Station, 2, 1543-1572.
Edil, T.B., and Fox, P.J. (1994). "Field Test of Thermal Precompression." *Specialty Conf. on Vertical and Horizontal Deformations of Foundations and Embankments,* ASCE, Settlement '94, College Station, 2, 1274-1286.
Edil, T.B., Fox, P.J., and Lan, L.T. (1994). "An assessment of one-dimensional peat compression." *Proc. 13th ICSMFE,* New Delhi, 1, 229-232.
Eriksson, L.G. (1989). "Temperature effects on consolidation properties of sulphide clays." *Proc. 12th ICSMFE,* Rio de Janeiro, 3, 2087-2090.
Feijó, R.L., and Martins I.S.M. (1993). "Relação entre compressão secundaria, OCR e K_0." *COPPEGEO' 93. Simpósio Geotécnico Comemorativo dos 30 anos da COPPE - UFRJ,* UFRJ, Rio de Janeiro, 27-40.
Graham, J., Crooks, J.H.A., and Bell, A.L. (1983). "Time effects on the stress-strain behaviour of soft natural clays." *Géotechnique,* 33(3), 327-340.
Hamouche, K.K. (1995). "Comportement des argiles Champlain sollicitées horizontalement." Ph.D. Thesis, Université Laval, Ste-Foy, Canada.
Hanzawa, H., Fukaya, T., and Suzuki, K. (1990). "Evaluation of engineering properties for an Ariake clay". *Soils and Foundations,* 30(4), 11-24.
Hicher, P.Y. (1988). "The viscoplastic behaviour of bentonite." *Int. Conf. on Rheology and Soil Mechanics,* Coventry, 89-107.
Hight, D.W. (1983). "Laboratory investigations of sea bed clays." Ph.D. Thesis, University of London, London, U.K.
Hight, D.W., Jardine, R.J., and Gens, A. (1987). "The behaviour of soft clays." Chapter 2 of *Embankments on Soft Clays, Bulletin of the Public Works Research Center of Greece,* Athens, 33-38.
Holtz, R.D., Jamiolkowski, M.B., and Lancellotta, R. (1986). "Lessons from oedometer tests on high quality samples." *J. Geotech. Engrg. Div.*, ASCE, 112(8), 768-776.
Holzer, T.L., Hoeg, K., and Arulanandan, K. (1973). "Excess pore pressures during undrained clay creep." *Canadian Geotechnical J.*, 10(1), 12-24.
Houston, S.L., Houston, W.N., and Williams, N.D. (1985). "Thermo-mechanical behavior of seafloor sediments." *J. Geotech. Engrg. Div.*, ASCE, 111(11), 1249-1263.

Hueckel, T., and Peano, A. (1987). "Some geotechnical aspects of radioactive waste isolation in continental clays." *Computers and Geotech.*, 3(2, 3), 157-182.

Hueckel, T., and Pellegrini, R. (1989). "Modeling of thermal failure of saturated clays." *Numerical Models in Geomechanics*, Elsevier, New York, 81-90.

Hueckel, T., and Baldi, G. (1990). "Thermoplasticity of saturated clays: Experimental constitutive study." *J. Geotech. Engrg. Div.*, ASCE, 116(12), 1778-1796.

Hungr, O. and Morgenstern, N.R. (1984). "High velocity ring shear tests on sand." *Géotechnique*, 34(3), 415-421.

Imai, G., and Tang, Y. (1992). "A constitutive equation of one-dimensional consolidation derived from inter-connected tests." *Soils and Foundations*, 32(2), 83-96.

Imai, G. (1995). "Analytical examinations of the foundations to formulate consolidation phenomena with inherent time-dependence." *Int. Symp. on Compression and Consolidation of Clayey Soils - IS-Hiroshima's 95*, Hiroshima, 2, 891-935.

Isenhower, W.M., and Stokoe, K.H. (1981). "Strain-rate dependent shear modulus of San Francisco Bay Mud." *Int. Conf. on Recent Advances in Geotechnical Earthquake Engineering and Soil Dynamics*, St. Louis, 2, 597-602.

Ishihara, K. (1985). "Stability of natural deposits during earthquakes." *Proc. 11th ICSMFE*, San Francisco, 1, 321-376.

Iwasaki, T., Tatsuoka, F., and Yoshikazu, T. (1978). "Shear moduli of sands under cyclic torsional shear loading." *Soils and Foundations*, 18(1), 39-56.

Jamiolkowski, M., Ladd, C.C., Germaine, J.T., and Lancellota, R. (1985). "New developments in field and laboratory testing of soils." *Proc. 11th ICSMFE*, San Francisco, 1, 57-153.

Jardine, R.J., St.John, H.D., Hight, D.W., and Potts, D.M. (1991). "Some practical applications of a non-linear ground model." *10th European Conf. on Soil Mechanics and Found. Engrg.*, Florence, 1, 223-228.

Kabbaj, M. (1985). "Aspects rhéologiques des argiles naturelles en consolidation." Ph.D. Thesis, Université Laval, Ste-Foy, Canada.

Kabbaj, M., Tavenas, F., and Leroueil, S. (1988). "In situ and laboratory stress-strain relations." *Géotechnique*, 38(1), 83-100.

Kavazanjian, E., and Mitchell, J.K. (1984). "Time dependance of lateral earth pressure." *J. Geotech. Engrg. Div.*, ASCE, 110(4), 530-533.

Kenney, T.C. (1966). "Shearing resistance of natural quick clays." Ph.D. Thesis, University of London, London, U.K.

Kenney, T.C. (1967). "The influence of mineral composition on the residual strength of natural soils." *Geotechnical Conf. on Shear Strength Properties of Natural Soils and Rocks*, Oslo, 1, 123-129.

Konrad, J.M., and Wagg, B.T. (1993). "Undrained cyclic loading of anisotropically consolidated clayey silts." *J. Geotech. Engrg. Div.*, ASCE, 119(5), 929-947.

Kulhawy, F.H., and Mayne, P.W. (1990). "Manual of estimating soil properties for foundation design." *Geotechnical Engineering Group*, Cornell University, Ithaca.

Lacasse, S., Jamiolkowski, M., Lancellotta, R., and Lunne, T. (1981). "In situ characteristics of two Norwegian clays." *Proc.10th ICSMFE*, Stockholm, 2, 507-511.

Lacerda, W.A., and Houston, W.N. (1973). "Stress relaxation in soils." *Proc. 8th ICSMFE*, 1, Moscow, 221-227.

Lacerda, W.A. (1976). "Stress-relaxation and creep effects on soil deformation." Ph.D. Thesis, University of California, Berkeley, U.S.A..

Ladanyi, B. (1982). "Determination of geotechnical parameters of frozen soils by means of the cone penetration test." *2nd European Symp. on Penetration Testing*, Amsterdam, 671-678.

Ladanyi, B. (1985). "Use of the cone penetration test for the design of piles in permafrost." *Journal of Energy Resources Technology*, 107, 183-187.
Ladd, C.C. (1991). "Stability evaluation during staged construction." *J. Geotech. Engrg. Division*, ASCE, 117(4), 540-615.
Larsson, R. (1977). "Basic behaviour of Scandinavian soft clays." Swedish Geotechnical Institute, Linköping, Report No. 4.
Larsson, R. (1987). "Long-term behaviour of two test fills in Sweden." *Int. Symp. on Geotechnical Engineering of Soft Soils*, Mexico City, 1, 239-247.
Lefebvre, G. (1981). "Fourth Canadian Geotechnical Colloquium: Strength and slope stability in Canadian soft clay deposits." *Canadian Geotechnical J.*, 18, 420-442.
Lefebvre, G., and Leboeuf, D. (1987). "Rate effects and cyclic loading of sensitive clays." *J. Geotech. Engrg. Div.*, ASCE, 113(GT5), 476-489.
Leidwanger, C., Flavigny, E., Giafferi, J.L., Catel, P., and Bufi, G. (1994). "Tassements différés et 'Diflupress L.D.'." *Proc. 13th ICSMFE*, New Delhi, 1, 233-236.
Lemos, L.J.L. (1986). "The effect of rate of shear on residual shear strength of soil." Ph.D. Thesis, University of London, London, U.K.
Leroueil, S. (1988). "Tenth Canadian Geotechnical Colloquium: Recent developments in consolidation of natural clays." *Canadian Geotechnical J.*, 25(1), 85-107.
Leroueil, S. (1994-96). "Compressibility of clays: fundamental and practical aspects." *J. Geotech. Engrg. Div.*, ASCE, 122(7), 534-543. Also partially published in the *Proc. of the ASCE Conf. on Vertical and Horizontal Deformations of Foundations and Embankments*, Settlement's 94, College Station, 1, 57-76.
Leroueil, S., and Tavenas, F. (1979). "Strain rate behaviour of Saint-Jean-Vianney clay. Discussion." *Canadian Geotechnical J.*, 16(3), 616-620.
Leroueil, S., Tavenas, F., Samson, L., and Morin, P. (1983). "Preconsolidation pressure of Champlain clays - Part II - Laboratory determination." *Canadian Geotechnical J.*, 20(4), 803-816.
Leroueil, S., Kabbaj, M., and Tavenas, F. (1985a). "Discussion on Theme Lecture No. 2-B on Laboratory Testing." *Proc. 11th ICSMFE*, San Francisco, 5, 2691-2692.
Leroueil, S., Kabbaj, M., Tavenas, F., and Bouchard, R. (1985b). "Stress-strain-strain rate relation for the compressibility of natural sensitive clays." *Géotechnique*, 35(2), 159-180.
Leroueil, S., Kabbaj, M., Tavenas, F., and Bouchard, R. (1986). "Stress-strain-strain rate relation for the compressibility of natural sensitive clays." Reply. *Géotechnique*, 36(2), 288-290.
Leroueil, S., and Vaughan, P.R. (1990). "The general and congruent effects of structure in natural soils and weak rocks."*Géotechnique*, 40(3), 467-488.
Leroueil, S., Boudali, M., Tavenas, F., Ouarzidini, A., and Boudaa, R. (1996a). "General stress-strain-strain rate behaviour of a natural soft clay." Submitted to *Géotechnique* for publication
Leroueil, S., Perret, D., and Locat, J. (1996b). "Strain rate and structuring effects on the compressibility of a young clay." *This volume*.
Leroueil, S., Vaunat, J., Picarelli, L., Locat, J., Lee, H., and Faure, R. (1996c). "Geotechnical characterization of slope movements." *7th Int. Symp. on Landslides*, Trondheim. 1, 53-74.
Lingnau, B.E., Graham, J., and Tanaka, N. (1995). "Isothermal modeling of sand-bentonite mixtures at elevated temperatures."*Canadian Geotechnical J.*, 31(1), 78-88.
Lo, K.Y., and Morin, J.P. (1972). "Strength anisotropy and time effects of two sensitive clays." *Canadian Geotechnical J.*, 9(3), 261-277.

Lowe, J.L.III, (1974). "New concepts in consolidation and settlement analysis." *J. Geotech. Engrg. Div.*, ASCE, 100(GT6), 574-612.
Lupini, J.F. (1980). "The residual strength of soils." Ph.D. Thesis, University of London, London, U.K.
Magnan, J.P. (1992). "Le rôle du fluage dans les calculs de consolidation et de tassement des sols compressibles." *Bulletin de liaison des laboratoires des ponts et chaussées*, 180, 19-24.
Marchand, G. (1982). "Quelques considérations sur le comportement avant rupture des pentes argileuses naturelles." M.Sc. Thesis, Université Laval, Ste-Foy, Canada.
Marques, M.E.S. (1996). "Influência da velocidade de deformação e da temperatura no adensamento de argilas naturais." M.Sc. Thesis. Research performed at Université Laval, Ste-Foy, Canada in cooperation with COPPE, Federal University of Rio de Janeiro, Brazil.
Martins, I.S.M. (1992). "Fundamentos de um modelo de comportamento de solos argilosos." D.Sc. Thesis, COPPE- Universidade Federal do Rio de Janeiro, Rio de Janeiro, Brazil.
Mayne, P.W., and Kulhawy, F.H. (1982). "K_0 - OCR relationship in soil." *J. Geotech. Engrg. Div.*, ASCE, 108(6), 851-872.
Mesri, G. (1987). "The fourth law of soil mechanics: the law of compressibility." *Int. Symp. on Geotechnical Engineering of Soft Soils*, Mexico City, 2, 179-187.
Mesri, G., and Godlewski, P.M. (1977). "Time- and stress-compressibility interrelationship." *J. Geotech. Engrg. Div.*, ASCE, 103(5), 417-430.
Mesri, G., Ullrich, C.R. and Choi, Y.K. (1978). "The rate of swelling of overconsolidated clays subjected to unloading." *Géotechnique*, 28(3), 281-307.
Mesri, G., and Choi, Y.K. (1985). "The uniqueness of the end-of-primary (EOP) void ratio-effective stress relationship." *Proc. 11th ICSMFE*, San Francisco, 2, 587-590.
Mesri, G., and Feng, T.W. (1986). "Stress-strain-strain rate relation for the compressibility of sensitive natural clays. Discussion." *Géotechnique*, 36(2), 283-287.
Mesri, G., and Castro, A. (1987). "C_α/C_c concept and K_0 during secondary compression. *J. Geotech. Engrg. Div.*, ASCE, 113(3), 230-247.
Mesri, G., and Feng, T.W. (1991). "Surcharging to reduce secondary settlements." *Int. Conf. on Geotechnical Engng. for Coastal Development - Theory and Practice on Soft Ground -Geo-coast' 91*, Yokohama, 1, 359-364.
Mesri, G., and Hayat, T.M. (1993). "The coefficient of earth pressure at rest." *Canadian Geotechnical J.*, 30(4), 647-666.
Mesri, G., Lo, D.O.K., and Feng, T.W. (1994). "Settlement of embankments on soft clays." *Specialty Conf. on Vertical and Horizontal Deformations of Foundations and Embankments*, ASCE, Settlement '94, College Station, 1, 8-56.
Mesri, G., Shahien, M., and Feng, T.W. (1995). "Compressibility parameters during primary consolidation." *Int. Symp. on Compression and Consolidation of Clayey Soils - IS-Hiroshima's 95*, Hiroshima, 2, 1021-1037.
Miliziano, S. (1992). "Effetti della temperatura sul comportamento meccanico delle terre coesive." Ph.D. Thesis, Università di Roma "La Sapienza," Rome, Italy.
Mitchell, J.K. (1964). "Shearing resistance of soils as a rate process." *J. Soil Mech. and Found. Engrg. Div.*, ASCE, 90(SM1), 29-61.
Mitchell, J.K. (1993). *Fundamentals of soil behavior*. 2nd Edition, John Wiley & Sons, Inc.
Mitchell, J.K., Campanella, R.G., and Singh, A. (1968). "Soil creep as a rate process." *J. Soil Mech. and Found. Engrg. Div.*, ASCE, 94(SM1), 231-253.
Mitchell, R.J., and Eden, W.J. (1972). "Measured movements of clay slopes in the Ottawa area." *Canadian J. of Earth Sciences*, 9, 1001-1013.

Moller, B., Rankka, K., Sällfors, G., and Ahnberg, H. (1989). "Horizontal stresses and deformations in slopes - Case histories." *Royal Swedish Academy of Engineering Sciences*. Commission on Slope Stability, Report 2-89, Linköping, Sweden

Moritz, L. (1995a). "Geotechnical properties of clay at elevated temperatures." Swedish Geotechnical Institute, Linköping, Report No. 47.

Moritz, L. (1995b). "Geotechnical properties of clay at elevated temperatures." *Int. Symp. on Compression and Consolidation of Clayey Soils - IS-Hiroshima's 95*, Hiroshima, 1, 267-272.

Murayama, S., and Shibata, T. (1961). "Rheological properties of clays." *Proc. 5th ICSMFE*, Paris, 269-273.

Murayama, S., and Shibata, T. (1964). "Flow and stress relaxation of clays." *Symp. on Rheology and Soil Mechanics*, Grenoble, 99-129.

Murayama, S. (1969). "Effect of temperature on elasticity of clays." *Effects of Temperature and heat on engineering behavior of soils*, Highway Research Board Special Report 103, 194-203.

Murayama, S., Michihiro, K. and Sakagami, T. (1984). "Creep characteristics of sands." *Soils and Foundations*, 24(2), 1-15.

Nutt, N.R.F., and Houlsby, G.T. (1995). "Time-dependent behaviour of sand from pressuremeter tests." *Int. Conf. on the Pressuremeter and its New Avenues*, Sherbrooke, 95-100.

Ottosson, E., and Johansson, L. (1995). "Behaviour of a natural slope close to failure." *11th European Conf. on Soil Mechanics and Foundation Engrg.*, Copenhagen, 4, 95-100.

Petley, D.J. (1966). "The shear strength of soils at large strains." Ph.D. Thesis, University of London, London, U.K.

Plum, R.L., and Esrig, M.I. (1969). "Some temperature effects on soil compressibility and pore water pressure." *Effects of Temperature and Heat on Engineering Behavior of Soils*, Highway Research Board Special Report 103, 231-242.

Powell, J.J.M., and Quatermann, R.S.T. (1988). "The interpretation of cone penetration tests in clays, with particular reference to rate effects." *Int. Symp. on Penetration Testing I*, Orlando, 2, 903-909.

Rampello, S., and Silvestri, F. (1993). "The stress-strain behaviour of natural and reconstituted samples of two overconsolidated clays." *Int. Symp. on Geotechnical Engineering of Hard Soils-Soft Rocks*, Athens, 1, 769-778.

Richardson, A.M., and Whitman, R.V. (1963). "Effect of strain rate upon undrained shear resistance of a saturated remoulded fat clay." *Géotechnique*, 13(3), 310-324.

Roscoe, K.H., Schofield, A.N., and Wroth, C.P. (1958). "On the yielding of soils." *Géotechnique*, 8, 22-53.

Roscoe, K.H., and Burland, J.B. (1968). "On the generalized stress-strain behaviour of 'wet' clay." *Symp. on Plasticity*, Cambridge, 535-610.

Roy, M., Tremblay, M., Tavenas, F., and La Rochelle, P. (1982). "Development of quasi-static piezocone apparatus." *Canadian Geotechnical J.*, 19(2), 180-188.

Roy, M., and Leblanc, A. (1987). "Factors affecting the measurements and interpretation of the vane strength in soft sensitive clays." *Int. Symp. on Laboratory and Field Vane Shear Strength Testing*, ASTM, Tampa, STP 1014, 117-128.

Ruddy, I., Andersen, M.A., Pattillo, P.D., Bishlawi, M., and Foged, N. (1989). "Rock compressibility, compaction, and subsidence in a high-porosity chalk reservoir: a case study of Valhall Field." *J. of Petroleum Tech.*, 41(7), 741-746.

Sällfors, G. (1975). "Preconsolidation pressure of soft high plastic clays." Ph.D. Thesis, Chalmers University of Technology, Gothenburg, Sweden.

Santucci de Magistris, F. (1996). "Comportamento di una sabbia limosa e argillosa costipata ed addizionata con bentonite." Ph.D. Thesis, Università di Napoli "Federico II", Italy.
Schmertmann, J.H. (1983). "A simple question about consolidation." *J. Geotech. Engrg. Div.*, ASCE, 109(1), 119-122.
Schofield, A.N. and Wroth, C.P. (1968). *Critical State Soil Mechanics*. McGraw Hill, London.
Sekiguchi, H. (1977). "Rheological characteristics of clays." *Proc. 9th ICSMFE*, Tokyo, 1, 289-292.
Sheahan, T.C. (1995). "Interpretation of undrained creep tests in terms of effective stresses." *Canadian Geotechnical J.*, 32(2), 373-379.
Sheahan, T.C., Ladd, C.C., and Germaine, J.T. (1994). "Time-dependent triaxial relaxation behavior of a resedimented clay." *Geotechnical Testing J.*, 17(4), 444-452.
Sheahan, T.C., Ladd, C.C., and Germaine, J.T. (1996). "Rate dependent undrained behavior of saturated clay." *J. Geotech. Engrg. Div.*, ASCE, 122(2), 99-108.
Sherif, M.A., and Burrous, C.M. (1969). "Temperature effects on the unconfined shear strength of saturated, cohesive soil." *Effects of Temperature and Heat on Engineering Behavior of Soils*, Highway Research Board Special Report 103, 267-272.
Shen, C. K., Arulanandan, K. and Smith, W. S. (1973). "Secondary consolidation and strength of a clay." *J. Soil Mech. and Found. Engrg. Div.*, ASCE, 95(1), 95-110.
Shibuya, S., Mitachi, T., Fukuda, F., and Degoshi, T. (1995). "Strain rate effects on shear modulus and damping of normally consolidated clay." *Geotechnical Testing J.*, 18(3), 365-375.
Sills, G.C. (1995). "Time-dependent processes in soil consolidation." *Int. Symp. on Compression and Consolidation of Clayey Soils - IS-Hiroshima's 95*, Hiroshima, 2, 875-890.
Singh, A.W., and Mitchell, J.K. (1968). "General stress-strain-time function for soils." *J. Soil Mech. and Found. Engrg. Div.*, ASCE, 94(1), 21-46.
Smith, R.E., and Wahls, H.E. (1969). "Consolidation under constant rates of strain." *J. Soil Mech. Found. Engrg. Div.*, ASCE, 95(SM2), 519-539.
St-Arnaud, G., Morel, R., and Lavallée, J.G. (1992). "Comportement de la fondation argileuse traitée avec des drains synthétiques sous le remblai Olga-C." *Internal Report*, Hydro-Québec, Service géologie et structures, Montréal, Québec, Canada.
Stokoe, K.H. II, Hwang, S.K., Lee, J.N.K., and Andrus, R.D. (1995). "Effects of various parameters on the stiffness and damping of soils at small to medium strains." *First Int. Conf. on Pre-failure Deformation Characteristics of Geomaterials*, Sapporo, 2, 785-816.
Suklje, L. (1957). "The analysis of the consolidation process by the isotaches method." *Proc. 4th ICSMFE*, London, 1, 200-206.
Tatsuoka, F., and Kohata, Y. (1995)."Stiffness of hard soils and soft rocks in engineering applications." *First Int. Conf. on Pre-failure Deformation Characteristics of Geomaterials*, Sapporo, 2, 947-1063.
Tatsuoka, F., Lo Presti, D., and Kohata, Y. (1995). "Deformation characteristics of soils and soft rocks under monotonic and cyclic loads and their relationships." *3rd Int. Conf. on Recent Advances in Geotechnical Earthquake Engineering and Soil Dynamics*, St. Louis.
Tavenas, F. (1984). "Landslides in Canadian sensitive clays. - A State-of-the-Art." *4th Int. Symp. on Landslides*, Toronto, 1, 141-153.

Tavenas, F., and Leroueil, S. (1977). "Effects of stresses and time on yielding of clays." *Proc. 9th ICSMFE*, Tokyo, 1, 319-326.
Tavenas, F., Leroueil, S., La Rochelle, P., and Roy, M. (1978). "Creep behaviour of an undisturbed lightly overconsolidated clay." *Canadian Geotechnical J.*, 15(3), 402-423.
Tavenas, F., and Leroueil, S. (1981). "Creep and failure of slopes in clays." *Canadian Geotechnical J.*, 18, 106-120.
Tavenas, F., and Leroueil, S. (1987). State-of-the-Art on "Laboratory and in situ stress-strain-time behavior of soft clays." *Int. Symp. on Geotechnical Engineering of Soft Soils*, Mexico City, 2, 1-46.
Tidfors, M., and Sällfors, G. (1989). "Temperature effect on the preconsolidation pressure." *Geotechnical Testing J.*, 12(1), 93-97.
Torstensson, B.A. (1977). "Time-dependent effects in the field vane test." *Int. Symp. on Soft Clays*, Bangkok, 387-397.
Urciuoli, G. (1992). "Rigonfiamento di un'argilla di alta plasticità e modellazione dei fenomeni erosivi del colle di Bisaccia." Ph.D. Thesis, Università di Napoli "Frederico II," Italy.
Vaid, Y.P., and Campanella, R.G. (1977). "Time-dependent behaviour of undisturbed clay." *J. Geotech. Engrg. Div.*, ASCE, 103(7), 693-709.
Vaid, Y.P., Robertson, P.K., and Campanella, R.G. (1979). "Strain rate behaviour of Saint-Jean-Vianney clay." *Canadian Geotechnical J.*, 16(1), 34-42.
Varnes, D.J. (1978). "Slope movements types and processes." Transportation Research Board Report 176. Landslides. Analyses and Control, 11-33.
Virely, D. (1993). "Le piézocône - Développement et usages." Ph.D. Thesis, Université Laval, Ste-Foy, Canada.
Vulliet, L. (1986). "Modélisation des pentes naturelles en mouvement." D.Sc. Thesis, École Polytechnique Fédérale de Lausanne, Switzerland.
Wiesel, C.E. (1973). "Some factors influencing in situ vane test results." *Proc. 8th ICSMFE*, Moscow, Vol.1.2, 475-479.
Wu, T.H., Resendiz, D., and Neukirchner, R.J. (1966). "Analysis of consolidation by rate process theory." *J. Soil Mech. and Found. Engrg. Div.*, ASCE, 92(SM6), 229-248.
Yoshikuni, H., Hirao, T., Nishiumi, H., and Ikegami, S. (1993). "The behavior of swelling and recompression due to unloading." *48th Annual Conf. of the Japan Society of Civil Engineers*, 48(3), 1010-1011. (In Japanese, ref. by Imai, 1995).
Yoshikuni, H., Nishiumi, H., Ikegami, S., and Seto, K. (1994). "The creep and effective stress-relaxation behavior on one-dimensional consolidation (in Japanese)." *29th Japan National Conf. on Soil Mechanics and Found. Engrg.*, Vol.29, 269-270.
Yoshikuni, H., Kusakabe, O., Okada, M., and Tajima, S. (1995a). "Mechanism of one-dimensional consolidation." *Int. Symp. on Compression and Consolidation of Clayey Soils - IS-Hiroshima's 95*, Hiroshima, 1, 497-504.
Yoshikuni, H., Moriwaki, T., Ikegami, S., and Nishiumi, H. (1995b). "Rebound due to partial unloading and subsequent recompression behaviour in 1-D consolidation." *Int. Symp. on Compression and Consolidation of Clayey Soils - IS-Hiroshima's 95*, Hiroshima, 1, 233-238.
Zavoral, D.Z., and Campanella, R.G. (1994). "Frequency effects on damping/ modulus of cohesive soil." *Conf. on Dynamic Geotechnical Testing*, ASTM, STP 1213, 191-201.

Appendix A

Basic properties of the main geotechnical materials discussed in the paper

		Depth (m)	W_O (%)	W_P (%)	W_L (%)	σ'_{Pconv} (kPa)		
Batiscan clay	I*		71-88	22	49		Mesri et al.	(1995)
Boston Blue clay	R			24	45		Sheahan et al.	(1996)
Berthierville clay	I	3.3	62	21	45	52	Boudali et al. Boudali Kabbaj	(1994) (1995) (1985)
B6 clay	I	6.8 10.1	50 48	24 22.3	38 32.5	145 175	Lefebvre & Leboeuf	(1987)
Illite	R		36-37				Mitchell et al.	(1963)
Kaolinite	R		33-42				Sherif & Burrous	(1969)
Louiseville clay	I	5.8	77	24	67	140	Hamouche	(1995)
Luleå clay	I	4	110	50	110	52	Eriksson	(1989)
Mascouche clay	I	≈ 5	60	27	56	300	Marchand	(1982)
Pasquasia clay	I		13				Hueckel & Baldi	(1990)
San Francisco Bay mud	I		96-103	36 45	88 93		Lacerda & Houston Arulanandan et al.	(1973) (1971)
Sarapui clay	I	2.8	163			15-32	Feijo & Martins	(1993)
Silty clay sand	C			22	35		Santucci de Magistris	(1996)
St-Alban clay	I	3	90	27	50	50	Tavenas et al.	(1978)
St-Hilaire clay	I		62-84	23	55		Mesri et al.	(1995)
St-Polycarpe clay	I	8.2	56	27	51		Marques	(1996)
St-Vallier clay	I		59	23	60	186	Lo & Morin	(1972)
Vallericca clay	I		29	28	54	2200	d'Onofrio	(1996)

* I = Intact; R = Reconstituted; C = Compacted

Modeling Aspects Associated with Time Dependent Behavior of Soils

Toshihisa Adachi[1]
Fusao Oka[2]
Mamoru Mimura[3]

INTRODUCTION

The effect of time on the loading process is a salient feature of soils, in particular, for clayey soils. This time effect is also observed even for sands (e.g. di Preisco, 1996). In general, there are two types of time-dependent behavior. One is due to the interaction of free pore water and soil skeleton called consolidation phenomena of soils with small permeability and the other is brought about by the inherent viscous characteristics of soil skeleton. In this review article, we will focus on the constitutive modeling of time dependent behavior due to the viscous nature of the materials that is known as creep, relaxation, rate sensitivity and secondary compression. The viscous properties are due to the microscopic structure in soils like clay. The structure of clayey soils is composed of small clay particles with high activity and ion-water system between particles that is mainly trapped in the micro-pores.

The very interesting and elegant description of the microscopic nature of the viscous behavior has been modeled by Murayama and Shibata (1956, 1964), Christensen and Wu (1964), Ter-Stepanian (1975), Mitchell (1976) using the rate process theory. The good review of this approach is discussed by Mitchell (1976). The microscopic aspects of viscous properties are not discussed herein because of the space limitations.

It is well known that there are three approaches for the modeling of time-dependent behavior from a macroscopic point of view. These are the empirical approach, the viscoelastic approach and the viscoplastic approach to construct a constitutive model taking into account the viscous behavior.

Empirical Approach

In the first step, the viscous or time-dependent behavior has been modeled by empirical relations on the basis of the observed or experimental results such as creep

---
[1]Professor, Dept. Civil Eng., Kyoto Univ., Sakyoku, Kyoto, 606, Japan
[2]Professor, Dept., Civil Eng., Gifu Univ., 1-1 Yanagito, Gifu, 501-11, Japan
[3]Associate Professor, D.P.R.I., Kyoto Univ., Gokasho, Uji, 611, Kyoto, Japan

and relaxation processes. The explicit relation between strain and logarithm of time during creep is commonly used. Garlanger (1972) proposed the compression model including secondary compression. This relation, which explicitly depends on time, however, is not a constitutive relation but the solution of the adequate constitutive relation except in the case where the inherent time dependency such as the attenuation of radioactivity. The constitutive relations are generally given by differential equations. The explicit introduction of time violates the principle of objectivity in continuum mechanics (Eringen, 1962). Therefore, this type of empirical relation is strictly limited to the specific boundary and loading conditions.

One of the typical relations was proposed by Singh and Mitchell (1968). In addition, this type of relation is one-dimensional and not applicable to general loading conditions. Saito and Uezawa (1961) found a useful empirical relation between failure time and minimum strain rate. This relation was confirmed by a viscoplastic constitutive model (Adachi et al. 1985).

Viscoelastic Approach

The most popular method of modeling time-dependent behavior of materials is the viscoelastic approach. This approach has been widely applied to many materials including metals, polymers, soils, concrete and rocks etc. In the linear viscoelastic model, the Maxwell model, the Voigt model and the three parameter model with Voigt element and elastic spring called the linear spring Voigt model are representative. For dynamic problems, it has been reported that the viscoelastic nature of cohesive soil can be approximately expressed by the linear spring-Voigt model (Kondner and Ho, 1965, Hori, 1974). By introducing the distribution of relaxation time into the viscoelastic model, the wide range of time-dependency with high and low strain rates can be modeled. Murayama (1983) proposed a non-linear viscoelastic and/or viscoelasticplastic model by modifying his original model including viscoelastic and viscoplastic elements. The linear viscoelastic approach is valid for the behavior in the range of small strains. In the large strain range, both plastic and viscous features are important. From this reason, viscoplastic modeling of soils is useful in the wide range of strain including failure.

Viscoplastic Approach

Based on the above discussions, viscoplastic modeling of soils is necessary to describe both the rate-sensitive and plastic behavior.

Clay is a strain hardening, rate sensitive material that has remarkable characteristics such as rate sensitivity of strength, secondary compression, creep and stress relaxation. Various elasto-viscoplastic constitutive models have been proposed to describe the rheological behavior of clay. Murayama and Shibata (1956) proposed a rheology model based on the rate process theory, the leading study in this field. Adachi and Okano (1974) proposed an elasto-viscoplastic constitutive model that extends the critical state energy theory (Roscoe et al., 1958, Schofield and Wroth , 1968). For this model, Perzyna's theory of an elasto-viscoplastic continuum (1963) was introduced to describe the rate sensitive behavior of normally consolidated clay.

Adachi and Oka (1982) generalized the Adachi and Okano model following the theory of Oka (1981), in which it is assumed that normally consolidated clay never reaches the static equilibrium state even at the end of primary consolidation and the viscoplastic volumetric strain is taken as a hardening parameter.

Sekiguchi (1977) proposed an elasto-viscoplastic constitutive model for normally consolidated clay based on a non-stationary flow surface. A viscoplastic potential has been introduced so that this model can describe universally rate sensitive behavior of clay, such as creep rupture.

Most elasto-viscoplastic constitutive models can be classified as overstress models or non-stationary flow surface models (Matsui et al., 1984). The Adachi and Oka model (1982), Dafalias model (1982), Katona model (1984), Baladi-Rohani model (1984) and Zienkiewicz-Humpheson-Lewis model (1975) belong to the former, whereas, the Sekiguchi model (1977), Dragon and Mroz model (1979), Nova model (1982) and the Matsui and Abe model (1985) belong to the latter class. Detailed theoretical structures of above-mentioned elasto-viscoplastic constitutive models are summarized by Sekiguchi in the report of ISSMFE subcommittee on constitutive laws of soils (1985) and Flavigny and Nova (1990).

CLASSIFICATION OF ELASTO-VISCOPLASTIC CONSTITUTIVE MODELS

An overstress elasto-viscoplastic constitutive model was introduced by Perzyna (1963). The key assumption in this model is that viscous effects become pronounced only after the material undergoes yielding, and that viscous effects are not essential in the elastic domain. Perzyna assumed that the strain rate is composed of the elastic and the viscoplastic part as follows:

$$\dot{\varepsilon}_{ij} = \dot{\varepsilon}_{ij}^{e} + \dot{\varepsilon}_{ij}^{vp} \tag{1}$$

where ε_{ij} denotes the total strain rate tensor, and the superscripts e and vp stand respectively for the elastic and viscoplastic component. Note that the viscoplastic part represents combined viscous and plastic effects. The elastic strain rate is assumed to obey the generalized Hooke's law as follows:

$$\dot{\varepsilon}_{ij}^{e} = C_{ijkl}\,\dot{\sigma}_{kl} \tag{2}$$

where C_{ijkl} is the elastic compliance and σ_{kl} is the effective stress rate.

The viscoplastic strain rate is assumed to obey the following associated flow rule:

$$\dot{\varepsilon}_{ij}^{vp} = \gamma \left\langle \Phi(F) \right\rangle \frac{\partial f}{\partial \sigma_{ij}} \tag{3}$$

Here $\Phi(F)$ is a scalar function and can be determined experimentally, and $\Phi(F)$ represents the strain rate effect on the yielding of the material. The function F, a static yield function, is expressed as follows:

$$F = \frac{f - \kappa_s}{\kappa_s} \tag{4}$$

where κ_s is a strain hardening parameter. The dynamic loading function, f in Eq. (4) is defined in the following form:

$$f = f\left(\sigma'_{kl}, \varepsilon^{vp}_{mn}, \kappa_s\right) \qquad (5)$$

As shown in the above equation, the overstress function, F is assumed to depend not only on the state of stress, but also on the viscoplastic strain as well as the amount of viscoplastic work, κ_s. In the case of work hardening, κ_s is defined in the following form:

$$\kappa_s = \int \sigma'_{ij} \, d\varepsilon^{vp}_{ij} \qquad (6)$$

In the case of strain hardening,

$$\kappa_s = \int d\kappa_s = \int g_{ijkl} \, d\varepsilon^{vp}_{kl} \qquad (7)$$

Here, g_{ijkl} is a 4th-order tensor.

The functional, $<\Phi(F)>$ is defined as follows:

$$\langle \Phi(F) \rangle = \left\{ \begin{array}{ll} \Phi(F) & \text{for } F > 0 \\ 0 & \text{for } F \leq 0 \end{array} \right\} \qquad (8)$$

The magnitude of the viscoplastic strain rate is evidently governed by the amount of overstress above the current static yield surface for a given state of viscoplastic straining. Based on the theoretical assumption stated above, it is required that the functional $<\Phi(F)>$ for the strain-hardening material should be a monotonously increasing function of F when F is positive, together with $<\Phi(0)> = 0$.

Non-stationary Flow Surface Model

The concept of stationary yield surface in the classical plasticity requires that the yield condition of a material does not change with time when plastic strains are held constant. Olszak and Perzyna (1966, 1970) extended this concept to a so-called non-stationary flow surface by introducing the time dependent yield condition. The time dependent yield condition is assumed in the following form:

$$F = F(\sigma'_{kl}, \varepsilon^{vp}_{mn}, \beta) \qquad (9)$$

where β is a scalar parameter which includes time dependent alteration of the material property. By identifying the above-mentioned non-stationary flow surface with a viscoplastic potential, the viscoplastic flow rule can be expressed as follows:

$$\dot{\varepsilon}^{vp}_{ij} = \Lambda \frac{\partial F}{\partial \sigma'_{ij}} \qquad (10)$$

where Λ is a non-negative multiplier. Note that the material undergoing viscoplastic flow should always satisfy F=0. If F is negative, the material is defined to be in the elastic domain. Then, the condition for continued viscoplastic flow can be expressed in the following form:

$$\dot{F} = \frac{\partial F}{\partial \sigma_{ij}'} \dot{\sigma}_{ij}' + \frac{\partial F}{\partial \varepsilon_{mn}^{vp}} \dot{\varepsilon}_{mn}^{vp} + \frac{\partial F}{\partial \beta} \dot{\beta} \qquad (11)$$

Substitution of Eq.(10) into Eq.(11) permits Λ to be specified form as follows:

$$\Lambda = -\frac{\dfrac{\partial F}{\partial \sigma_{kl}'} \dot{\sigma}_{kl}' + \dfrac{\partial F}{\partial \beta} \dot{\beta}}{\dfrac{\partial F}{\partial \varepsilon_{mn}^{vp}} \dfrac{\partial F}{\partial \sigma_{mn}'}} \qquad (12)$$

Thus, the viscoplastic rate equations of the non-stationary flow surface model are characterized by the stress rate terms, while the basic equations of the overstress model do not contain any such stress rate term (see Eq.(3)).

Almost all the possible elasto-viscoplastic models for clay are thought to have been proposed. Recently, however, there has been strong indications that such models should be applied to the practical problems conscious of their characteristics by clarifying the theoretical structure of the models. Many researchers have pointed out that overstress models cannot describe the acceleration creep process nor the creep rupture of normally consolidated clay. We here clarify the theoretical structures for the Adachi and Oka model, a typical overstress model, by comparing its features with those of the Sekiguchi model, a typical non-stationary flow surface model. We also discuss the effect of structural characteristics on the performances of the model.

In addition, we modified the Adachi and Oka model so that it can describe the acceleration creep process and undrained creep rupture without changing the framework of the constitutive model. By using the modified model to calculate the undrained behavior of normally consolidated clay, we are able to discuss the use of the modified model for determining the acceleration creep process and undrained creep rupture. After surveying the theoretical structures of Adachi and Oka model, we develop a theoretical modification of the model.

THEORETICAL STRUCTURE OF THE ELASTO-VISCOPLASTIC CONSTITUTIVE MODEL

General Remarks on the Adachi and Oka Model

The viscoplastic flow rule for an overstress type elasto-viscoplastic constitutive model is generally expressed as:

$$\dot{\varepsilon}_{ij}^{vp} = \gamma \langle \Phi(F) \rangle \frac{\partial f}{\partial \sigma_{ij}'} \qquad (13)$$

$$F = f - \kappa_s \tag{14}$$

in which, γ is the coefficient of viscosity, κ_s the static hardening parameter, f the static yield function, ε_{ij}^{vp} and $\dot{\varepsilon}_{ij}^{vp}$ the viscoplastic strain and strain rate, and F the functional of overstress, $f(\sigma'_{ij}) - \kappa_s(\varepsilon_{ij}^{vp})$. Here, according to the Schofield and Wroth (1968), the function f, a dynamic loading function, is defined as follows:

$$f = \frac{\sqrt{(2J_2)}}{M^*\sigma'_m} + \ln\left(\frac{\sigma'_m}{\sigma'_{m0}}\right) = \ln\left(\frac{\sigma'_{my}}{\sigma'_{m0}}\right) = \kappa_s \tag{15}$$

where J_2 is the second invariant of the deviatoric stress tensor s_{ij}, σ'_m is the mean effective stress, M^* is defined as the value of the stress ratio $\sqrt{2J_2}/\sigma'_m$ at the critical state, σ'_{my} is the work hardening parameter and σ'_{m0} is a unit effective stress. In Eq. (15), κ_s (= $\ln(\sigma'_{my}/\sigma'_{m0})$) is defined as the strain hardening parameter. Also, as the function, f is non-dimensional, the function F expressed by Eq. (14) is also non-dimensional. Following the Cam-clay model (Schofield and Wroth, 1968), the viscoplastic volumetric strain, ε_{kk}^{vp} is adopted as a hardening parameter as shown in the following equation:

$$\kappa_s = \int d\kappa_s = \int \frac{1+e}{\lambda - \kappa} d\varepsilon_{kk}^{vp} \tag{16}$$

Here, e denotes the void ratio, λ and κ denote the compression and swelling index respectively.

The Adachi and Oka model being a typical overstress model, as shown by Eq.(13), the stress strain relation is

$$\dot{\varepsilon}_{ij} = \dot{\varepsilon}_{ij}^e + \dot{\varepsilon}_{ij}^p = \frac{1}{2G}\dot{s}_{ij} + \frac{\kappa}{3(1+e_0)}\frac{\dot{\sigma}'_m}{\sigma'_m}\delta_{ij} + \frac{1}{M^*\sigma'_m}\Phi(F)\frac{s_{ij}}{\sqrt{2J_2}}$$
$$+ \frac{1}{3M^*\sigma'_m}\Phi(F)[M^* - \frac{\sqrt{2J_2}}{\sigma'_m}]\delta_{ij} \tag{17}$$

$$\gamma\Phi(F) = C \cdot M^* \cdot \sigma'_m \cdot \exp\left[m'\left\{\frac{\sqrt{2J_2}}{M^*\sigma'_m} + \ln\frac{\sigma'_m}{\sigma'_{me}} - \kappa_s\right\}\right] \tag{18}$$

Here, s_{ij} and J_2 are the deviatoric stress tensor and the second invariant of deviatoric stress, δ_{ij} is Kronecker's delta, σ_m' the mean effective stress, σ_{me}' the mean effective stress at the end of consolidation, G the elastic shear modulus and M^* the effective stress ratio in the critical state. The superscripts, e and p denote the elastic and viscoplastic components. There are six parameters for this model: the compression and swelling (recompression) indices, λ and κ, the critical stress ratio, M^*, the elastic shear modulus, G, the parameter that estimates the secondary compression, m', and the viscoplastic parameter, C. The parameters λ, κ, M^* and G can be determined by empirical methods, by consolidation swelling tests and by strain rate-controlled undrained compression tests.

Let us here determine m' and C. These parameters can be found from results of undrained triaxial compression tests for at least two different constant rates of strain. First, we determine m' from data on Osaka alluvial clay. Undrained triaxial compression tests were carried out in different constant rates of strain. The effective stress paths for these tests are shown in Fig. 1. Note the equi-mean effective stress, p'=p* given in the figure. The following relation is obtained from the constitutive model by taking into account the rate dependency of normally consolidated clay,

$$\ln\left\{\frac{\dot{\varepsilon}_{11}^{(1)}}{\dot{\varepsilon}_{11}^{(2)}}\right\} = \frac{m'}{M}\left\{\frac{q^{(1)}}{p^{*1}} - \frac{q^{(2)}}{p^{*1}}\right\} \tag{19}$$

There is a valid linear relation in Fig. 2 between the logarithm of the strain rate and the effective stress ratio as verified by Eq.(19). The parameter m' can be determined from the slope of the straight lines in this figure provided the $M(=(3/2)M^*)$ value is given.

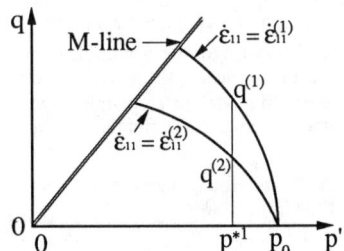

Fig. 1 Typical Effective Stress Paths for NC Clay Under Undrained Condition

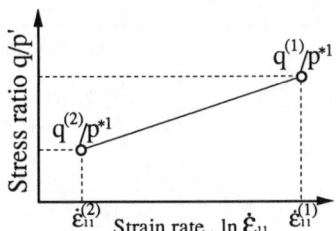

Fig. 2 Determination of the Viscoplastic Parameter, C

Next, the determination of the viscoplastic parameter C. Originally, C has been introduced as:

$$C = \frac{c_0}{M^*\sigma_m'} \exp\left[-m' \ln \frac{\sigma_{myi}'^{(s)}}{\sigma_{me}'}\right] \tag{20}$$

in which, $\sigma'^{(s)}_{myi}$ is the initial static hardening parameter and c_0 the material parameter. However, the static equilibrium state cannot be found experimentally. Because it is very difficult to identify the values of $\sigma'^{(s)}_{myi}$ and c_0, knowledge of parameter C from the relation expressed by Eq.(21) explained later is sufficient. Thus, all the parameters can be determined from the experimental results.

Comparison of Structures

In this section, we give a practical explanation of the theoretical structures of the Adachi and Oka model (overstress model) by comparing its features with those of a non-stationary flow surface model (the Sekiguchi model).

A comparison of the theoretical structures of the Adachi and Oka model with those of the Sekiguchi model is shown in Table 1. The comparison is made for strain rate-controlled undrained triaxial compression as proposed by Mimura and Sekiguchi (1985 a). The variables X_4, X_5, X_6 derived from the condition of the constant rate of strain are the same for both models. There are, however, some differences in the variables X_1, X_2, X_3 derived from the undrained condition (v=0). Because the effect of the stress rate is ignored in an overstress model, $X_1=\kappa/(1+e_0)$, $X_2=0$ can be derived automatically, but, the main factors that make the performances of the Adachi and Oka model and the Sekiguchi model different are the terms, $\dot{v}_0$ for the Sekiguchi model and $C(M-q/p')$ for the Adachi and Oka model. During the shearing process, in

Table 1 Comparison of the Theoretical Structures for Overstress Model and Non-stationary Flow Surface Model

Adachi and Oka Model (Overstress Type)	Sekiguchi Model (Non-stationary Flow Surface Type)
$\frac{\dot{p}}{p}X_1 + \frac{\dot{q}}{p}X_2 - X_3 = 0$ (Undrained Condition)	$\frac{\dot{p}}{p}X_4 + \frac{\dot{q}}{p}X_5 - X_6 = 0$ (Constant Strain Rate Condition)
$X_1 = \frac{\kappa}{1+e_0}$, $X_2 = 0$ $X_3 = -C \cdot (M-q/p') \cdot \exp\left\{\frac{f-v^p}{*\alpha}\right\}$ $X_4 = -\frac{\kappa}{1+e_0}$, $X_5 = \frac{M-q/p'}{3G/p'}$, $X_6 = \dot{\varepsilon}_{11}(M-q/p')$	$X_1 = **A(t) \cdot D\left(M - \frac{q}{p'}\right) + \frac{\kappa}{1+e_0}$ $X_2 = D \cdot **A(t)$, $X_3 = -\dot{v}_0 \cdot \exp\left\{\frac{f-v^p}{\alpha}\right\}$, $X_4 = -\frac{\kappa}{1+e_0}$ $X_5 = \frac{M-q/p'}{3G/p'}$, $X_6 = \dot{\varepsilon}_{11}(M-q/p')$

$*\alpha = \frac{\lambda - \kappa}{1+e_0} \cdot \frac{1}{m'}$, $**A(t) = 1 - \exp\left[-\frac{v^p}{\alpha}\right]$

which $\dot{v}_0$ in Sekiguchi model is assumed to be constant, whereas $C(M-q/p')$ is not. Although the viscoplastic parameter C is assumed to be constant, $(M-q/p')$ decreases as the shear deformation advances in an undrained condition. Therefore the term

C(M-q/p') decreases during undrained shear. The effect of this difference between $\dot{v}_0$ and C(M-q/p') on the performances of constant strain rate triaxial compression has been shown elsewhere by Adachi et al. (1985).

Description of Acceleration Creep by the Overstress Model

Katona (1984), Oka (1985), Mimura and Sekiguchi (1985 b) have shown that because of its theoretical structure, the overstress type model cannot describe the acceleration creep process. We here prove this phenomenon mathematically.

Three characteristic phases, the primary, steady and acceleration creep phases appear in the undrained creep process (Fig. 3). At the same time, there are relations between the creep strain rate and the time elapsed shown schematically in Fig. 4.

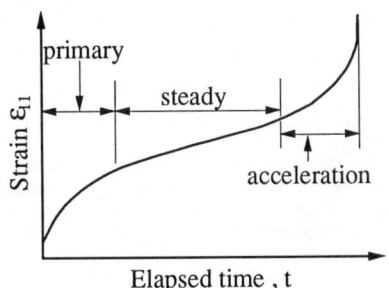

Fig. 3 Typical Creep Curves for NC Clay : Primary, Steady and Acceleration Phases

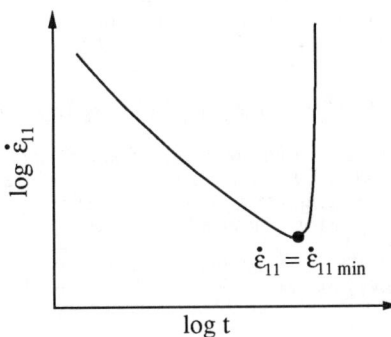

Fig. 4 Schematic log $\dot{\varepsilon}_{11}$ - log t Relation for Undrained Creep

Clearly, when there is creep rupture, the creep strain rate, $\dot{\varepsilon}_{11}^{vp}$, diverges to infinity just after its minimum stage. In other words, the condition necessary for the onset of acceleration creep is that there be a solution for the equation, $\ddot{\varepsilon}_{11}^{vp}=0$, as has been pointed out by Sekiguchi (1984). The stress strain relation for the Adachi and Oka model is

$$\ddot{\varepsilon}_{11}^{vp} = \sqrt{\frac{2}{3}} C \cdot \exp\left[m'\left\{\frac{q}{Mp'} + \frac{\lambda}{\lambda - \kappa} \ln \frac{p'}{p_0}\right\}\right] \quad (21)$$

By differentiating Eq.(21) under the condition of undrained creep with q=constant, the following relation can be derived (Adachi et al., 1987 a).

$$\ddot{\varepsilon}_{11}^{vp} = -\left(M - \frac{q}{p'}\right) \cdot \{\dot{\varepsilon}_{11}^{vp}\}^2 \frac{m'(1+e_0)}{(\lambda - \kappa)} \quad (22)$$

Here, $\{\dot{\varepsilon}_{11}^{vp}\}^2 > 0$ and $m'(1+e_0)/(\lambda-\kappa)$ is positive definite. Therefore, the variation in $\ddot{\varepsilon}_{11}^{vp}$ depends only on the value of $(M-q/p')$. Because $(M-q/p')$ is a monotonically decreasing, positive function through the undrained shearing process, $\ddot{\varepsilon}_{11}^{vp}$ is always negative and the condition for $\ddot{\varepsilon}_{11}^{vp} = 0$ is satisfied only at failure, $(q/p'=M)$.

Thus, the Adachi and Oka model is proved to have the characteristics of the creep strain rate decreasing gradually in the undrained creep process that converges to a certain value at creep rupture. As many researchers have pointed out, this is the mathematical proof that an overstress type elasto-viscoplastic constitutive model cannot describe the acceleration creep process.

MODIFICATION OF THE CONSTITUTIVE MODEL TAKING INTO ACCOUNT THE VARIATION IN THE VISCOPLASTIC PARAMETER

Because the overstress type model cannot describe the undrained creep rupture phenomenon that follows the acceleration creep process, we have modified the constitutive model to describe undrained creep rupture, as well as strain-controlled undrained shearing, without changing the structure of the theoretical framework of the model.

Interrelation of Undrained Strength and the Viscoplastic Parameter

First, we introduce the theoretical solution for undrained strength, q_f and discuss the background for the modification of the constitutive model by clarifying the interrelation of the strain rate with the viscoplastic parameter. The following relation is derived from the basic equation for the model

$$\dot{q} = 3G\left[\dot{\varepsilon}_{11} - C \cdot \exp\left\{m'\left(\frac{q}{Mp'} + \ln\frac{p'}{p_0} - \frac{1+e_0}{\lambda - \kappa}v^p\right)\right\}\sqrt{\frac{2}{3}}\right] \quad (23)$$

Taking into account the undrained condition, $v^p = -v^e = -\frac{\kappa}{1+e_0}\ln(p'/p_0')$ and $\dot{q} = 0$, $q = q_f = M\ p_f$ at failure, Eq.(23) can be expressed as

$$\dot{\varepsilon}_{11} = \sqrt{\frac{2}{3}}\ C \cdot \exp\left[m'\left\{1 + \ln\frac{q_f}{Mp_0}\left(1 + \frac{\lambda}{\lambda - \kappa}\right)\right\}\right] \quad (24)$$

After some calculations, the following relation is derived.

$$\frac{\dot{\varepsilon}_{11}}{\sqrt{2/3}\, C} = \exp(m') \cdot \left[\frac{q_f}{Mp_0}\right]^{\frac{\lambda-\kappa}{m'\lambda}} \tag{25}$$

Therefore, the undrained strength, q_f can be introduced as:

$$q_f = Mp_0 \left[\exp\left\{-\left(1-\frac{\kappa}{\lambda}\right)\right\}\right] \cdot \left[\frac{\dot{\varepsilon}_{11}}{\sqrt{2/3}\,C}\right]^{m'\lambda/(\lambda-\kappa)} \tag{26}$$

Because $M p_0 \exp\{-(1-\kappa/\lambda)\}$ denotes the undrained strength for the Cam clay model in Eq.(26), the undrained strength for the Adachi and Oka model depends on the strain rate as in the Cam clay model. q_f also is a monotonic function for the term $(\dot{\varepsilon}_{11}/\sqrt{2/3}C)$, there being such relations as $\lim \dot{\varepsilon}_{11} \to 0, q_f \to 0$ and $\lim \dot{\varepsilon}_{11} \to \infty, q_f \to \infty$. The relation derived as the analytical solution (Eq.(26)) is shown in Fig. 5. The material parameters used were determined for Osaka alluvial clay (Adachi et al., 1985). Experimental data also are plotted in Fig. 5 for comparison. Clearly, calculations made with the model predict the experimental data accurately.

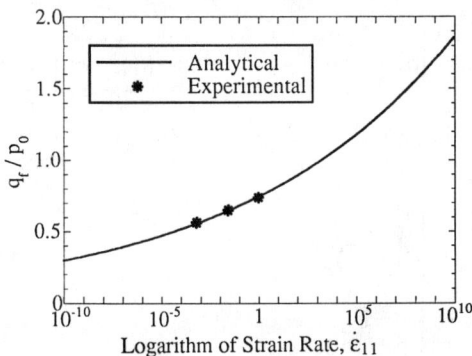

Fig. 5 Theoretical Solution for Undrained Strength with the Adachi & Oka Model

As is evident from Eq.(26), the effects of the strain rate, $\dot{\varepsilon}_{11}$, and the viscoplastic parameter, C, on the undrained strength, q_f, are equivalent, the increase in the strain rate corresponds to the decrease of the viscoplastic parameter, and the decrease in the strain rate is equivalent to the increase of viscoplastic parameter. Taking into account all the characteristics given above, the overstress type constitutive model can be modified to describe undrained creep rupture.

Variation in the Viscoplastic Parameter

As shown above, the Adachi and Oka model accurately predicts the undrained strength for shear under a constant rate of strain assuming that the viscoplastic parameter C is constant. But because the viscoplastic parameter C contains uncertain factors, closer consideration must be given for this parameter. Variation in the viscoplastic parameter, C, during an undrained creep process is discussed here, by comparing its values with those of the experimental results.

The strain rate varies under constant creep stress during the undrained creep process. A decrease in the mean effective stress caused by excess pore water pressure generation contributes decisively to the undrained creep rupture of normally consolidated clay. Therefore, we checked the relation between the strain rate, $\dot{\varepsilon}_{11}$, and the viscoplastic parameter, C, by estimating the effective stress state, (M-q/p').

Sekiguchi (1984) showed that undrained creep behavior, including the creep rupture of normally consolidated clay, can be explained universally by the non-stationary flow surface type elasto-viscoplastic constitutive model that he proposed. Therefore, the characteristics of the viscoplastic parameter, C, will be considered based on the experimental data for Umeda clay reported by Sekiguchi (1984).

A lnC - (M-q/p') relation for some creep stress levels is shown in Fig. 6. Whereas the value of the viscoplastic parameter, C, increases slightly in the early stages of the undrained creep process, independent of the creep stress. However, as

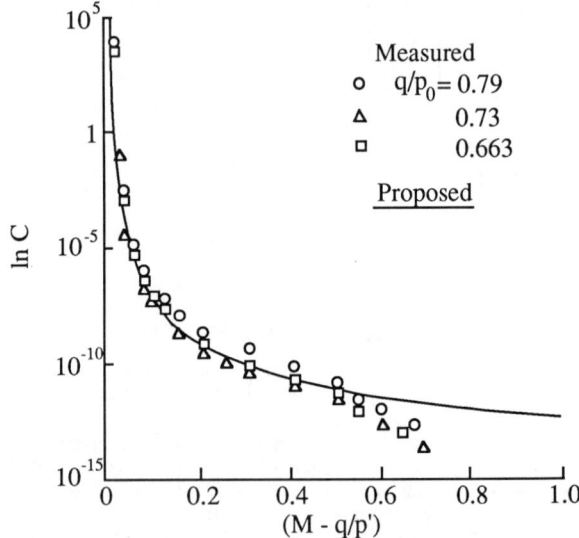

Fig. 6 Variation in the Viscoplastic Parameter, C with the Effective Stress State, (M-q/p')

the critical state draws near, it increases rapidly and diverges to infinity. It is evident from Eq.(26) that the strain rate implicitly contributes to this relation. Fig.7 shows the $\dot{\varepsilon}_{11}/C$ - $\dot{\varepsilon}_{11}$ relations; a unique relation exists between $\dot{\varepsilon}_{11}/C$ and $\dot{\varepsilon}_{11}$ except in the acceleration creep stage in which viscoplastic flow occurs. Although the relation is slightly convex, it can be regarded as almost linear in the range of the strain rate normally used for undrained triaxial compression tests. Therefore, it is natural that the C value determined from strain-controlled undrained triaxial compression tests should be constant. But, in cases such as the acceleration creep process, in which the strain rate changes drastically, the value of the viscoplastic parameter, C, also changes accordingly.

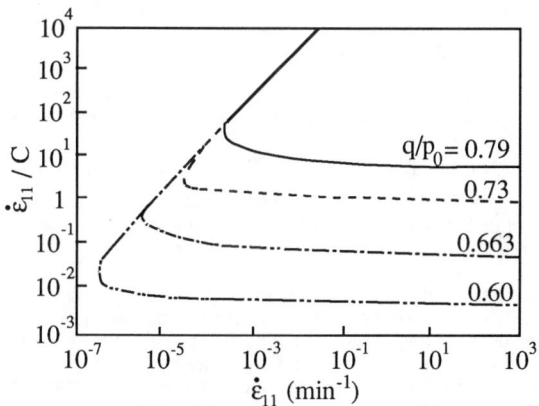

Fig. 7 Relation Between Strain Rate, $\dot{\varepsilon}_{11}$ and the Viscoplastic Parameter, C

Formulation Taking into Account Variation in the Viscoplastic Parameter

The viscoplastic parameter, C, has been shown to vary under constant creep stress (Figs. 6 and 7). Here, the change of the viscoplastic parameter C, with the stress state is calculated by solving Eq.(21) for C. The proposed curve shown in Fig. 6 is determined using the method of least square using the experimental data for Umeda clay. In the undrained creep process, the effective stress is reduced because of the generation of excess pore water pressure. Because the clay structure deteriorates with the decrease in effective stress, the variation in viscoplastic parameter C, can be considered equivalent to this deterioration of the clay structure during undrained creep. From Sekiguchi's experimental results, the relation between the viscoplastic parameter C and the stress ratio (M-q/p') is

$$C = \exp\left[\frac{\delta}{(M - q/p')} - \xi\right] \quad (27)$$

in which, δ and ξ are material constants determined experimentally. When the sustained load is prescribed, the viscoplastic parameter C, is determined by Eq.(27), and the strain rate $\dot{\varepsilon}_{11}$ is calculated as the function of the viscoplastic parameter and the mean effective stress.

The modified stress strain relation for undrained condition is formulated as:

$$\dot{\varepsilon}_{11} = \dot{\varepsilon}_{11}^e + \dot{\varepsilon}_{11}^{vp} = \frac{\dot{q}}{3G} + \sqrt{\frac{2}{3}}\ C\ (\delta,\xi)\cdot\exp\left\{m'\left(\frac{q}{Mp'} + \frac{\lambda}{\lambda - \kappa}\ln\frac{p'}{p_0}\right)\right\} \quad (28)$$

Calculated Performance of Undrained Creep

Calculations for undrained creep were made with the modified overstress elasto-viscoplastic constitutive model defined by Eq.(28). The material constants used are those for Umeda clay reported by Sekiguchi (1984). Two more material constants in Eq.(27), i.e., δ and ξ, are required to describe the proposed relation between the viscoplastic parameter C and the stress state (M-q/p'). Values of δ and ξ were determined as 0.3 and 13.0 from the experimental results shown in Fig. 6 by using the method of least squares.

The calculated creep strain rate - elapsed time relations are shown in Fig. 8. This rate decreases in the primary stage and, after the minimum rate, it diverges to infinity, independent of creep stress. The modified constitutive model based on the assumption that the viscoplastic parameter, C, varies with the function of the stress ratio (M-q/p'), thus has been adapted so that it can describe the undrained creep process including acceleration creep and creep rupture (Adachi et al., 1987 a).

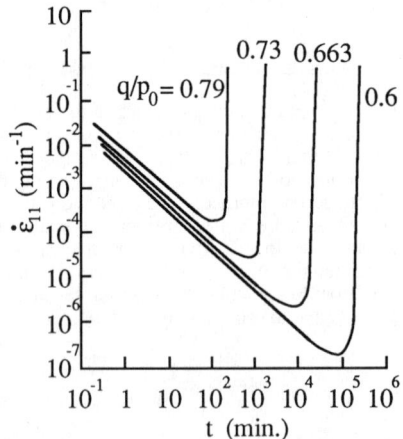

Fig. 8 Calculated Results for $\dot{\varepsilon}_{11}$ - log t Relations with the Modified Overstress Model

The creep strain - elapsed time relations are shown in Fig. 9. The calculated values show that undrained creep rupture follows acceleration creep independent of the magnitude of the creep stress. The values for calculated creep rupture life, t_f, are shown in Table 2, along with the experimental values. Calculated creep rupture life is known to be extremely sensitive to the sustained load, therefore, no estimation of the model's ability to predict undrained creep rupture life is possible with the limited amount of data available. But, we did ascertain that the modified overstress type model can describe the process of acceleration creep and can predict creep rupture life to some extent.

The next point to consider is the physical meaning of viscoplastic parameter C. Because some earlier models often contain physically ambiguous parameters, it is impossible to determine those parameters when applying the models to practical problems. The viscoplastic parameter, C, is closely related to the coefficient of viscosity. A definitive definition of C has yet to be made. It is certain, however, that the deterioration of the soil structure is a main factor of this parameter. Aubry et al. (1985) introduced this concept into constitutive equations mathematically, but, to apply the damage theory to soil mechanics, clarification of the physical meaning of damage or deterioration of the soil structure is needed.

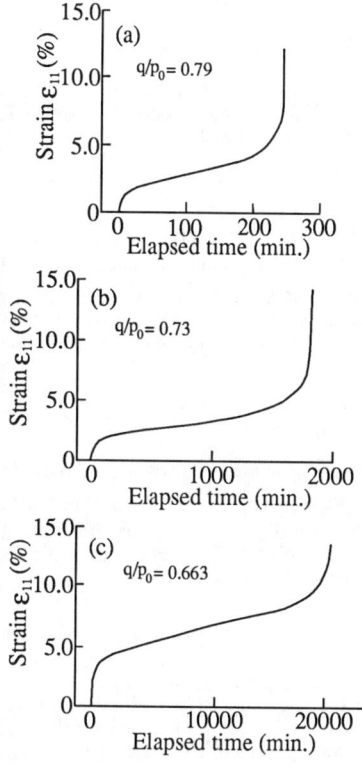

Fig. 9 Calculated Strain - Time Relations for Triaxial Undrained Creep with the Modified Overstress Model

Table 2 Comparison of Creep Rupture Life

Creep Stress (q/p_0)	t_f (min.) (Calculated)	t_f (min.) (Measured)
0.790	253	154
0.730	1840	2210
0.660	20700	18000

Lastly, how can the model represent the viscoplastic behavior of the clay mass? Structural characteristics greatly affect the performance of the model when the behavior of a clay mass that shows complicated responses, such as stress redistribution or migration of pore water, is simulated (Mimura and Sekiguchi, 1986). The next task is to confirm the viscoplastic algorithm that represents various initial and boundary problems for such a clay mass.

RECENT DEVELOPMENTS

Bounding Surface Model

Kaliakin (1988) proposed an elastoplastic-viscoplastic bounding surface model for isotropic cohesive soils. The bounding surface in stress space is analytically defined as follows:

$$F(\bar{\sigma}_{ij}, q_n) = 0 \qquad (29)$$

where a bar over a stress quantity indicates an image point on the bounding surface. The actual stress point σ_{ij} always lies within or on the surface. To each σ_{ij}, a unique image stress point $\bar{\sigma}_{ij}$ is assigned by a properly defined mapping rule. Here, the radial mapping rule (Dafalias, 1979) is used. The dependence of the bounding surface on s_{ij} is expressed in terms of the following three stress invariants:

$$I = \sigma_{kk}, \quad J = \sqrt{\frac{1}{2} s_{ij} s_{ij}}, \quad \alpha = \frac{1}{3}\sin^{-1}\left[\frac{\sqrt{3}}{2}\left\{\frac{s_{ij} s_{jk} s_{ki}}{J^3}\right\}\right] \qquad (30)$$

where s_{ij} and α ($-\pi/6 \leq \alpha \leq \pi/6$) represent the deviatoric part of σ_{ij} and the Lode angle respectively. A section of the surface for a given value of α is schematically shown in Fig. 10. The actual stress point (I,J) is related to its image value $(\bar{I},\bar{J})$ on the bounding surface through the radial mapping rule which is analytically expressed in the following form:

$$\bar{I} = b(I - CI_0) + CI_0, \quad \bar{s}_{ij} = b\, s_{ij} \rightarrow \bar{J} = bJ, \quad \bar{\alpha} = \alpha \qquad (31)$$

where C represents a model parameter ($0 \leq C < 1$) and I_0 represents the intersection of the bounding surface with the positive I-axis. Using CI_0 as the projection center, the image stress is obtained by the radial projection of the actual stress on the bounding surface. The strain rate is decomposed into an elastic and inelastic part. The inelastic part consists of plastic and viscoplastic components. Denoting the total strain tensor by ε_{ij} and its elastic, plastic and viscoplastic components by the superscripts e, p and v, respectively, the following linear decomposition is assumed:

$$\dot{\varepsilon}_{ij} = \dot{\varepsilon}^e_{ij} + \dot{\varepsilon}^p_{ij} + \dot{\varepsilon}^v_{ij}$$
$$= C_{ijkl}\,\dot{\sigma}_{kl} + \langle L \rangle \frac{\partial F}{\partial \bar{\sigma}_{ij}} + \langle \phi \rangle \frac{\partial F}{\partial \bar{\sigma}_{ij}} \qquad (32)$$

where the symbol $\langle \rangle$ denotes the Macaulay bracket, and C_{ijkl} is the fourth order tensor of elastic compliance. F=0 represents the analytical expression of the boundary surface. The overstress function, ϕ is defined as follows (Kalialin and Dafalias, 1990 a):

$$\phi = \frac{1}{V}\exp\left(\frac{J}{NI}\right)\left[\frac{\hat{\delta}}{r - \frac{r}{s_v}}\right]^n \qquad (33)$$

The distances $\hat{\delta}$ and r and the critical state line slope, N are shown in Fig. 10. The parameter, s_v denotes a model parameter which defines the size of elastic nucleus and V and n denote model parameters which control the viscous effect, such as tertiary creep rupture. The calculated performance was verified (Kaliakin and Dafalias, 1990 b) by comparing with the experimental data on San Francisco Bay Mud by Lacerda (1976) and Arulanandan et al. (1971).

Comparison is made between the experimental and numerical generated results in the variation in strain due to undrained creep for various stress levels on San Francisco Bay Mud (Fig. 11 (a)). It can be seen that the calculated performance can well predict the advance in axial strain during undrained creep at various stress levels. Fig. 11 (b) compares the stress relaxation response obtained on San Francisco Bay Mud, here it is seen that the performance of the stress drops when strain keep constant as well as the gain in stress at reloading process can be qualitatively predicted by the numerical calculation.

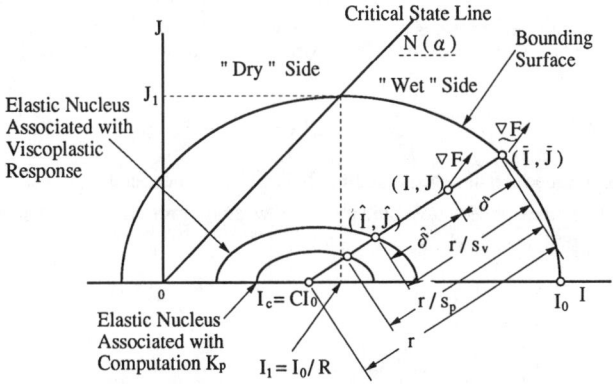

Fig. 10 Schematic Illustration of the Radial Mapping Rule and of the Bounding Surface in Stress Invariant Space (Kaliakin, 1988)

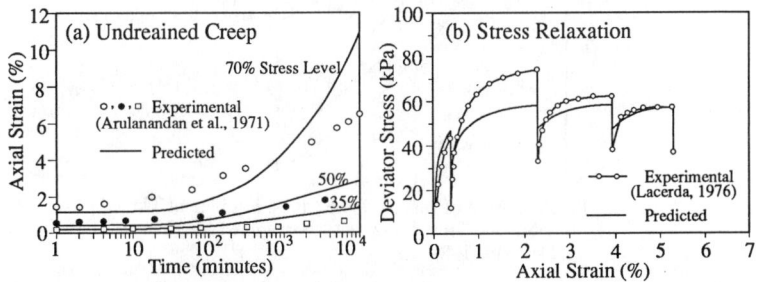

Fig. 11 Undrained Creep Axial Strain With Time Relations (a) and Undrained Stress Relaxation Response (b) on San Francisco Bay Mud (Kaliakin and Dafalias, 1990 b)

Yin and Graham Model

Yin and Graham (1989 a) proposed a one-dimensional viscous-elastic-plastic model using a new concept for establishing "equivalent times" during time-dependent straining. The equivalent times, t_e is defined as the time needed to creep from a reference time line to the current value of the vertical strain, ε_z and the effective vertical stress, σ_z' under constant effective stress. To count equivalent times, a reference time line should be defined along which equivalent times have zero value as shown in Fig. 13 (Yin and Graham, 1989 b). Equivalent times below the reference time line are positive in the range $0<t_e<\infty$. Above the reference time line, equivalent times are negative, in the range $-t_0<t_e<0$, where t_0 is a material parameter. At the same time, a limit line is also defined beyond which the behavior is time independent even for viscous clayey soils. As shown in Fig. 12, the limit line is defined as the time line that has an equivalent time $t_e=\infty$ and creep rate equal to zero.

Strains and deformations are both assumed small so that the usual definition of engineering strain can be used. Vertical strains in the oedometer are divided into three components:

$$\varepsilon_z = \varepsilon_z^e + \varepsilon_z^{sp} + \varepsilon_z^{tp} \qquad (34)$$

where ε_z^e is the recoverable elastic strain, ε_z^{sp} is the time independent plastic strain induced only by the applied stress, and ε_z^{tp} is the viscous strain, i.e., time dependent plastic strain. Both ε_z^{sp} and ε_z^{tp} are irreversible.

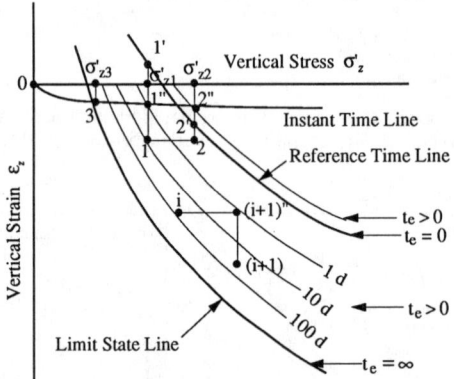

Fig. 12 Illustration of Instant Time Lines, Equivalent Times, Reference Time Lines and Limit Time Line (Yin & Graham, 1989, b)

The time independent elastic strain is assumed linear in $V - \sigma_z$ space in the following form:

$$\varepsilon_z^e = \varepsilon_{z0}^e + \frac{\kappa}{V} \ln\left(\frac{\sigma_z'}{\sigma_{z0}'}\right) \tag{35}$$

where V (=1+e) is the specific volume of the clay, κ is the swelling index, the subscripts, z and 0 denote the vertical component and initial value, resulting in σ_{z0}' being the initial vertical effective stress. The time independent plastic strain, ε_z^{sp} can be obtained from the slope λ of the V - ln σ_z' relationship at stress higher than the preconsolidation stresses, σ_{zc}' from the following equation:

$$\varepsilon_z^{sp} = \varepsilon_{z0}^{sp} + \frac{\lambda - \kappa}{V} \ln\left(\frac{\sigma_z'}{\sigma_{z0}'}\right) \tag{36}$$

Creep compression at constant σ_z' is assumed to be the usual secondary consolidation, i.e., to vary with logarithm of time, and given by:

$$\varepsilon_z^{tp} = \varepsilon_{z0}^{tp} + \frac{\psi}{V} \ln\left(\frac{t}{t_0}\right) \tag{37}$$

where ψ is defined as $\psi = \Delta V/\ln (t/t_0)$.

Then, the general stress - strain - strain rate relationship has been derived as follows:

$$\dot{\varepsilon}_z = \frac{\kappa}{V} \frac{\dot{\sigma}_z'}{\sigma_z'} + \frac{\psi}{V t_0} \exp\left[-\varepsilon_z \frac{V}{\psi}\right]\left[\frac{\sigma_z'}{\sigma_{z0}'}\right]^{\frac{\lambda}{\psi}} \tag{38}$$

Calculated performance for constant rate of strain tests is shown in Figs. 13 (Yin and Graham, 1994). Fig.13 (a) shows predictions for three tests with different strain rates and creep parameter $\psi/V = 0.007$, while Fig. 13 (b) shows predictions using the same initial values and strain rates, but using a smaller creep parameter. The effect of strain rate on stress strain behavior is larger when a larger value is used

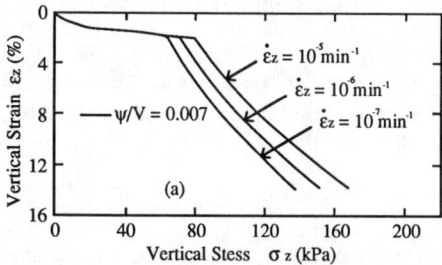

Fig. 13 (a) Calculated Performance for Constant Rate of Strain Consolidation Tests for $\psi/V = 0.007$ (Yin & Graham, 1994)

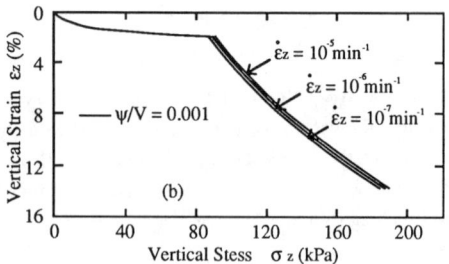

Fig. 13 (b) Calculated Performance for Constant Rate of Strain Consolidation Tests for $\psi/V = 0.001$ (Yin & Graham, 1994)

for the creep parameter ψ/V.

Liang and Ma Model

Liang and Ma (1992) proposed a unified elasto-viscoplastic model based on Perzyna's viscoplastic theory. The limit surface and the conjugated static yield surface form the basic framework from which the time and rate effects can be accounted for by using a single internal state variable. The internal variable, the preconsolidation pressure in this model is affected by the aging effect.

Cyclic Model

Oka (1988) proposed an elasto-viscoplastic constitutive model for saturated clay under the general cyclic loading conditions by extending the preceding works (Oka, 1982, Adachi and Oka, 1986). This cyclic elasto-viscoplastic constitutive model is based on the theory of overstress viscoplasticity. The overconsolidated boundary surface is introduced to define the boundary between the normally consolidated region ($f_b > 0$) and the overconsolidated region ($f_b < 0$) by the following equation:

$$f_b = \overline{\eta}^*(0) + M_m^*(\theta) \ln\left[\frac{\sigma_m'}{\sigma_{md}'}\right] \tag{39}$$

in which σ_m' is the mean effective stress, θ is the Lode angle, M_m^* is a material parameter, σ_{md}' is a material parameter related to deformation history and $\eta^*(0)$ is a stress parameter defined as follows:

$$\overline{\eta}^*(0) = \left[(\eta_{ij}^* - \eta_{ij(0)}^*)(\eta_{ij}^* - \eta_{ij(0)}^*)\right]^{1/2} \tag{40}$$

in which $\eta_{ij}^* = s_{ij}/\sigma_m'$, the subscript (0) represents the value at the end of anisotropic consolidation and s_{ij} is a deviatoric stress tensor. In the normally consolidated region, the behavior of saturated clay is assumed to obey the elasto-viscoplastic constitutive model proposed by Adachi and Oka (1982).

The plastic yield function is given by

$$\overline{\eta}_n^* - \kappa_s = 0 \qquad (41)$$

in which κ_s is a hardening parameter and $\eta_{(0)}^*$ is a relative stress parameter defined as

$$\overline{\eta}_n^* = \left[(\eta_{ij}^* - \eta_{ij(n)}^*)(\eta_{ij}^* - \eta_{ij(n)}^*)\right]^{1/2} \qquad (42)$$

$$\eta_{ij}^* = \frac{s_{ij}}{\sigma'_m} \qquad (43)$$

in which σ_{mo}' is a unit mean effective stress, c and d are the material parameters and $\eta_{ij(n)}^*$ denotes the value of η_{ij}^* at the nth times turning over of the loading direction. $\eta_{ij(n)}^*$ is updated whenever the loading direction on the π plane is changed.

Based on the failure stress components (σ_{1f}, σ_{2f}, σ_{3f}), the relative failure stress ratio is expressed as follows (Oka et al., 1988):

$$\eta_f^* = \left[(\eta_{ij(n)}^* - \eta_{ij(f)})(\eta_{ij(n)}^* - \eta_{ij(f)})\right]^{1/2} \qquad (44)$$

$$\eta_{ij(f)} = \frac{s_{ij(f)}}{\left[c\left(\sigma'_{m(f)}/\sigma'_{m0}\right)^d\right]} \qquad (45)$$

where $s_{ij(f)}$ and $\sigma_m'(f)$ are the values of s_{ij} and σ_m' derived from the failure stress components. The hardening function is assumed as follows:

$$\gamma^{p*} = \frac{\overline{\eta}_{(n)}^* \eta_{(f)}^*}{G'\left(\eta_{(f)}^* - \overline{\eta}_{(n)}^*\right)} \qquad (46)$$

where γ^{p*} is the relative deviatoric strain which is given by the following form:

$$\gamma^{p*} = \left[(e_{ij}^p - e_{ij(n)}^p)(e_{ij}^p - e_{ij(n)}^p)\right]^{1/2} \qquad (47)$$

where $e_{ij(n)}^p$ is the value of plastic deviatoric strain tensor at the nth turn over of the loading direction.

The plastic potential function f_p is assumed in the following expression:

$$f_p = \overline{\eta}_{(n)}^* + \widetilde{M}^* \ln\left(\sigma_m'/\sigma_{ma(n)}'\right) = 0 \qquad (48)$$

in which the parameter $\widetilde{M}^*$ is given by

$$\widetilde{M}^* = -\frac{\eta^*}{\ln(\sigma'_m/\sigma'_{mc})} \qquad (49)$$

where η^* is a stress parameter defined by

$$\eta^* = \left[\eta^*_{ij}\eta^*_{ij}\right]^{1/2}, \qquad \eta^*_{ij} = s_{ij}/\sigma'_m \qquad (50)$$

In the normally consolidated region, $\widetilde{M}^*$ is equal to M_m^*.

The viscoplastic constitutive equations are given as follows:

$$\dot{\varepsilon}^{vp}_{ij} = \langle\Phi_{ijkl}(F)\rangle\frac{\partial f_p}{\partial \sigma_{kl}} \qquad (51)$$

$$F = (f - \kappa_s)/\kappa_s$$

$$\langle\Phi_{ijkl}(F)\rangle = 0 \qquad (F \leq 0)$$
$$= \Phi_{ijkl}(F) \qquad (F > 0)$$

In this theory, $\Phi_{ijkl}(F)$ is assumed to be the 4th order isotropic tensor expressed by

$$\Phi_{ijkl}(F) = C_{ijkl}\,\Phi'(F), \quad C_{ijkl} = A\,\delta_{ij}\delta_{kl} + B(\delta_{ik}\delta_{jl} + \delta_{il}\delta_{jk}) \qquad (52)$$

where δ_{ij} is Kronecker's delta, A and B are the material constants and $\Phi'(F)$ is a material function whose concrete form is as follows:

$$\Phi'(F) = \sigma'_m \exp\left[m'_0\,(\overrightarrow{\eta}^*_{(n)} - \kappa_s)\right] \qquad (53)$$

where, m_0' is a material parameter.

Oka (1992) pointed out that the rotational hardening model (Oka, 1988) has two shortcomings, namely, (1) the rotational hardening model can not describe the retardation of changes in the strain rate direction after the stress direction is changed, and (2) a singularity arises in the constitutive equations at the stress turning point due to renewal of the kinematical hardening parameters. In order to overcome these shortcomings, the elasto-viscoplastic constitutive model was proposed (Oka, 1992) by introducing the non-linear kinematical and cyclic isotropic hardening rule, which was initially introduced by Armstrong and Frederick (1966) and developed by Chaboche (1977, 1983).

The over consolidation boundary surface is the same that is used in the rotational hardening model (Oka, 1988). However, considering the non-linear kinematical hardening rule, static yield functions are assumed as follows:

$$f_{y1} = \left\{(\eta^*_{ij} - x^*_{ij})(\eta^*_{ij} - x^*_{ij})\right\}^{1/2} - R_{D1} = 0 \qquad (54)$$

where x_{ij}^* is the kinematic hardening tensor and R_{D1} is a scalar variable. The evolution equation for kinematical hardening tensor x_{ij}^* is given by

$$dx_{ij}^* = B_1^* (A_1^* de_{ij}^{vp} - x_{ij}^* d\gamma^{vp*}) \tag{55}$$

where de_{ij}^{vp} is the deviatoric viscoplastic strain increment tensor, A_1^* and B_1^* are material constants and $d\gamma^{vp*}$ is the second invariant of the viscoplastic deviatoric strain increment tensor, expressed in the following form:

$$d\gamma^{vp*} = \left[de_{ij}^{vp} de_{ij}^{vp}\right]^{1/2} \tag{56}$$

For changes in the mean effective stress, the following yield function is applied:

$$f_{y2} = M_m^* \left| \ln\left(\frac{\sigma_m'}{\sigma_{mo}'}\right) - y_m^* \right| - R_{D2} = 0 \tag{57}$$

where y_m^* is a scalar kinematical hardening parameter, M_m^* is the value of $\sqrt{\eta_{ij}^* \eta_{ij}^*}$ at the maximum compression and σ_{mo}' is the unit value of the mean effective stress. The kinematical hardening parameter, y_m^* can be decomposed into y_{m1}^* and y_{m2}^*. The evolution equations for the kinematical hardening scalar parameters are assumed to be non-linear for y_{m1}^* and linear for y_{m2}^* expressed as follows:

$$dy_{m1}^* = B_2^* (A_2^* dv^{vp} - y_{m1}^* |dv^{vp}|) \tag{58}$$

$$dy_{m2}^* = H_2 \, dv^{vp} \tag{59}$$

where A_2^* and B_2^* and H_2 are the material constants for kinematical hardening and σ_{mo}' is the unit mean effective stress. The evolution equation for dx_{11}^* can be rewritten under the triaxial stress condition as follows:

$$dx_{11}^* = B_1^*(A_1^* de_{11}^{vp} - \sqrt{\frac{2}{3}} B_1^* e_{11}^{vp}) \tag{60}$$

When $de_{11}^{vp} > 0$, integrating Equation (60) following relation can be obtained:

$$x_{11}^* = \sqrt{\frac{2}{3}} A_1^* \left\{ 1 - \exp\left(-\sqrt{\frac{2}{3}} B_1^* e_{11}^{vp}\right) \right\} \tag{61}$$

Parameter A_1^* corresponds to stress ratio M_f^* at failure and B_1^* is the material parameter related to the rate of kinematical hardening. The value of A_1^* is changed only when the sign for the direction of the principal stress is changed.

For the first yield function, the plastic potential function is assumed as follows:

$$f_p = \left\{(\eta_{ij}^* - x_{ij}^*)(\eta_{ij}^* - x_{ij}^*)\right\}^{1/2} + \widetilde{M}^* \ln (\sigma_m'/\sigma_{ma}') = 0 \tag{62}$$

$\widetilde{M}^*$ is determined by

$$\widetilde{M}^* = -\frac{\eta^*}{\ln(\sigma_m'/\sigma_{mc}')} \tag{63}$$

where

$$\sigma_{mc}' = \sigma_{mb}' \exp\left(\frac{\vec{\eta_0}}{M_m^*}\right) \tag{64}$$

in which $\eta^* = \sqrt{\eta_{ij}^* \eta_{ij}^*}$, $\vec{\eta_0}$ is the relative stress ratio. In the case of isotropic consolidation, $\vec{\eta_0} = 0$. $\widetilde{M}^*$ can be determined by the current stress and σ_{mc}'. In addition, it is assumed that $\widetilde{M}^*$ becomes equal to M_m^* once $\widetilde{M}^*$ reaches M_m^* and $\widetilde{M}^* = M_m^*$ in the region where $f_b \geq 0$.

For the second yield function, the following plastic potential function is introduced.

$$f_{p(2)} = \vec{\eta_x^*} + M_m^* \ln \frac{\sigma_m'}{\sigma_{ma(2)}'} \tag{65}$$

where $\sigma_{ma(2)}'$ is a constant.

The viscoplastic strain rate tensor, $\dot{\varepsilon}_{ij(1)}^{vp}$ is assumed as follows:

$$\dot{\varepsilon}_{ij(1)}^{vp} = \langle \Phi_{1\ ijkl}(F_1) \rangle \Phi_2(\xi) \frac{\partial f_p}{\partial \sigma_{ij}'} \tag{66}$$

where $\langle \Phi_{1\ ijkl}(F_1) \rangle$ is defined as

$$\langle \Phi_{1\ ijkl}(F_1) \rangle = \left\{ \begin{array}{c} \Phi_{1\ ijkl}(F_1)\ ;\ F_1 > 0 \\ 0\ ;\ F_1 \leq 0 \end{array} \right\} \tag{67}$$

In Equation (67), $\Phi_{1\ ijkl}(F_1)$ is the functional of F_1 which shows rate sensitivity and can be experimentally determined, and is given by:

$$F_1 = f_{y1} = \left[(\eta_{ij}^* - x_{ij}^*)(\eta_{ij}^* - x_{ij}^*)\right]^{1/2} - R_{D1} \tag{68}$$

in which $F_1=0$ denotes the static yield function. While in Perzyna's theory, F_1 is dealt with as the scalar function, the first material function for the present model is assumed to be the fourth order isotropic tensor function, which is:

$$\Phi_{1\ ijkl}(F_1) = C_{ijkl}\ \Phi_1'(F_1) \tag{69}$$

The concrete form of the material function is given by Adachi and Oka (1982) and takes the following form:

$$\frac{\Phi_{1\,ijkl}(F_1)}{\sigma_m'} = \exp\left[m_0'\left\{(\eta_{ij}^* - x_{ij}^*)(\eta_{ij}^* - x_{ij}^*)\right\}^{1/2}\right] \quad (70)$$

where m_0' is the viscoplastic parameter and C_{ijkl} is the fourth order isotropic tensor,

$$C_{ijkl} = A\,\delta_{ij}\delta_{kl} + B(\delta_{ik}\delta_{jl} + \delta_{il}\delta_{jk}) \quad (71)$$

For convenience in numerical application, the equations are set as follows (Oka, 1982):

$$C_{01} = 2B,\ C_{02} = 3A + 2B \quad (72)$$

In order to take the rate independence at failure into account, the second material function, $\Phi_2(\xi)$, was introduced (Adachi et al., 1987 b). This function, $\Phi_2(\xi)$, is extended to include the effect of cyclic loading and is given by:

$$\Phi_2(\xi) = 1 + \xi \quad (73)$$

where ξ is the internal variable. The evolution equation for the material variable, ξ, is a general ordinary differential equation. When ξ is zero at the initial state, the following equation is obtained by integration

$$\xi = \frac{\overline{\eta}_x^* M_f^*}{G_2^*\left[M_f^* - \dfrac{\eta_{mn}^*(\eta_{mn}^* - x_{mn}^*)}{\overline{\eta}_x^*}\right]} \quad (74)$$

where G_2^* is a material constant and $\overline{\eta}_x^*$ is defined as

$$\eta_x^* = \left[(\eta_{mn}^* - x_{mn}^*)(\eta_{mn}^* - x_{mn}^*)\right]^{1/2} \quad (75)$$

$$\dot{\varepsilon}_{ij(2)}^{vp} = \langle\Phi(F_2)\rangle\frac{\partial f_{p(2)}}{\partial \sigma_{ij}} \quad (76)$$

On the basis of the above-mentioned discussion, the total viscoplastic strain rate is obtained by the summation of the first and second viscoplastic strain rate.

$$\dot{\varepsilon}_{ij}^{vp} = \dot{\varepsilon}_{ij(1)}^{vp} + \dot{\varepsilon}_{ij(2)}^{vp} \quad (77)$$

Calculated performance of stress strain relation for cyclic triaxial test on Eastern Osaka Clay is shown in Fig. 14 (a). The material parameters are summarized by Oka (1992). For comparison, the experimental results of cyclic loading test is shown in Fig. 14 (b). Deterioration of the clay rigidity with the increasing number of cycles is well simulated by this model. Similarly, calculated and experimental effective stress paths is shown in Fig. 15 (a) and (b). The decrease in effective stress caused by the generation of excess pore water pressure during cyclic loading can be simulated by this model qualitatively.

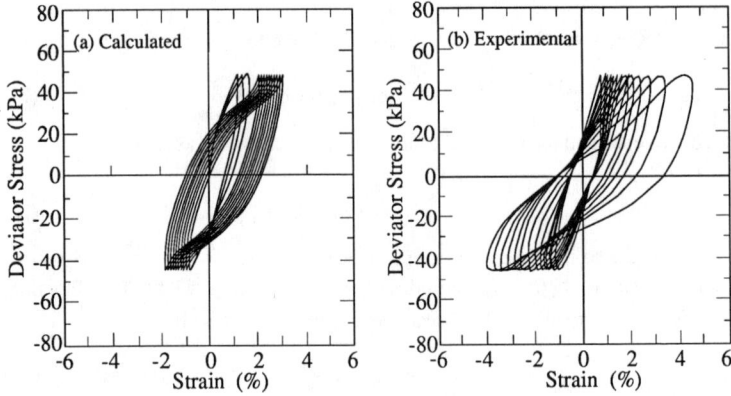

Fig. 14 Calculated Stress-strain Relations of Cyclic Triaxial Test for Eastern Osaka Clay Under Undrained Condition (Oka, 1992)

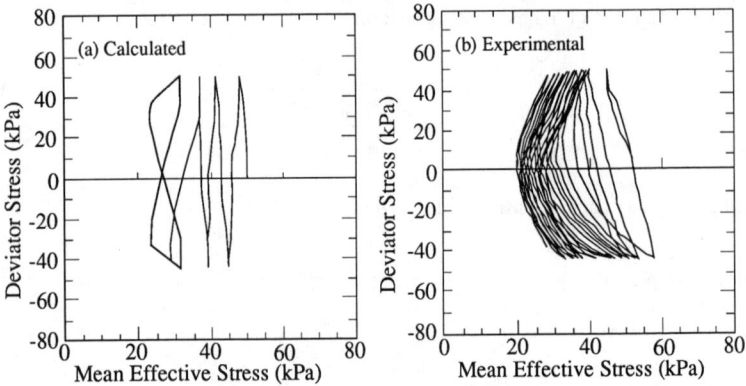

Fig. 15 Calculated Effective Stress Path of Cyclic Triaxial Test for Eastern Osaka Clay Under Undrained Condition (Oka, 1992)

Anisotropy

Oka (1993 a) constructed an anisotropic constitutive model for over-consolidated clay introducing an initial structural tensor into the viscoplastic model for overconsolidated clay (Oka et al., 1988). The proposed model is capable of describing the triaxial behavior of clay specimen with different angles between sedimentation plane and principal stress direction.

Pseudo Anisotropy

Oka (1993 b) also developed viscoplastic model accounting for a current stress induced pseudo anisotric behavior using a transformed stress tensor based on the current stress dependent structural tensor. The current stress induced pseudo anisotropic behavior is brought about by the intermediate stress dependency of the soil. Using this model, the deviatoric flow rule for soil which means the dependency of the direction of the strain rate tensor on the intermediate stress (b value) could be well reproduced. In addition, it was found that the Matsuoka and Nakai's yield criterion(1977) is equivalent to the generalized extended von-Mises criterion obtained by the transformed stress tensor.

Elasto-viscoplastic Model Based on the Stress History Tensor

Oka (1985) proposed an elasto-viscoplastic model based on the specific strain measure and stress history tensor, σ^*_{ij}, in order to continuously describe the transition and the viscoplastic and plastic behavior following the similar concept of endochronic theory by Valanis (1971). The proposed model can continuously describe the transition between viscoplastic and plastic states and the strain softening behavior.

The total strain rate, $\dot{\varepsilon}_{ij}$, is composed of the elastic strain rate, $\dot{\varepsilon}^e_{ij}$, and the viscoplastic strain rate, $\dot{\varepsilon}^{vp}_{ij}$, as shown in the following form:

$$\dot{\varepsilon}_{ij} = \dot{\varepsilon}^e_{ij} + \dot{\varepsilon}^{vp}_{ij} \tag{78}$$

In this model, a new definition of time measure is used and is given by the following expression:

$$dz = g\left(\dot{\varepsilon}_{ij}\right) dt \tag{79}$$

where t is the real time and g is the rate dependent function. The total stress history, σ^z_{ij}, is assumed to be dependent on both current stress and reduced stress history, $\sigma^z_{r\,ij}$.

Following Eringen's approach and the aforementioned time measure, the full history of the stress tensor is replaced by the present stress and the reduced stress tensor as follows:

$$\sigma^z_{ij} = \left[\sigma_{ij}(z), \sigma^z_{r\,ij}(z-z'), 0 < z' \le z\right] \tag{80}$$

$$\sigma^z_{r\,ij} = \left[\sigma_{ij}(z-z'), 0 < z' \le z\right] \tag{81}$$

The total stress history tensor is a function of the reduced stress tensor with respect to the time measure dz shown in the following expression:

$$\sigma^*_{ij} = \sigma^*_{ij}\left(\sigma^z_{r\,ij}(z-z')\right) \tag{82}$$

By using a kernel function, the stress history tensor is expressed as follows:

$$\sigma_{ij}^* = \int_0^z K(z - z') \sigma_{ij}(z') \, dz' \tag{83}$$

where K is a continuous scalar function and is assumed to satisfy $\partial K / \partial z < 0$. From this definition, the stress history tensor satisfies the principle of fading memory. The influence of the old stress becomes small. The single exponential function is a typical example of the kernel function and is used for the modeling of the behavior of frozen sand (Adachi et al., 1990). From the theorem of an integral in the wide sense, the interval of integration is defined by

$$0 \leq z' \leq z \tag{84}$$

The viscoplastic flow rule is expressed as follows:

$$d\varepsilon_{ij}^{vp} = H \frac{\partial f_p}{\partial \sigma_{ij}} df_y \tag{85}$$

where f_P is the viscoplastic potential function, f_y is the yield function and H is the hardening-softening function. The yield function depends on the stress history tensor and on the hardening-softening parameter κ as shown in the following form:

$$f_y(\sigma_{ij}^*, \kappa) = 0 \tag{86}$$

The loading criteria are given by

$$d\varepsilon_{ij}^{vp} \neq 0 \tag{87}$$

when $f_y = 0$ and $df_y = \left[\left(\frac{\partial f_y}{\partial \sigma_{ij}^*}\right) d\sigma_{ij}^*\right] > 0$ are satisfied,

$$d\varepsilon_{ij}^{vp} = 0 \tag{88}$$

when $f_y = 0$ and $df_y = 0$ are satisfied,

$$d\varepsilon_{ij}^{vp} = 0 \tag{89}$$

when $f_y = 0$ and $df_y < 0$ are satisfied,

$$f_y = \eta^* - \kappa = 0 \tag{90}$$

The viscoplastic potential function depends on the current stress and the stress history parameter, L; i.e.,

$$f_p(\sigma_{ij}, L) = 0 \tag{91}$$

Adachi et al. (1990) and Oka et al. (1994) successfully applied this model to the triaxial behavior of frozen sand and discussed the effect of strain rate, temperature, confining stress and soil concentration. Calculated performance of stress strain relations as well as the volumetric strain with the progress of deviatoric strain are shown in Fig. 16 together with the experimental results. The calculated results can predict the experimental data at different rate of strain well.

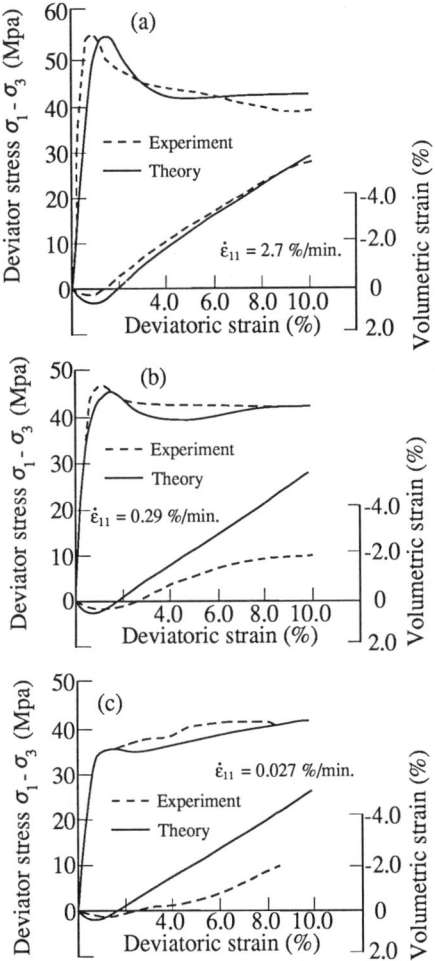

Fig. 16 Stress-strain Relations of Frozen Sand (Oka et al., 1994)

CONCLUSION

The aspect of modeling of rate dependent behavior of soils has been discussed in the present review paper, emphasizing the formulation of the viscoplastic constitutive models. The viscoelastic modeling of the behavior is more important in the dynamic loading conditions. On the other hand, the viscoplastic modeling is necessary to simulate the quasi-static and large strain problems such as creep and progressive failure. The combination of viscoelastic and viscoplastic model is effective to obtain the full formulation of the time-dependent behavior of soils. The experimental identification of the viscoplastic and viscoelastic strains, however, is not so easy.

The other salient properties of soils is a temperature dependency of the mechanical behavior. It is well-known that the viscosity of the materials is inherently dependent on the temperature. We need the thermo-viscoplastic modeling in the future.

The validity of the many proposed models has been evaluated using rather simple laboratory test results such as the triaxial test. The more elaborate models taking account of other effects such as anisotropy and cyclic loading could be constructed only by use of the relevant experimental results under complex loading conditions and field observations. In that sense, in order to generalize the present models, we need the more experimental or measured data.

In the application of the constitutive models to geomaterials, the time-dependent models are more useful than the rate-independent models to accurately simulate the deformation of ground. In particular, the deformation of clayey soil deposits should be analyzed by the time-dependent models in the framework of soil-water coupling formulation. The many applications of the rate dependent models to the consolidation analysis have been successfully reported although many analyses are done using the small strain formulation. In recent studies, the simulation of shear banding that is a precursor to failure has been tried in relation to the progressive failure and localization. For that purpose, the models should be formulated in the context of finite strain theory.

In order to elucidate many of the points that remain to be solved and discussed in the paper, more experimental and theoretical studies are necessary in the future.

REFERENCES

Adachi, T. and Okano, M. (1974), "A Constitutive Equation for Normally Consolidated Clay," *Soils and Foundations*, Vol.14, No.4, pp.53-73.

Adachi, T. and Oka, F. (1982), "Constitutive Equations for Normally Consolidated Clay Based on Elasto-Viscoplasticity," *Soils and Foundations*, Vol.22, No.4, pp.57-70.

Adachi, T., Oka, F. and Mimura, M. (1985), "Descriptive Accuracy of Several Existing Constitutive Models for Normally Consolidated Clays," *Proc. 5th ICONMG*, Vol.1, pp.259-266.

Adachi, T., Oka, F. and Mimura, M. (1987), "An Elasto-viscoplastic Theory for Clay Failure," *Proc. 8th Asian Regional Conf. on SMFE*, Vol.1, pp.5-8.

Adachi, T., Oka, F. and Mimura, M. (1987), "Mathematical Structure of An Overstress Elasto-Viscoplastic Model for Clay," *Soils and Foundations*, Vol. 27, No. 3, pp.31-42.

Adachi, T., Oka, F. and Poorooshasb, H.B. (1990), "A Constitutive Model for Frozen Sand," *Journal of Energy and Resources Technology*, ASME,112, pp.208-212.

Adachi, T. and Oka, F. (1995), "Elasto-Viscoplastic Model with Strain Softening," *Proc. 10th Asian Regional Conf. on SMFE*, Beijing, China, International Academic Press, Vol. 1, pp.1-4.

Armstrong, P. J. and Frederick, C. O. (1966), "A Mathematical Representation of the Multiaxial Bauschinger Effect," *C.E.G.B., Report RD/B/N 731*.

Arulanandan, K., Shen, C.K. and Young, R.B. (1971), "Undrained Creep Behaviour of a Coastal Organic Silty Clay," *Geotechnique*, Vol.1, No.4, pp.359-375.

Aubry, D., Kodaissi, E. and Meimon, E. (1985), "A Viscoplastic Constitutive Equations for Clays Including Damage Law, *Proc. 5th ICONMG*, Vol.1, pp.421-428.

Baladi, G.Y. and Rohani, B. (1984), "Development of An Elasto-Viscoplastic Constitutive Relationship for Earth Materals,"*Mechanics of Engineering Materials*, eds., C.S. Desai and R.H. Gallagher, John Wiley & Wiley Ltd., pp.23-43.

Chaboche, J. L. (1977), "Viscoplastic Constitutive Equations for the Description of Cyclic and Anisotropic Behavior of Metals," *Bull. de l'Acad. Polonaise des Sciences, Serie Sc. et Tech.*, Vol.25, No.1, pp. 33-42.

Chaboche, J. L. and Rousselier, G. (1983), "On the Plastic and Viscoplastic Constitutive Equations, Part 1 : Rules Developed with Internal Variable Concept," *Journal Pressure Vessel Technology*, ASME, Vol.105, pp.153-158.

Cheney, J.A. and Arulanandan, K. (1977), "Stress Relaxation in Sand and Clay, Constitutive Equations of Soils," *Proc. Specialty session 9, 9th ICSMFE*, Tokyo, pp. 279-285.

Christensen, R.W. and Wu, P.L. (1964), "Analysis of Clay Deformation as the Rate Process, " *Journal of the Soil mechanics and Foundation Division*, ASCE, Vol. 90, No. SM6, pp. 125-157.

Dafalias, Y. (1982), "Bounding Surface Elastoplasticity-Viscoplasticity for Particulate Cohesive Media," *Proc. IUTAM Symp. on Deformation and Failure of Granular Materials*, pp.97-107.

di Prisco, C. and Imposimato, S. (1996), "Time Dependent Mechanical Behaviour of Loose Sands," *Mechanics of Cohesive-Frictional Materials*, Vol. 1, No. 1, pp. 45-74.

Dragon, A. and Mroz, Z. (1979), "A Model for Plastic Creep of Rock-Like Materials Accounting for the Kinetics of Fracture, *Int. Journal Rock Mech. Min. Sci. Geomech.*, Abstr., Vol.16, pp.253-259.

Eringen, C. (1962), "Non-Linear Theory of Continuous Media," *McGraw-Hill.*

Flavigny, L.E. and Nova, R. (1990), "Viscous Properties of Geomaterials," *GEOMATERIALS*, Constitutive equations and modelling, ed. by Felix Darve, Elsevier Appl. Sci. Pub. pp.27-54.

Garlanger, J.E. (1972), "The Consolidation of Soils Exhibiting Creep Under Constant Effective Stress," *Geotechnique*, Vol. 22, No. 1, pp. 71-78.

Hori, M. (1974), "Fundamental Study of Wave Propagation Characteristics Through Soils," *Thesis for Dr. Eng.*, Kyoto University.

Kaliakin, V.N. (1988), "An Elastoplastic- Viscoplastic Bounding Surface Model for Isotropic Cohesive Soils," *Proc. Int. Conf. on Rheology and Soil Mech.*, pp.147-163.

Kaliakin, V.N. and Dafalias, Y.F. (1990,a) , "Theoretical Aspects of the Elastoplastic- Viscoplastic Bounding Surface Model for Cohesive Soils, *Soils and Foundations,* Vol.30, No.3, pp.11-24.

Kaliakin, V.N. and Dafalias, Y.F. (1990,b), "Verification of the Elastoplastic-Viscoplastic Bounding Surface Model for Cohesive Soils, *Soils and Foundations*, Vol.30, No.3, pp.25-36.

Katona, M. G. (1984), "Evaluation of Viscoplastic Cap Model," *Journal Geotechnical Engineering*, ASCE, Vol. 110, No. 8, pp.1106-1125.

Kondner, R.L. and Ho, M.M.K. (1965), "Viscoelastic Response of a Cohesive Soil in the Frequency Domain," *Trans. Soc. of Rheology*, Vol. 9, No. 2, pp. 329-342.

Lacerda, W.A. (1976) , "Stress-Relaxation and Creep Effect on Soil Deformation," *Ph.D Dissertation*, University of California, Berkeley.

Liang, R.Y. and Ma, F.G. (1992), "A Unified Elasto-Viscoplastic Model for Clays, Part 1 : Theory" *Computers and Geotechnics*, Vol. 13, No.2, 71-87.

Matsui, T. and Abe, N. (1984), "Introduction to Constitutive Equations of Soils, *Tsuchi-to-Kiso*, JSSMFE, Vol. 32, No. 1, pp.71-79 (in Japanese).

Matsui, T. and Abe, N. (1985), "Elasto/Viscoplastic Constitutive Equation of Normally Consolidated Clays Based on Flow Surface Theory, *Proc. 5th ICONMG*, Vol.1, pp.407-413.

Matsuoka, H. and Nakai, T. (1977), "Stress-Strain Relationship of Soil Based on SMP, *Proc. Specialty session 9, 9th ICSMFE*, pp. 153-162.

Mitchell, J. K. (1976), "Fundamental of Soil Behavior," *Wiely & Sons.*

Mimura, M. and Sekiguchi, H. (1985, a), "A Review of Elasto-Viscoplastic Models with Particular Emphasis on Stress-Rate Effect," *Proc. 20th Japan Nat. Conf. SMFE*, pp.403-406 (in Japanese).

Mimura, M. and Sekiguchi, H. (1985, b), "A Review of Existing Viscoplastic Constitutive Models Regarding the Performance of Creep Rupture Prediction," *Proc. 40th Japan Nat. Conf. JSCE*, pp.461-462 (in Japanese).

Mimura, M. and Sekiguchi, H. (1986), "Some Remarks on the Delayed Failure of Clay Foundations," *Proc. 21th Japan Nat. Conf. SMFE*, pp.1097-1100 (in Japanese).

Murayama S. and Shibata, T. (1956), "On the Rheological Characters of Clay, *Transaction of JSCE*, No.40, pp.1-31 (in Japanese).

Murayama, S. and Shibata, T. (1964), "Flow and Stress Relaxation of Clays," *Proc. IUTAM Symp. on Rheology and Soil Mechanics*, Grenoble, pp. 99-129.

Murayama, S. (1983), "Formulation of Stress-Strain-Time Behavior of Soils Under Deviatoric Stress Condition," *Soils and Foundations*, Vol. 23, No. 2, pp. 41-57.

Nova, R. (1982), "A Viscoplastic Constitutive Model for Normally Consolidated Clay," *Proc. IUTAM Symp. on Deformation and Failure of Granular Materials*, pp.287-295.

Oka, F. (1981), "Prediction of Time Dependent Behavior of Clay," *Proc. 10th ICSMFE*, Vol.1, pp.215-218.

Oka, F. (1982), "Elasto-viscoplastic Constitutive Equation for Overconsolidated Clay," *Proc. Int. Symp. on Numerical Models in Geomechanics*, Vol.1, pp.147-156.

Oka, F. (1985), "Elasto-Viscoplastic Constitutive Equations with Memory and Internal Variables," *Computer and Geomechanics*, Vol.1, pp.59-69.

Oka, F (1988), "A Cyclic Viscoplastic Constitutive Model for Clay," *Proc. 6th ICONMG*, Vol.1, pp.293-298.

Oka, F., Adachi, T. and Mimura, M (1988), "Elasto-Viscoplastic Constitutive Models for Clays," *Proc. Int. Conf. Rheology and Soil Mechanics*, Elsevier, pp.12-28.

Oka, F. (1992), "A Cyclic Elasto-Vscoplastic Constitutive Model for Clay Based on the Non-linear Hardening Rule," *Proc. 4th Numerical Models in Geomechanics*, Balkema, pp. 105-114.

Oka, F. (1993 a), "An Elasto-Viscoplastic Constitutive Model for Clay Using a Transformed Stress Tensor," *Mechanics of Materials*, Vol. 16, pp.47-53.

Oka, F. (1993 b), "Anisotropic and Pseudo-Anisotropic Elasto-Viscoplastic Constitutive Models for Clay," *Modern Approaches to Plasticity, Proc. of Int. Workshop in Horton*, June, Elsevier, pp.505-526.

Oka, F., Adachi, T. and Yashima, A. (1994), "Strain Localization Analysis by Elasto-Viscoplastic Softening Model for Frozen Sand," *Int. Journal Numerical and Analytical Methods in Geomechanics*, Vol. 18, pp.813-832.

Oka, F., Adachi T. and Yashima, A. (1995), "A Strain Localization Analysis of Clay Using a Strain Softening Viscoplastic Model," *Int. Journal of Plasticity*, Vol. 11, No. 5, pp.523-545.

Oka, F. and Yashima, A. (1995), "A Cyclic Elasto-Viscoplastic Model for Cohesive Soil," *Proc. 11th ECSMFE*, Copenhagen, The Danish Geotechnical Society, Vol. 6, pp.145-150.

Olszak, W. and Perzyna, P. (1966), "The Constitutive Equations of the Flow Theory for A Non-stationary Yield Condition, " *Proc. 11th Int. Congress of Applied Mechanics*, pp. 545-553.

Olszak, W. and Perzyna, P. (1970), "Stationary and Non-stationary Viscoplasticity, "*Inelastic Behaviour of Solids*, eds., M.F. Kanninen et al., McGraw-Hill, pp. 53-75.

Perzyna, P. (1963), "The Constitutive Equations for Workhardening and Rate Sensitive Plastic Materials," *Proc. of Vibrational Problems*, Warsaw, Vol.4, No.3, pp.281-290.

Roscoe, K. H., Schofield, A. N. and Wroth, C. P. (1958), "On the Yielding of Soils," *Geotechnique*, Vol. 8, pp.22-53.

Saito, M. and Uezawa, H. (1961), "Failure of Soil Due to Creep," *Proc. 5th ICSMFE*, Paris, Vol. 1, pp. 315-318.

Schofield, A. N. and Wroth, C. P. (1968), "Critical State Soil Mechanics," *McGraw-Hill*.

Sekiguchi, H. (1977), "Rheological Characteristics of Clays," *Proc. 9th ICSMFE*, Vol.1, pp.289-292.

Sekiguchi, H. (1984), "Theory of Undrained Creep Rupture of Normally Consolidated Clay Based on Elasto-Viscoplasticity," *Soils and Foundations*, Vol.24, No.1, pp.129-147.

Sekiguchi, H. (1985), "Macrometric Approaches -Static- Intrinsically Time-Dependent," Constitutive Laws of Soils, *Report of ISSMFE Subcommittee on Constitutive Laws of Soils and Proc. of Discussion Session 1A*, ISSMFE Sub Committee on Constitutive Laws of Soils,11th ICFMFE, San Francisco, pp.66-98.

Singh, A. and Mitchell, J.K. (1968), "General Stress-Strain-Time Function for Soils," *Journal of Soil Mechanics and Foundation Engineering*, ASCE, Vol. 94, No.SM1, pp. 21-46.

Ter-Stepanian, G (1975), "Creep of Clays During Shear and Its Rheological Model," *Geotechnique*, Vol. 25, No. 2, pp. 299-320.

Ting, W.J. (1983), "On the Nature of the Minimum Creep Rate-Time Correlation for Soil, Ice and Frozen Soil," *Canadian Geotechnical Journal*, Vol. 20, No. 1, pp.176-182.

Valanis, K.C. (1971), "A Theory of Viscoplasticity Without a Yield Surface-1," *Archiv of Mechanics*, Vol. 23, No. 4, pp. 517-533.

Yin, J.-H. and Graham, J. (1989, a), "Viscous-Elastic-Plastic Modelling of One-Dimensional Time-Dependent Behaviour of Clays," *Canadian Geotechnical Journal*, Vol. 26, No.2, pp.199-209.

Yin, J.-H. and Graham, J (1989, b), "General Elastic Viscous Plastic Constitutive Relations for 1-D Straining in Clays," *Proc. 3rd NUMOG*, pp.108-117.

Yin J.-H. and Graham, J (1994), "Equivalent Times and One-Dimensional Elastic Viscoplastic Modelling of Time-Dependent Stress-Strain Behaviour of Clays," *Canadian Geotechnical Journal*, Vol. 31, No.1, pp.42-52.

Zienkiewicz, O.C., Humpheson, C. and Lewis, R.W. (1975), "Associated and Non-Associated Viscoplasticity and Plasticity in Soil Mechanics,"*Geotechnique*, Vol. 25, No. 4, pp. 671-689.

Simulation of Pore Pressures in Triaxial Creep Tests
Horst G. Brandes[1] and Armand J. Silva[2]

Abstract

The finite element method is used to investigate the pore pressure response of a fine-grained, low permeability deep sea clay in triaxial creep tests, where the sample is subjected to compression and shearing with axisymmetric drainage. Following loading, the pore pressures dissipate rapidly near the drainage boundaries, but initially continue to increase at internal locations due to the Mandel-Cryer effect. Maximum excess pore pressures fall within the theoretical range of values predicted by Cryer (1963) for an elastic material. Overall, pore pressures and deformations during primary consolidation are affected slightly by the secondary compression parameter in Taylor's (1942) creep equation. Creep is found to decrease the rate at which pore pressure dissipation takes place compared to the no-creep case. Deformation of the triaxial specimen, particularly during the shearing phase, agrees with test results and further supports the constitutive model chosen. This type of modeling is useful in preparing loading schedules for drained tests that minimize excess pore pressures and preserve uniform effective stresses.

Introduction

Triaxial stress-strain-time tests are used extensively in the development and verification of continuum-type constitutive models. The implicit assumption is that the behavior of the finite-size cylindrical specimen is representative of an infinitesimal region in the continuum. Therefore, a desirable condition during the progress of triaxial tests is to maintain uniform conditions throughout the specimen so that properties determined from the test can be related to uniform effective stresses. Inhomogeneities are introduced through complex boundary conditions and axisymmetric drainage. We focus on the effect that drainage has on the pore pressure distribution, and hence effective stresses, in triaxial creep specimens that are allowed to drain vertically through porous stones at the top and bottom and radially through filter strips along the length of the specimen.

Of interest is the response of triaxial specimens in drained creep tests that are subjected to compression and shearing. These test conditions are encountered in long-term tests conducted at the University of Rhode Island's Marine Geomechanics

---

[1] Assistant Professor of Civil Engineering, Department of Civil Engineering,
University of Hawaii, Honolulu, HI 96822, USA
[2] Professor of Ocean and Civil Engineering, Director of Marine Geomechanics Laboratory,
University of Rhode Island, Narragansett, RI 02882, USA

Laboratory that are part of an overall research program that is investigating the behavior of fine-grained sediments on submarine slopes (Silva et al., 1989; Brandes, 1992; Silva and Brandes, 1996). The type of modeling discussed here is useful in preparing loading schedules that minimize excess pore pressures and hence maintain reasonably uniform effective stresses.

The pore pressure dissipation in triaxial tests has been considered by several investigators, including Gibson and Henkel (1954), Gibson (1963) and Savage (1988), who present theoretical solutions to Biot's equations for an imposed constant rate of axial strain. However, their expressions were derived for elastic materials and assume one-dimensional flow, neither of which apply in our case. Consideration of a realistic elasto-plastic constitutive model that includes time-dependency does not allow for a closed form solution. Instead, the computations are carried out using the finite element code GEO-CP (Brandes, 1992; Brandes et al., 1996), which uses Biot's consolidation equations in combination with Borja and Kavazanjian's (1984) critical state stress-strain-time model. Additional numerical and experimental studies have been conducted to investigate inhomogeneities induced by axisymmetric drainage (Carter, 1982; Atkinson et al., 1985; Airey, 1991), although none of them considers the effects of creep. In the analyses presented below we pay particular attention to the influence that the creep parameter ψ in Taylor's (1942) creep equation has on pore pressures and deformations. A similar parametric study that investigates the predictive capabilities of the three-parameter Singh-Mitchell (1968) creep equation is presented by Brandes et al.(1996).

We center our attention on modeling the pore pressure response during a particular creep test, DS-1, conducted on an undisturbed deep sea clay from the North Central Pacific ocean, 900 km north of Hawaii in water depth of 5800m. Several piston and large-diameter gravity cores from this location have been analyzed and an extensive data base exists on strength, compressibility, permeability, and creep properties (Akers and Silva, 1980; Silva and Jordan, 1984; Siciliano, 1984; Silva and Brandes, 1996). This deep sea 'red clay' consists of 67% (<2μm) clay and 33% silt and has liquid and plastic limits of 86 and 40 respectively. It has a highly flocculated structure that was formed by very slow sedimentation of windblown materials that have settled through the water column (1mm per 1000 years). The upper 4 to 5 m show evidence of apparent overconsolidation, attributed to high interparticle bonding and perhaps some cementation. This type of microstructure is characterized by

Table 1. Material Parameters

Parameter		Input Value
Poisson's ratio,	ν	0.33
Compression index,	λ	0.40
Recompression index,	κ	0.04
Critical state intercept,	e_c	3.86
Critical state slope,	M	1.24
Permeability,	k	10^{-8} m/s
Hyperbolic constants,	a	0.0072
	R_f	0.95
Secondary compression,	ψ	0.010
Singh-Mitchell constants,	A	0.53×10^{-5}
	m	0.80
	α	4.4

relatively high void ratios but low permeabilities. A summary of average material parameters used as input for this study is shown in Table 1. All the analysis presented below use these values unless indicated otherwise.

Constitutive and Numerical Formulation

The constitutive model and finite element formulation implemented in the code GEO-CP are described in detail elsewhere (Borja and Kavazanjian, 1984; Brandes, 1992; Brandes et al., 1996) and only the main features and equations are presented here. Rates of total strains $\dot{\varepsilon}_{kl}$ are composed of elastic ($\dot{\varepsilon}_{kl}^e$), plastic ($\dot{\varepsilon}_{kl}^p$), and creep portions ($\dot{\varepsilon}_{kl}^t$):

$$\dot{\varepsilon}_{kl} = \dot{\varepsilon}_{kl}^e + \dot{\varepsilon}_{kl}^p + \dot{\varepsilon}_{kl}^t \tag{1}$$

The creep-inclusive constitutive equation is given by Borja and Kavazanjian (1984):

$$\dot{\sigma}_{ij} = C_{ijkl}^{ep}\dot{\varepsilon}_{kl} - C_{ijkl}^{ep}\dot{\varepsilon}_{kl}^t - \frac{\dfrac{\partial F}{\partial p_c}\left(\dfrac{\psi}{\lambda-\kappa}\right)\dfrac{p_c}{t_v}C_{ijkl}^e}{\dfrac{\partial F}{\partial \sigma_{mn}}C_{mnkl}^e\dfrac{\partial F}{\partial \sigma_{kl}} - \dfrac{\partial F}{\partial p_c}\dfrac{\partial F}{\partial p}\left(\dfrac{1+e}{\lambda-\kappa}\right)p_c}\frac{\partial F}{\partial \sigma_{kl}} \tag{2}$$

where C_{ijkl}^e and C_{ijkl}^{ep} are the elastic and the modified Cam Clay elasto-plastic consitutive tensors respectively, σ_{ij} is the effective stress tensor, p_c is the preconsolidation stress, e is the void ratio, λ and κ are the critical state compression parameters, ψ is the critical state coefficient of secondary compression ($C_\alpha = \psi \ln 10$), t_v is a state variable that is a measure of void ratio reduction due to creep ($t_v = t_{v_i} \exp(\Delta e/\psi)$), and F is the modified Cam clay volumetric yield function:

$$F = \frac{q^2}{M^2} + p(p - p_c) \tag{3}$$

where p and q are the volumetric and deviatoric stresses respectively, and M is the slope of the critical state line.

Creep strains are calculated using Taylor's (1942) equation for secondary compression:

$$\dot{\varepsilon}_{kl}^t = \frac{\psi}{(1+e)(2p-p_o)t_v}\frac{\partial F}{\partial \sigma_{kl}} \tag{4}$$

Hardening is due to plastic volumetric straining, $\dot{\varepsilon}_v^p$, and time-dependent effects, and is measured by the rate of change of the preconsolidation stress:

$$\dot{p}_c = \frac{1+e}{\lambda-\kappa}p_c\dot{\varepsilon}_v^p + \frac{\psi}{\lambda-\kappa}\frac{p_c}{t_v} \tag{5}$$

Standard finite element discretization procedures lead to a set of non-linear, incremental equations:

$$K\Delta u + L\Delta w = \Delta F$$
$$L^T\Delta u - \Phi\beta\Delta t\Delta w = \Delta H + \Phi\Delta t w\big|_{t=t_n} \quad (6)$$

where the primary unknowns are the displacements **u** and the pore pressures **w**. **K**, **L** and Φ are the displacement and pore pressure stiffnesses. These equations are solved iteratively using the Newton-Rahpson technique with the time integration parameter β set to 0.5 for optimum accuracy and unconditional stability (Brandes, 1992). Variable time increments were used ranging from as little as 0.01 minutes at the start to as much as 1500 minutes at the end. A number of time step schemes were tried and optimum increments were selected for each analysis that resulted in stable predictions that did not vary with further refinement in the size of the time steps. As expected, the larger the applied load, the smaller the time steps necessary during the ensuing pore pressure dissipation.

In the finite element analysis that follows the upper quadrant of the cylindrical specimen is divided into a number of axisymmetric isoparametric elements with four corner pore pressures, four displacement nodes, and four integration points. Three mesh densities were considered consisting of 48, 112 and 231 elements. As discussed below, the 48-element mesh gave results that were deemed unsatisfactory, whereas the 112 and 231-element meshes gave essentially identical results. The analyses were therefore conducted using the 112-element arrangement shown in Figure 1. Drainage occurs through the sides, the top, and the bottom of the sample by specifying zero excess pore pressures along those boundaries. The loading sequence is typical of triaxial creep tests conducted at URI and involves backpressuring to 413 kPa (60 psi), followed by a two-stage isotropic consolidation process, the first one to 41 kPa (6 psi) and the second one to 93 kPa (13.5 psi). The sample is then subjected to gradual shearing by increasing the axial stress while maintaining the lateral stress constant, followed by consolidation and creep under a constant applied stress.

Isotropic Compression

Sample DS-1 was in an overconsolidated state at the beginning of the isotropic consolidation phase due to unloading and apparent overconsolidation. Assuming an average buoyant density of 0.5 gm/cm^3, the vertical overburden stress at the depth interval of the sample of 235-245 cm is about 12 kPa. Since the OCR at that depth is approximately 2 (Silva and Jordan, 1984), the preconsolidation stress is 24 kPa. Therefore, the first isotropic consolidation step of 41 kPa brings the sample into the normally consolidated region. The material is assumed to be in a normally consolidated condition at the beginning of the second consolidation step where we begin our analysis. At that point we close the drainage valve, increase the cell pressure by 52 kPa, and then open it to allow drainage. Numerically, the 52 kPa stress is applied undrained at $t = 0$, followed by full drainage along the top and bottom of the sample at $t = 1$ minute.

The distribution of excess pore pressures w, normalized by the applied stress $w_{o,}$ is shown for two different times in Figure 2. Because of isotropic undrained loading during the first minute, excess pore pressures initially rise uniformly to 52 kPa. As soon as drainage is allowed along the outside of the sample, large pore pressure gradients develop near the boundaries. Due to the low permeability of the material, the interior of the sample has not undergone any consolidation and in fact first sees an increase in excess pore pressure beyond the applied 52 kPa. This is known as the Mandel-Cryer effect (Cryer, 1963) and is due to the coupled nature of the consolidation process. It results from the inward propagation of the effects of volume reduction that take place near the drainage boundaries. Internal locations, which have not had a chance to consolidate, undergo a temporary increase in total stress.

As expected, pore pressures take the longest to dissipate at the center of the sample (Figure 3). The 112 and 231-element meshes result in essentially identical predictions, suggesting that the 112-element mesh density and the element type chosen can be expected to give accurate results for this particular problem. Although all the analysis were conducted by restraining the top of the sample in the horizontal direction to simulate the top cap, Figure 3 indicates that this type of restraint does not affect the pore pressure predictions at the center of the sample and has a minimal effect at node 121 (Figure 1) which is located near the top boundary.

The effect of creep on the pore pressure history at this location is illustrated in Figure 4, which compares predictions for three different values of the creep parameter ψ. There is little difference in the maximum excess pore pressure and the time at which it occurs for the cases with $\psi=0.010$ and $\psi=0.015$, although for $\psi=0.030$ the peak pore pressure is somewhat higher. Maximum pore pressures are 30%, 32% and 42% larger than the applied stress for $\psi=0.010$, $\psi=0.015$, and $\psi=0.030$ respectively. These values fall within the theoretical range of 0% for $v=0.5$ to 57% for $v=0.5$ found for an elastic sphere by Cryer (1963). Once the pore pressures begin to dissipate (Figure 4), they decrease at a slightly higher rate for lower values of ψ. The additional volumetric deformation caused by creep acts to increase the fluid pressure, which in turn delays pore pressure dissipation. Pore pressure throughout the sample has essentially dissipated after 100 minutes, with less than 10% remaining at the center of the sample. A permeability decrease of half an order of magnitude, to $k = 5 \times 10^{-9}$ m/s, results in a similar peak pore pressure of 32% above the applied stress, but drastically reduces the rate of dissipation (Figure 5). Changes in permeability of this order are not uncommon in fine-grained sediments.

Volumetric deformation of the triaxial specimen is affected by creep as shown in Figure 6. During primary consolidation larger values of the creep parameter ψ result in somewhat smaller strains as excess pore pressures dissipate more slowly. After dissipation of the pore pressures, the rate of deformation becomes dependent on the value of ψ. The predictions in Figure 6 that use the average input parameters in Table 1 underestimate the measured values, which were collected with a burette system connected to the sample. However, reasonable agreement is obtained if we assume different permeability and compression parameters (Figure 6). Whereas a permeability of $k = 5 \times 10^{-9}$ m/s falls within the range of values determined for this type of material (10^{-8} to 3×10^{-10} m/s), a compressibility of $\lambda = 0.70$ is slightly larger than the measured range (0.20 to 0.65). It is possible that the measured volume changes were affected by leakage across the sample membrane, which over time can lead to an overestimate of volume change, and hence an apparently larger compressibility. Recent improvements (Tian, 1992) have essentially eliminated this problem resulting in generally lower measured volume changes for similar conditions. Note that the predictions are slightly higher than the test data during the first 50 minutes. This is attributed to the assumption that excess pore pressures readily drop to zero along the outside of the sample when the drainage valve is opened. Given the conventional triaixal set up, where thin filter strips are placed along the length of the sample, pore pressures likely dissipate at a slower rate and therefore volumetric strains are somewhat smaller than the predictions during early consolidation.

Shearing

Following the second isotropic consolidation stage, the axial load was increased while maintaining the lateral pressure constant until the deviatoric load reached 35 kPa, which was 20% of the estimated failure strength. The load was applied gradually over a period of 1600 minutes (Figure 7) in an effort to keep excess pore pressures to a minimum. For the remainder of the test the loads were kept constant while axial and volumetric deformations were monitored for a period of 60,000 minutes (42 days). The vertical load was adjusted daily to account for changes in sample area. Numerically, the analysis assumed initial effective stresses equal to the applied consolidation pressure (93 kPa) and a larger preconsolidation stress (116 kPa) to account for 2928 minutes of creep during the previous isotropic compression stage (see Brandes et al., 1996). The simulated loading followed the same stress-time schedule shown in Figure 7.

The observed general pore pressure pattern was similar to that for the isotropic loading case with sharp pore pressure gradients near the drainage boundaries and slower pore pressure dissipation at the center of the sample. At the center the effect of creep on the pore pressure history is as shown in Figure 8. During the first 440 minutes the pore pressure is significantly affected by changes in the creep parameter ψ as illustrated, with a moderate rise in pore pressure inmediately following the first load for the case with $\psi=0.015$. This initial rise is due to the creep component in the constitutive model as can be seen by comparison with the creep *turn-on* case where no external loads are applied (Figure 8). During the last five load increments there is little difference in the pore pressure response between the analyses with $\psi=0.010$ and $\psi=0.015$.

We can estimate the expected rise in pore pressure w in a triaxial test during undrained loading by the equation (Skempton, 1954):

$$\Delta w = B\Delta\sigma_3 + A(\Delta\sigma_1 - \Delta\sigma_3) \tag{7}$$

where $\Delta\sigma_1$ and $\Delta\sigma_3$ are the applied stresses in the vertical and lateral directions respectively, and A and B are pore pressure parameters that depend on the compressibility of the pore fluid, C_w, and the compressibility of the sediment structure, C_c. In our creep test only the axial stress is increased and therefore:

$$\Delta w = A\Delta\sigma_1 \tag{8}$$

where

$$A = \frac{1}{n\frac{C_w}{C_c} + 3} \tag{9}$$

where n is the porosity. For most soils $C_w/C_c \approx 0$, and therefore $\Delta w \approx \Delta\sigma_1/3$. Note that the increase in pore pressure resulting from the last five deviatoric loads are indeed equal to or slightly larger than 1/3 of the applied $\Delta\sigma_1$, with the excess probably due to the Mandel-Cryer effect (Figure 8). For example, after increasing the load by 3.1 kPa at 1265 minutes the pore pressure rises 1.1 kPa, compared to a predicted value of 1.0 kPa from Equation (8). Similarly, the pore pressure increases 2.1 kPa for an applied load of 6.1 kPa versus a predicted value of 2.0 kPa.

A decrease in the permeability to $k = 5\times10^{-9}$ m/s again reduces the rate of pore pressure dissipation (Figure 9). The reduced permeability does not allow for full dissipation of pore pressures in between the applied loads and pore pressures reach higher total levels as compared to the case with $k = 1\times10^{-8}$ m/s, particularly during the last five loads. Such higher pore pressures lead to less uniform effective stresses throughout the sample. Based on the predictions shown in Figures 8 and 9 we would propose a loading rate that allows for about 50 minutes of drainage between each 1 kPa of load application for a permeability of $k = 1\times10^{-8}$ m/s, and 100 minutes of drainage between each 1 kPa of load application for a permeability of $k = 5\times10^{-9}$ m/s. In general, large load increments should be avoided because of the possibility of initiating premature failure and to increase uniformity of stresses and displacements.

The effect of the creep parameter ψ on axial strain is as shown in Figure 10. A value of $\psi = 0.030$ appears to best match the rate of deformation of the test data after primary consolidation, although the levels of axial strain reached do not match the experimental data very well for the average compressibility in Table 1. There is better agreement with the test data for a compressibility of $\lambda = 0.58$, which still is within the range of values determined for this deep sea clay. The comparisons in Figure 10 indicate that creep is an important component of total axial strain. The differences among the individual curves become most apparent in the long-term, after excess pore pressures have dissipated. This is of special importance to long-term applications such as deformations of slopes since different creep rates can lead to significantly different accumulated deformations over long periods of time (Silva and Brandes, 1996).

Summary

Long-term drained triaxial creep tests conducted at URI involve back-pressuring, isotropic (or anisotropic) consolidation in one or several steps, and step-wise deviatoric loading, followed by consolidation and creep under constant loads for periods ranging from a few weeks to over a year. The effects of this sequence on the pore pressure distribution throughout the cylindrical specimen was investigated in a particular test conducted on a deep sea clay from the North Central Pacific. The analysis was started from a normally consolidated condition and included one isotropic consolidation step and step-wise deviatoric loading to 20% of the estimated failure strength A realistic constitutive model, based on critical state principles, was used to address the more important aspects of fine-grained marine sediment behavior, including elasto-plasticity, creep, and volumetric and time-dependent hardening. Creep strains were calculated using the Taylor (1942) creep equation.

Once drainage is allowed, pore pressures decrease rapidly near the boundaries. Internally, pore pressures are seen to first increase beyond the applied stress before dissipation takes place, which is attributed to the Mandel-Cryer effect. For isotropic loading, maximum excess pore pressures at internal locations agree with the theoretical calculations by Cryer (1963) for an elastic material. For deviatoric loading the peak excess pore pressures are only slightly larger than the values predicted by Skempton's (1954) equation, with the excess again due to the Mandel-Cryer effect. Predicted volumetric strains during isotropic compression and axial strains during shearing match the experimental data for the compressibility parameters near the upper end of values measured for this deep sea clay. A secondary compression parameter ψ of 0.030 best matches the rate of axial strain after primary consolidation.

The effect of creep during pore pressure dissipation is minimal. Creep acts to delay the rate at which both dissipation and strains take place during primary consolidation. Permeability changes of as little as half an order of magnitude have a dramatic effect on the pore pressure history. For this particular material and test configuration, a loading rate that allows for approximately 50 minutes of drainage for every 1 kPa of applied deviatoric load is found to be appropriate for a permeability of $k = 1\times10^{-8}$ m/s. Such a gradual loading rate will minimize excess pore pressures and help maintain nearly uniform effective stresses throughout the specimen.

References

Airey, D.W. (1991). Finite element analyses of triaxial tests with different end and drainage conditions. *Computer Methods and Advances in Geomechanics*, G. Beer, J.R. Booker and J.P. Carter Eds. Rotterdam: A.A. Balkema, 1:225-230.

Atkinson, J.H., Evans, J.S. and Ho, E.W.L. (1985). Non-uniformity of triaxial samples due to consolidation with radial drainage. *Geotechnique*, 35(3):353-356.

Akers, S.A. and Silva, A.J. (1980). Stress-strain and strength properties of marine sediments from the North Pacific. *Technical Report*, Department of Ocean Engineering, University of Rhode Island.

Borja, R.I. and Kavazanjian, E. (1984). Finite element analysis of time-dependent behavior of soft clays. *Geotechnical Engineering Research Report No. GT1* Department of Civil Engineering, Stanford University, CA.

Brandes, H.G., Sadd, M.H. and Silva, A.J. (1996). Finite element modeling of a deep sea clay in long-term laboratory creep tests. To be published by: *International Journal for Numerical and Analytical Methods in Geomechanics*.

Brandes, H.G. (1992). Finite element modeling of time-dependent deformations in marine clays. *Ph.D. Dissertation*, Department of Ocean Engineering, University of Rhode Island, Kingston, RI, 136 p.

Carter, J.P. (1982). Predictions of non-homogeneous behavior of clay in the triaxial test. *Geotechnique*, 32(2):55-58.

Cryer, C.W. (1963). A comparison of the three-dimensional consolidation theories of Biot and Terzaghi. *Quarterly Journal of Mechanical and Applied Mathematics*, XVI(4):401-412.

Gibson, R.B. and Henkel, D.J. (1954). Influence of duration of tests at constant rate of strain on measured drained strength. *Geotechnique*, 4:6-15.

Gibson, R.E. (1963). Analysis of system flexibility and its effect on time lag in pore water pressure measurements. *Geotechnique*, 13:1-11.

Savage, W.Z. (1988). Pore-pressure distributions in constant strain-rate triaxial tests. In: *Advanced Triaxial Testing of Soil and Rock*; R.T. Donaghe, R.C. Chaney and M.L. Silver Eds. ASTM STP 977; Philadelphia: American Society for Testing and Materials, pp. 582-591.

Siciliano, R. (1984). Constitutive behavior of isotropically and Ko consolidated marine sediments in undrained shear. *M. S. Thesis*, University of Rhode Island, Kingston, RI, 319p.

Silva, A.J. and Brandes, H.G. (1996). Drained creep behavior of undisturbed and reconstituted marine clay. ASCE Washington Convention Session, *Measuring and Modeling Time-Dependent Soil Behavior* (this volume).

Silva, A.J., Brandes, H.G., Sadd, M.H., Karamanlidis, D., Tian, W.-M. and Laine, E.P. (1989). Experimental and analytical study of creep deformations of submarine slopes. Proceedings, *Oceans '89 Conference*, 5:1530-1535.

Silva, A.J. and Jordan, S.A. (1984). Consolidation properties and stress history of some deep sea sediments. In: *Seabed Mechanics*; B. Denness Ed., Proceedings of IUTAM and IUGG Symposium; London: Graham & Trotman, pp. 25-40.

Singh, A. and Mitchell, J.K. (1968). General stress-strain-time functions for soils. *Journal of the Soil Mechanics and Foundation Division*, ASCE, SM1:21-46.

Skempton, A.W. (1954). The pore-pressure coefficients A and B. *Geotechnique*, IV:143-147.

Tian, W.-M. (1992). Long-term creep behavior of cohesive sediments under drained conditions. *Ph.D. Dissertation*, Department of Ocean Engineering, University of Rhode Island, Kingston, RI, 269 p.

Taylor, D.W. (1942). Research on consolidation of clays. *Report Serial 82*. Department of Civil and Sanitary Engineering, Massachusetts Institute of Technology, Cambridge, MA.

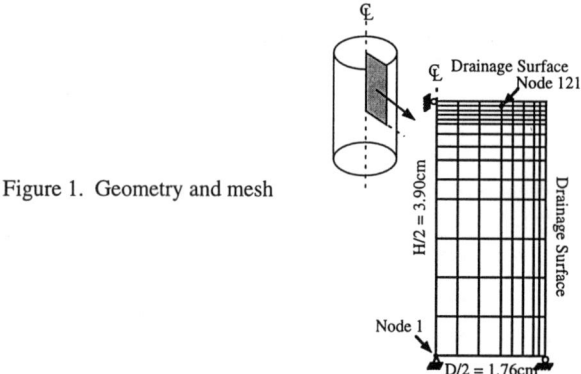

Figure 1. Geometry and mesh

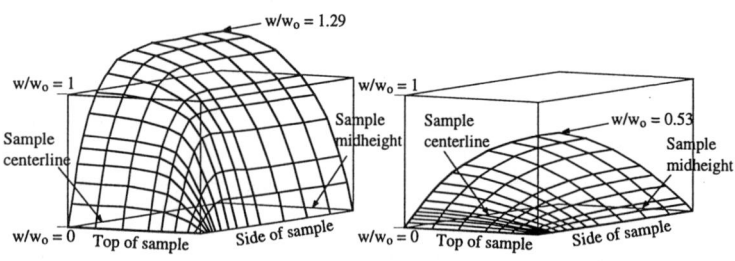

Figure 2. Pore pressure distribution in triaxial specimen

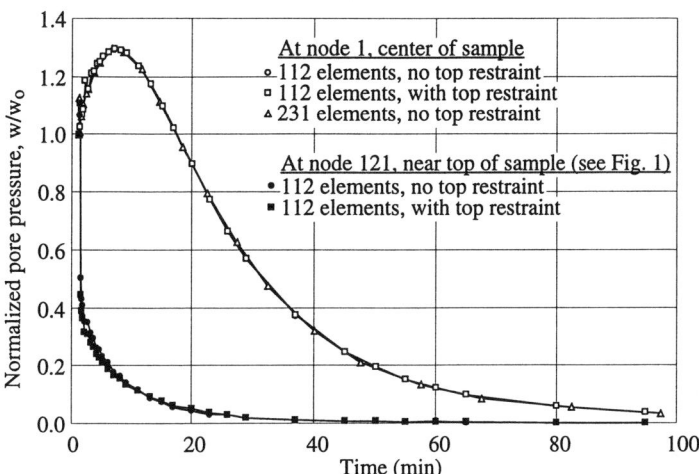

Figure 3. Mesh density and top restraint effects on pore pressure

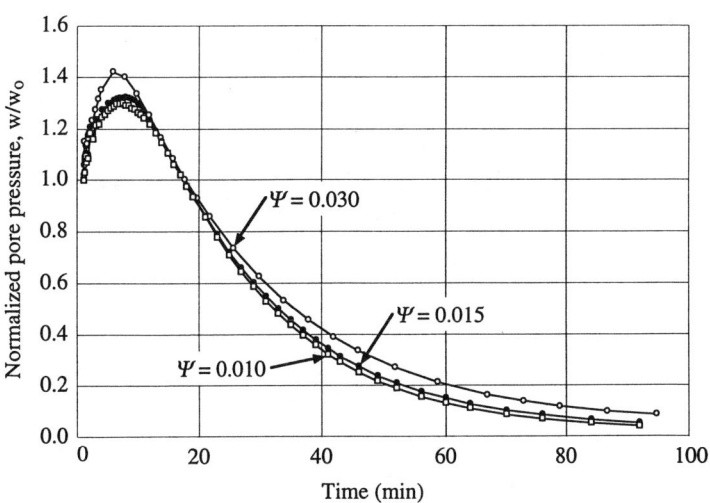

Figure 4. Effect of creep on pore pressure at center of sample (other parameters in Table 1)

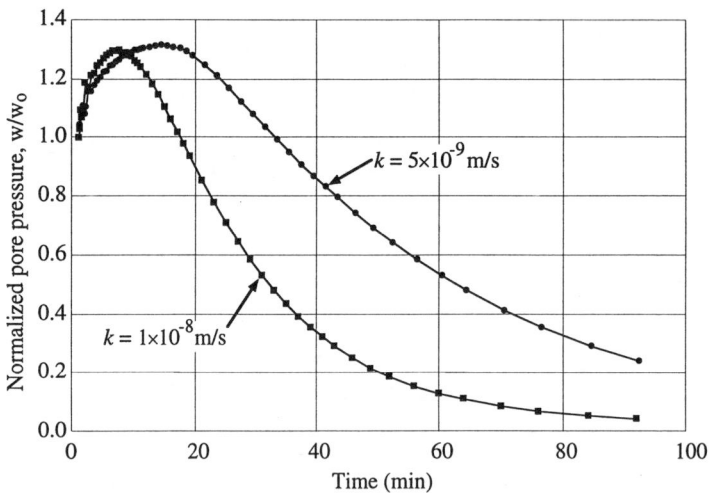

Figure 5. Effect of permeability on pore pressure at center of sample (other parameters in Table 1)

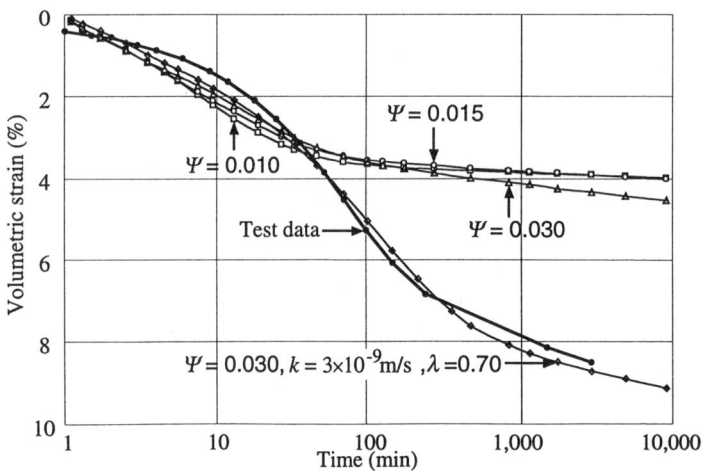

Figure 6. Isotropic compression - effect of secondary compression coefficient (other parameters in Table 1)

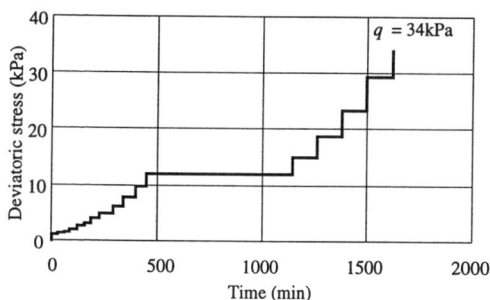

Figure 7. Deviatoric loading schedule

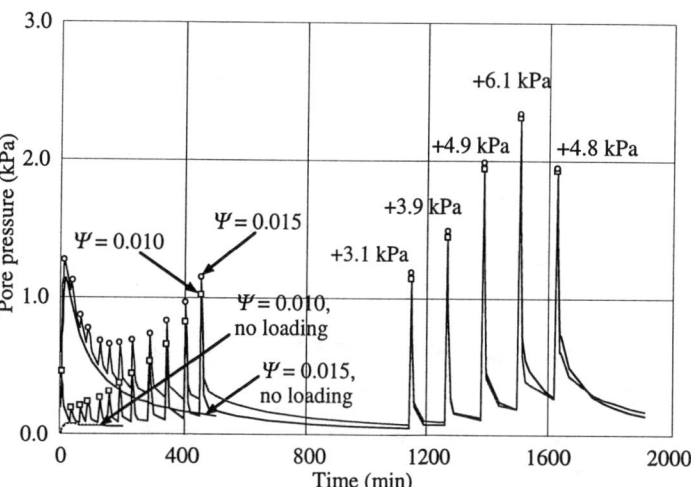

Figure 8. Effect of creep on pore pressure at center - shearing (other parameters in Table 1)

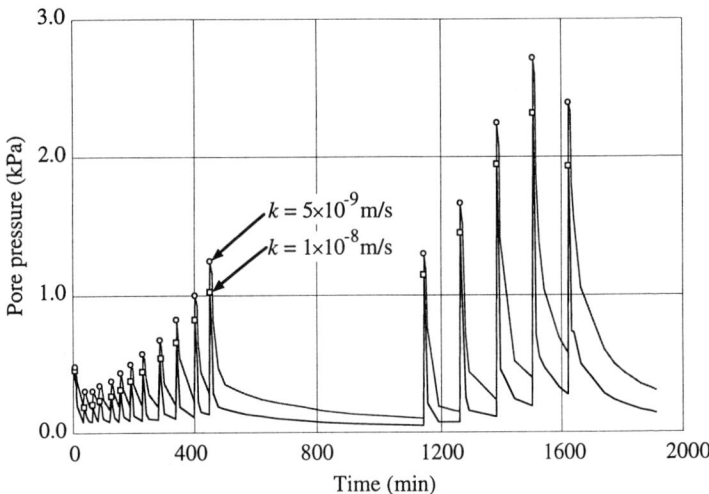

Figure 9. Effect of permeability on pore pressure at center - shearing (other parameters in Table 1)

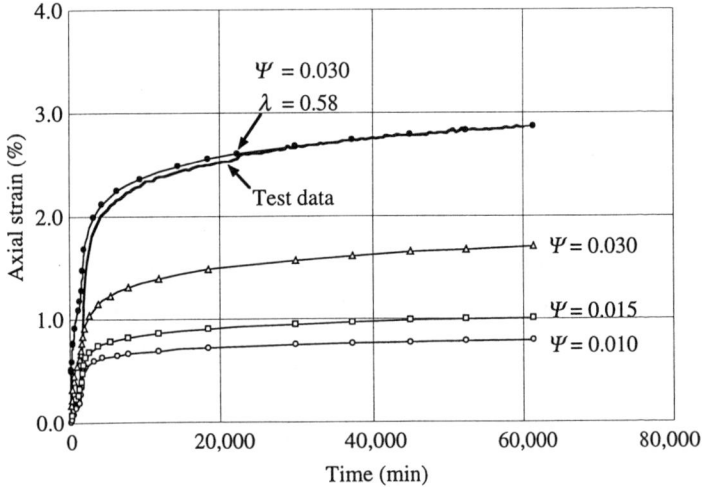

Figure 10. Effect of secondary compression coefficient - shearing (other parameters in Table 1)

Stabilization of a Creeping Slope Using Soil Nails

By Peter R. Cali[1], Member ASCE

Abstract

Soil nails are being used in an attempt to stabilize a creeping bank along the Mississippi River in Baton Rouge, Louisiana. Riverward creep of a 150 m section of the bank has been documented at the site for over 60 years. The movement has caused distress in an adjacent railroad and highway embankment, and has jeopardized plans for improvement of the flood protection in the vicinity. Slope inclinometers indicated the maximum rate of creep to be 7 cm per year, occurring along a well-defined arc within a Pleistocene clay formation. A single row of 126 steel H-piles were driven to pin the upper creeping zone to the lower stable zone. The design approach assumed that the reduction in mobilized soil resistance due to a tenfold reduction in strain rate in the creep zone would be transferred to the piles. Continued monitoring of the instrumentation will indicate their effectiveness.

Stability analysis of the bank, performed using residual strength values from unconsolidated-undrained triaxial compression tests (strain rate 1% per minute), yielded safety factors greater than 1.40, which did not model the creep mechanism. To estimate the shear strength more applicable to high strain, but low strain rates, new borings were made and tested. Strain rates were slowed and varied during unconsolidated-undrained triaxial compression testing, and consolidated-drained repeated direct shear tests were performed at strain rates of 0.00046 cm per minute to a displacement of 20-36 cm. From these data, the in-situ mobilized shear strength was estimated, and the effect of reduced in-situ strain rates on available resistance was approximated. The additional resisting force needed to achieve the design safety factor was the basis for the soil nail design.

---

[1]Geotechnical Engineer, U.S.Army Corps of Engineers, CELMN-ED-FD, P.O. Box 60267, New Orleans, LA 70160-0267

Introduction

This paper introduces a case history involving mechanical stabilization of a a creeping section of riverbank along the Mississippi River in Baton Rouge, Louisiana. Riverward creep of the bank is seasonal, fluctuating with the rise and fall of the Mississippi River. The bank required stabilization before the flood protection could be raised. The site was monitored using slope inclinometers and piezometers for several years prior to stabilization efforts. A stability berm was not economically or logistically feasible. Steel H-Pile inclusions, generically referred to as soil nails in this paper, were driven in November 1995 to slow the creep to an acceptable level.

Soil nails have been used since the 1950's in Europe, and with increasing frequency in the United States, to reinforce earth retaining structures, unstable slopes, and sliding earth masses. Two manuals provide design methods for most soil nailing applications: Mitchell and Vliet (1987) and an English language translation of the French soil nailing design manual, *Recommendations CLOUTERRE-1991*, published by the U.S. Federal Highway Administration (1993).

Background

Baton Rouge, Louisiana is located approximately 425 km north of the mouth of the Mississippi River at the Gulf of Mexico. This represents the northern extent of the man-made flood protection levee system on the left descending bank of the Mississippi River in Louisiana, beyond which higher bluffs provide flood protection. The river stage varies from the low water reference plane of el. 0.8 m above sea level, or National Geodetic Vertical Datum (N.G.V.D.), to project flood stage of el. 14.3 m NGVD. Flood protection at the site is scheduled to be upgraded from the existing el. 14.3 m NGVD to the required design grade of el. 15.2 m NGVD. A small levee or floodwall would provide the grade increase. Although the small levee represents a minor added load, the overall embankment stability was questioned when settlement and cracks were observed in the roadway just landward of the embankment and settlement of the railroad ballast became a significant maintenance problem. Figure 1 is a site plan showing the flood protection and the proximity of the roadway, an active railroad, and a cluster of oil transport pipelines parallell to the levee, none of which can be relocated.

Subsurface investigations provided a look at the soil stratification beneath the embankment. Soil borings revealed the presence of a large zone of gravel and rubble above an overconsolidated Pleistocene clay formation, all of which is overlain by accretionary deposits of silt and silty clay. The three-dimensional extent of the rubble lens was mapped using geophysical methods. Figure 2 is an isopach map, showing the thickness and lateral extent of the gravel deposit, which roughly coincided with the observed cracks in the landside roadway. Research into 18th and 19th century hydrographic survey maps revealed numerous riverbank scarps occurred in the vicinity, which had been the site of a river ferry landing since the mid 19th century. It was concluded these scarps were backfilled with gravel, bricks, and ballast stone for construction of the the railroad. A cross section cut through the embankment, Figure 3, details the stratification of the foundation and illustrates the proximity of the utilities affecting the design.

Instrumentation was installed in 1988 to monitor the bank movement, which was found to occur only during river stages below el. 4.6 m NGVD, characteristically from June-November. An unrelated shuffling of project priorities and funding delayed the schedule of the Baton Rouge flood protection enlargement project. This permitted several years of instrumentation monitoring before design, a rare luxury for geotechnical engineers, but invaluable in this case given the seasonal nature of the bank creep.

Slope inclinometers were monitored closely, particularly during periods of active movement. Instruments were replaced promptly after excessive deflection rendered them unreadable (approximately 10 cm) to maintain a continuity of data. The slope inclinometer data indicated movement perpindicular to the bank at a rate of 7 cm/year maximum during extreme low water periods. Figure 4 presents a time log of cumulative instrument deflection at the location of the series of inclinometers noted as SI-1 in Figures 1 and 3. Movement here occurred at el. -1.2 m NGVD, and clearly a falling river stage was the catalyst for bank creep. Along with field observation of the surface cracks and the results from other inclinometers, the failure was modeled as a flat circular arc through the Pleistocene clay formation. At the location selected for the soil nail row, 30.5 m riverward of the flood protection, the depth of failure was estimated to be el. -3.1 m NGVD.

Piezometers were read when the slope inclinometers were read, and when possible during high river stages. A time log for Piezometers P-1 and P-2, located as shown in Figures 1 and 3, and a correlation to river stage are presented in Figure 4. The piezometric data showed nearly total phreatic surface lag in the bank, indicating that a rapid drawdown condition existed during low river stages. Along with recharge from the river, runoff from the adjacent landside ground recharged the pool of water in the rubble deposit. Closer examination of the piezometric data indicated that the head did not dissipate with time due in part to the impervious nature of the accretionary deposits.

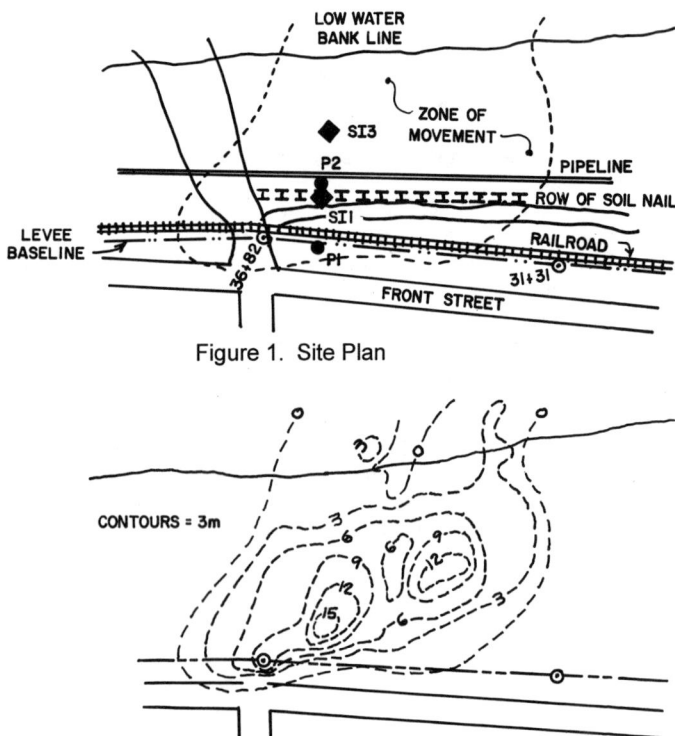

Figure 1. Site Plan

Figure 2. Gravel Isopach

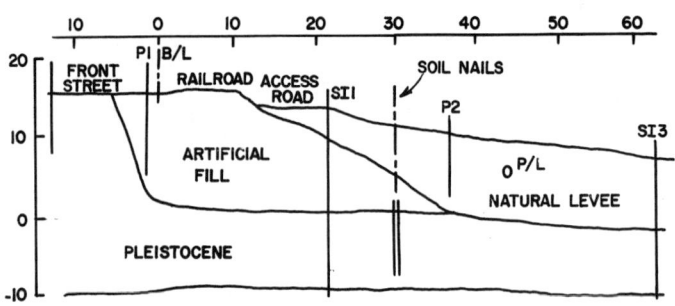

Figure 3. Stratigraph of the Riverbank at Baton Rouge

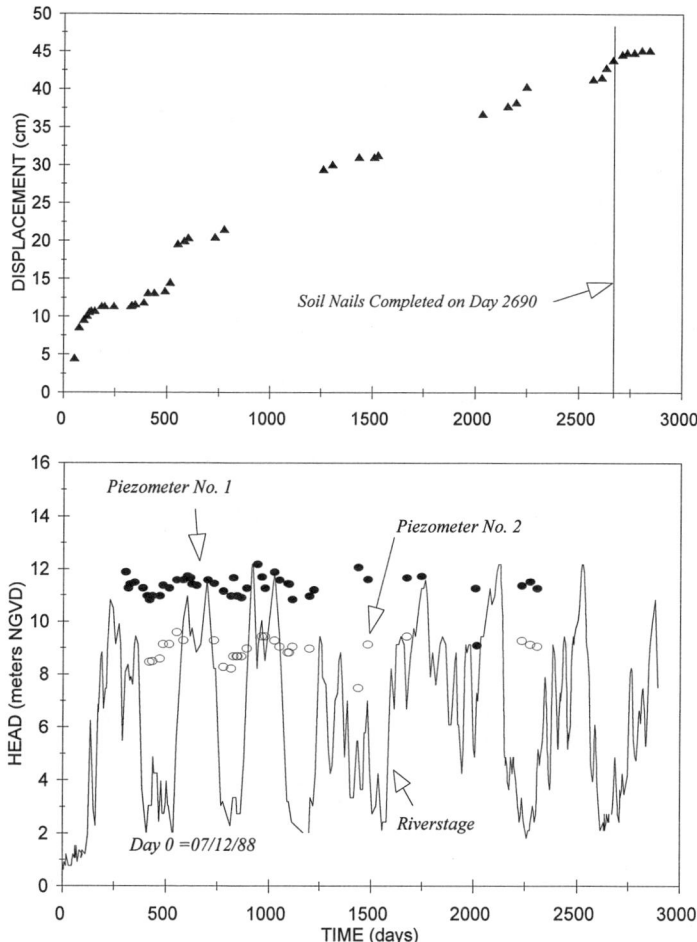

Figure 4. Correlation of Piezometer and Slope Inclinometer Data to River Stage

It was determined that the zone of creep was within the Pleistocene clay formation between el. +3.0 and el -9.0 m NGVD. This deposit is a stiff to very stiff fat clay, having generalized properties of water content = 20-50%; liquid limit = 40-84; dry density = 15-18 kN/m^3; maximum past consolidation pressure = 280-380 kPa; and undrained shear strength = 48-77 kPa. Undisturbed borings were made and continuous sampling was performed using a 12.7 cm diameter Shelby tube sampling barrel. The samples were extruded and classified at the New Orleans District soils laboratory and tested at the Waterways Experiment Station, Vicksburg, Mississippi. Particular attention was paid to the strength-strain rate dependence of the Pleistocene clay. Conventional unconsolidated-undrained (UU) tests performed at a uniform strain rate of 1.0% per minute, e.g., see Figure 5, showed a characteristic residual shear strength of 40-60% of the peak shear strength. To estimate the shear strength more applicable to high strain, but low strain rates, additional borings were made and tested. Strain rates were slowed and varied during unconsolidated-undrained triaxial compression testing according to procedures developed by Leinenkugel (1976) and reported by Winter *et al.*, (1983). In addition, consolidated-drained repeated direct shear tests were performed at a strain rate of 0.0009 cm/min and sheared to displacements of 20-36 cm by reversing the direction of shear after each 2.5 cm of displacement. Normalized values for this "ultimate" strength for a typical test are presented in Figure 6. The results of the UU tests with the variable strain rate technique and the CD-DS tests are summarized in Table 1. Profiles of peak and residual shear strength with depth are shown in Figure 7.

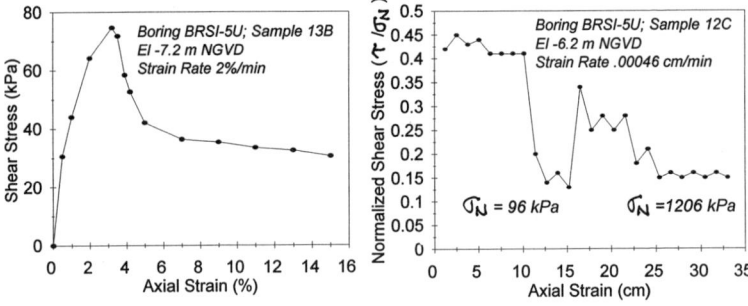

Figure 5. Stress-Strain Curve UU Test

Figure 6. Stress-Strain Curve CD Repeated Direct Shear Test

Table 1. Shear Test Data

Consolidated Drained Repeated Direct Shear Tests

Sample	Elev m NGVD	ε cm/min	τ/σ_N σ_N=147kPa	τ/σ_N σ_N=287kPa
12U-9D	+0.7	.0.0009	0.15	0.06
12U-13D	-4.1	.0.0009	0.22	0.24
13U-10C	-1.6	.0.0009	0.12	0.10
13U-11D	-4.8	.0.0009	0.54	0.46
13U-13C	-5.9	.0.0009	0.33	0.24
2.5U	-1.2	0.00046	0.12	0.15
2.5U	-2.5	0.00046	0.11	0.18
5U	-6.2	0.00046	0.13	0.15

Unconsolidated Undrained Triaxial Compression Test with Jump Technique

Sample	Elev m NGVD	ε % per min	c residual kPa	I_V
12U-13C	-3.8	2.0-0.067	43	0.05
12U-14B	-5.1	2.0-0.083	25	0.04
13U-10C	-1.4	2.0-0.083	61	0.03
13U-11D	-2.9	2.0-0.083	69	0.03
13U-13C	-5.0	2.0-0.083	48	0.07

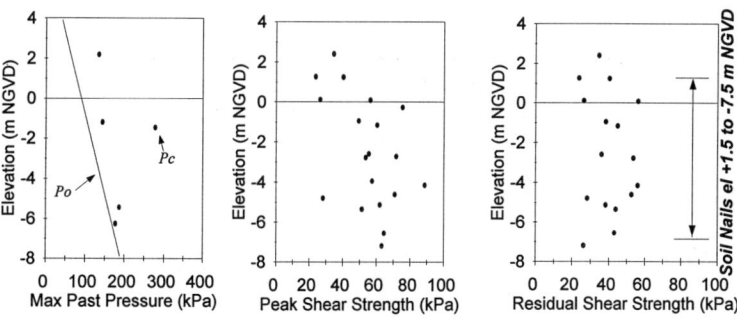

Figure 7. Shear Strength Profile

Slope stability analyses, as summarized in Figure 8, were performed to model the bank movement using residual undrained shear strengths in the clay, and the zone of movement defined by the inclinometers and observed cracks. The assumed failure surface which passed through the known points of movement yielded a factor of safety of 1.40, which did not model the failure. However, after repeating the analysis, using the average of the ultimate shear strength values obtained from the repeated direct shear tests (τ/σ_N = 0.20), a factor of safety of 0.93 was calculated. This represented a more accurate model of the creep mechanism. Conventional stability berms were designed, but dismissed due to cost and interference with adjacent commercial river traffic. Soil nailing was examined as an alternative that might reduce the magnitude of bank creep to an acceptable level, which was set as a tenfold reduction in the rate of creep.

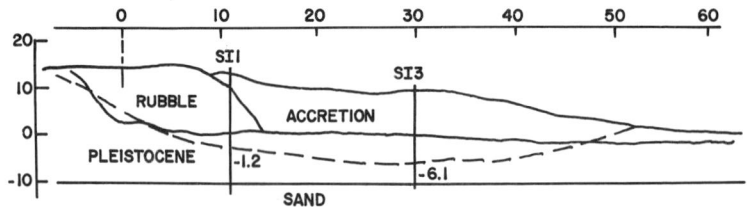

CASE	SOIL PROPERTIES									RIVER STAGE m NGVD	FACTOR OF SAFETY
	RUBBLE			ACCRETION			PLEISTOCENE				
	γ kN/m³	ϕ deg	c kPa	γ kN/m³	ϕ deg	c kPa	γ kN/m³	ϕ deg	c kPa		
1	20.4	11	0	18.4	15	9.6	17.3	0	38.3	0.8	1.40
2	20.4	11	0	18.4	15	9.6	17.3	0	0.20*	4.6	1.02
3	20.4	11	0	18.4	15	9.6	17.3	0	0.20*	0.8	0.93
4	20.4	11	0	18.4	15	9.6	17.3	0	0.20*	0.8	1.20

*τ/σ_N = 0.20 from consolidated-drained repeated direct shear tests
Case 1 mean low water and residual strengths from UU shear tests
Case 2 river stage at which movement begins and ultimate strengths from repeated direct shear tests
Case 3 mean low water and ultimate strengths from CD repeated direct shear tests
Case 4 design condition for Case 3 with mean low water and ultimate strengths from repeated direct shear tests; required resistance from soil nails = 219 N/m

Figure 8. Slope Stability Analysis

Soil Nail Design

The soil nail design method used is from a compilation of methods from Mitchell and Villet (1987), and is based on methods by Ito and Matsui (1975); Matsui et al. (1982); Brinch-Hansen (1961); Leinenkugel (1976); and Winter et al. (1983). The procedure makes several simplifying assumptions:

- The creeping slope acts as an isotropic, rigid plastic, flowing mass. The load imparted to the pile inclusions as a result of the flowing soil is described by Ito and Matsui (1975). This load is transferred to the lower, non-creeping soil according to the simplified pressure distribution proposed by Brinch-Hansen (1961). This can be taken as the maximum lateral load which can be transferred to the soil by the design pile.
- A reduction in the flow velocity, or strain rate, causes a corresponding reduction in the mobilized shear strength of the creeping zone. Conversely, if the mobilized shear strength is reduced, the creep rate will slow. The difference in the the mobilized shear strength needed to reduce the creep rate to an acceptable level over the contributary area of the failure zone becomes the design load on the soil nails. This approach was proposed by Winter et al. (1983).
- Unlike true soil nails, the steel H-piles are rigid inclusions. This implies that the main mechanism of soil-pile interaction is the passive soil resistance against the face of the pile. An assumption of flexible inclusions would require large displacements to mobilize the soil-pile friction as a means of developing the limit lateral earth pressure on the pile.

The designer further assumed that the slope inclinometer data accurately defined the slip surface. Additionally, the soil nail design had to satisfy two criteria: 1) they must provide 219 kN/m resistance for the 152 m section of creeping bank to bring the sliding mass into static equilibrium; and 2) they must be able to transfer the shear forces imparted by the creeping zone to the lower non-creeping zone, reducing the shear stress in the soil, therby reducing the creep.

Creep Rate Reduction

In order to slow the creep rate by a factor of 10, from 7 to 0.7 cm/year, the shear stress along the creep plane would have to be reduced by $\Delta\tau$. The corresponding change in mobilized shear strength is given as:

$$\Delta\tau = -s_u(\varepsilon_i)I_v \ln(\varepsilon/\varepsilon_i) \tag{1}$$

where ε_i = present strain rate; ε = reduced strain rate; $su(\varepsilon_i)$ = shear strength associated with strain rate ε_i; and Iv = viscosity index. The viscosity index can be determined in the laboratory according to the procedure outlined by Leinenkugel (1976) and applied by Winter (1983), in which a soil sample is allowed to reach its residual strength at high strain, then is subjected to positive and negative jumps in the applied strain rate. The resulting increase and decrease in shear strength give the stress-strain relationship at high strains. The average viscosity index was calculated to be 4.2 for five samples taken from the vicinity of the failure plane. An empirical relationship was presented by Winter (1983), which relates viscosity index to liquid limit, Figure 9. The laboratory determined values from this case history are in close agreement with Winter's correlation for the test liquid limits of 72-84, empirically estimated to be $I_v = 4.1$.

The required pile resistance (Q) was then calculated by multiplying the mobilized shear strength reduction for a ten-fold strain rate reduction ($\Delta\tau$) by the contributory area per pile

$$Q = S_{eq}\Delta\tau \qquad (2)$$

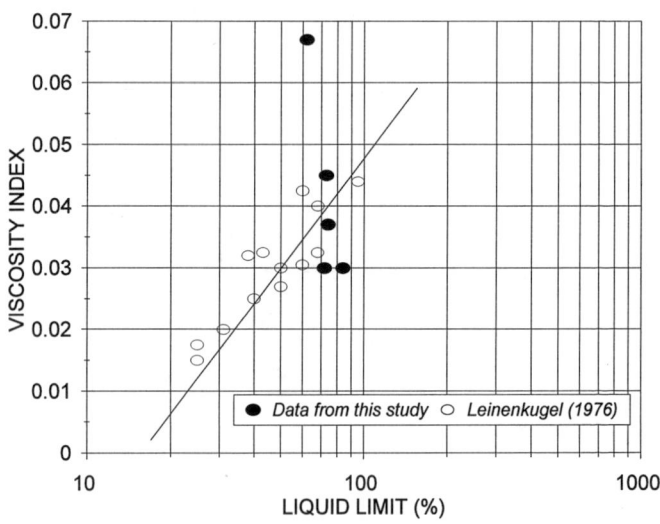

Figure 9. Relationship between Viscosity Index (I_v) and Liquid Limit (after Leinenkugel, 1976)

Lateral Load

The equations developed by Ito and Matsui (1975) describe the lateral load acting on a pile as a result of soil flow in the direction of the x-axis. It is assumed that the soil is in a plastic state, and that the Mohr-Coulomb yield criteria apply within the cross-hatched area shown in Figure 10.

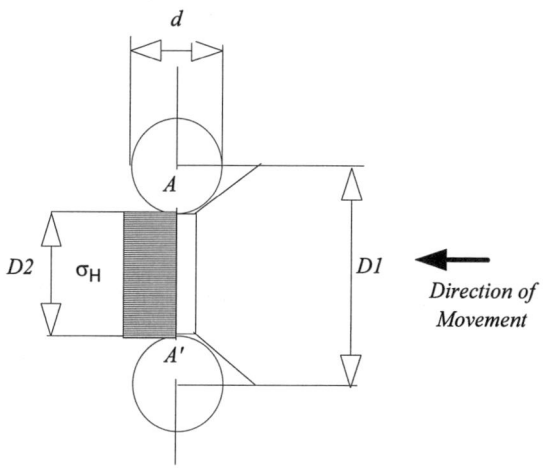

Figure 10. Pressure Distribution on Row of Piles
Due to Plastic Flow of Soil (after Ito and Matsui, 1975)

The equations, which were developed for circular piles, are being applied to H-Piles, introducing a slight error. The lateral pressure p acting in the direction of the x-axis on a cross-section of piles having spacing, D_1, is given by Matsui et al. (1982) as:

$$p = cD_1 \left(3 \ln \frac{D_1}{D_2} + \frac{D_1-D_2}{D_2} \tan \frac{\pi}{8} \right) + \sigma_H (D_1-D_2) \qquad (3)$$

where c is the cohesion of the clay; D_1 and D_2 are the center-to-center and clear intervals between piles, respectively; and σ_H is the lateral earth pressure acting on the Plane A-A'.

Using p as the uniform lateral pressure applied to the piles, length $2h$, as a result of the monolithic displacement of the creeping mass, and using the Brinch-Hansen pressure distribution, shown in Figure 11, the total load per pile is calculated as:

$$Q = 0.414 p h \qquad (4)$$

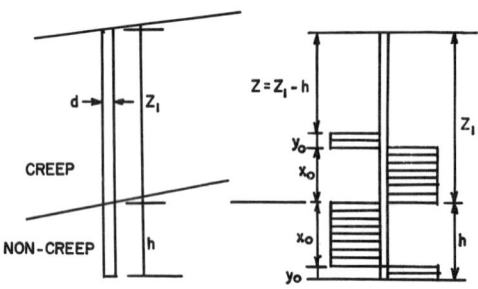

Figure 11. Idealized Pressure Distribution for Perfectly Plastic Flow between the Inclusions (after Brinch-Hansen, 1961)

Pile types, lengths, and spacings were tried until the lateral pile load required to slow the creep (Q from Eq 2) was equal to the capacity of the soil to transfer the load (Q from Eq 4). By trial and error, a pile spacing can quickly be determined. The piles were then sized for structural adequacy considering shear and bending.

Construction

In November 1995, 126 HP-14x73 piles were driven 1.2 m on center in a line paralleling the bank, oriented with the strong axis in the direction of bank movement. The location of the pile row was carefully positioned at 30.5 m riverward of the levee baseline to avoid interference with active utilities. The design piles are 9.1 m in length, with the butt at el. +1.5 m NGVD and the tip at el. -7.6 m NGVD, straddling the failure plane of el. -3.1 m NGVD at this location. Pilot holes were predrilled to full depth to insure plumbness and proper orientation. The holes were backfilled with a sand-cement grout mixture immediately after driving each pile. Pile driving was accomplished using a single-acting drop hammer fitted with a follower.

This paper serves as an introduction to the case history. Slope inclinometers are in-place and the effectiveness of the soil nails in reducing the creep rate is being monitored. Shortly after installation of the nails in November 1995, the Mississippi River rose to its high water level, and remained there until July 1996. No measurements were possible during this period, but as stages drop, the success or failure of the H-piles in slowing the creep will become apparent. A follow-up paper is planned, describing all the soil testing methods in detail, as well as comparisons between laboratory and observed field data, and hopefully a discussion between the techniques used here to evaluate creep potential and those proposed by Sheahan (1994).

References

Brinch-Hansen, J. (1961). "The Ultimate Resistance of Rigid Piles Against Transversal Forces," *Bull.*, Danish Geotechnical Institute, No. 12.

Ito, T., and Matsui, T. (1975). "Methods to Estimate Lateral Forces Acting on Stabilizing Piles," *J. JSSMFE*, Vol. 15, No. 4, pp 45-49.

Matsui, T., Hong, W.P., and Ito T.(1982). "Earth Pressures on Piles in a Row Due to Lateral Soil Movements," *J. JSSMFE*, Vol. 22, No. 2, pp 71-81.

Leinenkugel, H. J. (1976). "Deformations und Festigkeitsver Halten Bindiger Erdstoffe," Experimentelle ErgesBnisse und Ihre Physikalische Deutung Veroffentl. Inst. f. Bodenmech. u Felsmech. 66, Karlsruhe.

Winter, H., Schwarz, W., and Geudhus, G. (1983). "Stabilization of Clay Slopes by Piles," *Proc. 8th European Conference on Soil Mechanics and Foundations Engineering*, Helsinki, Vol. 2.

Mitchell J. K., and Villet, W.C.B. (1987). *Reinforcement of Earth Slopes and Embankments*, National Cooperative Highway Research Program Report 290, Transportation Research Board, National Research Council, Washington, D. C., pp 297-312.

Recommendations CLOUTERRE-1991, (English Translation) Soil Nailing Recommendations-1991, Report FHWA-SA-93-026, U.S. Federal Highway Administration, Washington D.C. (1993)

EVALUATION OF LONG-TERM TIME-RATE PARAMETERS OF SUBGLACIAL TILL

C. L. Ho[1], M. ASCE, J. C. Vela[2], S.M. ASCE, P. U. Clark[3], and J. W. Jenson[4]

Abstract

Geotechnical tests have been conducted to interpret laboratory tests to evaluate the rheological parameters of the clay-rich till deposited by the Lake Michigan Lobe of the Laurentide Ice Sheet. Yield strength, residual strength, frictional parameters, and time-rate parameters are established for implementation into a nonlinear numerical model. The model is based on a general constitutive law that can treat a wide range of sediment behavior. The effects of deforming subglacial till on ice sheet dynamics can then be evaluated.

Clay and till samples were collected from an open pit site in Wedron, Illinois. Consolidation, triaxial, and direct simple shear tests employed in a two test methodology are used to determine the parameters. The laboratory method utilizes a phenomenological rheological model based on a non-linear viscoplastic model of mass movement. Three different methods are used to evaluate the time-rate behavior based on stress controlled strain-rate test measurements taken at time intervals between one hour and 24 hours.

The long-term time-rate value being approached by the trend developed in plotting the parameters against time is evaluated. It was found that the long-term evaluation of the time-rate parameters is effective in establishing the rheologic parameters, as the test results indicated the existence of a limiting value over long term testing . Characterization of viscoplastic response over a time interval representative of continental glaciation can be based upon shorter term laboratory tests by mathematically evaluating the long term limiting value.

---

[1] Assoc. Prof., Dept. Civ. & Env. Engrg., U. Mass., Amherst, MA 01003.
[2] Res. Asst., Dept. Civ. and Env. Engrg., Wash. St. Univ., Pullman, WA 99164.
[3] Assoc. Prof., Dept. of Geosciences, Ore. St. Univ., Corvallis, OR 97331.
[4] Asst. Prof., Water and Energy Res. Inst. W. Pacific, Univ. of Guam, Mangilao, Guam 96923.

Introduction

Time-rate deformation of soil plays an important role in aspects varying from design considerations on and around slopes, to the influence a deforming subglacial sediment has on glacial ice-sheet growth and collapse resulting from viscous behavior of subglacial soil. In many geotechnical mass movement problems, the expense and difficulty of direct field observations of plastic and viscoplastic parameter prevent their use in mass-movement evaluation. Determination of time-rate parameters of a soil allows for full rheologic characterization of a material in a nonlinear numerical modeling application. Establishment of an experimental method to evaluate rheologic parameters of a soil, and incorporation of the mechanical properties into a nonlinear, rate-dependant, constitutive model, allows for the influence of the deforming soil under consideration to be determined.

Iverson (1985) derived a general constitutive relation for slow mass movement phenomena as

$$T = 2\mu D + \frac{\kappa + \alpha p}{\Pi_D^{\kappa}} D \tag{1}$$

with

$$\mu = \mu_0 \left(\frac{\Pi_D^{\kappa}}{D_0} \right)^{\frac{(1-n)}{n}} \tag{2}$$

where the deviator stress tensor, T, is defined by the apparent viscosity, μ, a rate of deformation, D, the Drucker-Präger parameters for cohesion and friction, κ and α, the mean confining pressure, p, and the second principal invariant of the deformation rate, Π_D. The parameters of equation 2 (μ_0, D_0, and n). Although developed for application to creeping landslides, this equation is appropriate for coupled ice sheet movement over deforming sediment because such a system is merely a special case of mass movement (Jenson et al, 1995). For n = 1 the equation reduces to the linearly viscous (Bingham) flow law.

Procedures described herein to evaluate results of time-rate laboratory experiments, provide a means to characterize rheologic behavior of a soil. Three methods to determine time-rate parameters are described. The first method evaluates time-rate parameters for constant stress conditions. The second method evaluates time-rate parameters for given time increments. The third method evaluates the trend of the time-rate parameters over time.

The rheologic behavior of a soil can be characterized by the established laboratory parameters. In site specific cases, a reference depth, D_o, is a function of slope geometry, shear stress, and flow regime. For observable mass movement phenomena such as landslides, D_o can be inferred from the field data (Iverson, 1984; Wong, 1992; Wong et al, 1995). In such cases where direct observation is financially or physically impossible, reference parameters need to be determined by calculating a velocity profile using the established test parameters. Newtonian reference parameter can be determined from graphic analysis of the velocity profile (Jenson, 1993).

The Batestown till used in this study was recovered from an open pit quarry site near Wedron, Illinois. The till is a sandy, clayey silt (ML) with liquid limit of 17 % and a plastic limit of 15 % (33% sand, 46% silt, 21% clay; clay fraction 73% illite, 20% kaolinite + chlorite, 7% expandable clay minerals). This study is part of a larger project to investigate the influence of subglacial till rheology on glacial movement (Jenson, 1993; Vela et al, 1996; Jenson, et al, 1995, 1996). The active behavior of the southern lobes of the Laurentide Ice Sheet, which overrode this site some 18,000 to 21,000 years ago (Clayton et al, 1985), along with the accessible and well-documented sediment record at the site made the subglacial till suitable for evaluation. The matrix of the till was predominantly fine-grained material with interspersed rock larger than 9.5 mm (3/8 in) diameter consisting of less than 2% of the total material by weight.

Test Methodology

Three types of tests were incorporated in the program to determine the nonlinear viscoplastic behavior of the soil. Consolidation tests were preformed on undisturbed block samples of lacustrine clay to establish an upper bound of possible confining stress on the till layer. The remainder of the program consisted of two types of tests on specimens from the till layer remolded to two different densities (2.08 and 2.39 g/cm$^{3)}$. The specimens were formed in cylindrical molds with a static vertical load. The specimens were molded to test size because of difficulties in trimming such dense samples. The two tests used were consolidated drained shear strength tests and consolidated drained time-rate tests conducted in triaxial and direct simple shear devices.

The consolidated drained strength tests established the plastic yield parameters. The time-rate tests established the time-rate or rheological parameters. In the isotropically consolidated triaxial testing sequence each of the two specimen densities were used in the strength tests and time-rate tests. In the simple shear testing sequence, only specimens from the lower densities were used in the strength tests and in the time-rate tests.

Determination of Rheologic Parameters

To establish an upper limit for the confining stress of the till overlying the lacustrine clay layer, consolidation tests were performed on the dense clay samples. Casagrande (1936) construction was used to determine a past preconsolidation pressure, σ_p', of 1500 kPa for the soil. This value is consistent with estimations of the ice thickness and the relatively thin layer of subglacial soil. Since the exact condition of the soil at the time of glaciation is not known, establishing the upper bound for confinement through consolidation and establishing a lower bound of confinement at normally consolidated conditions becomes important for the evaluation of the nonlinear viscoplastic parameters. Actual in-situ conditions would be enveloped between these upper and lower bounds.

The values of past preconsolidation pressures derived from the block clay samples suggest an ice thickness of 140-170 meters. These estimates are close to those suggested by Sauer and Christiansen (1991) for the Battleford formation in Saskatchewan, and the East-White sublobe of central Indiana studied by Harrison (1958). Consistency in the results gives an indication that the obtained values for past preconsolidation pressure are acceptable as an upper bound for effective confining stress.

Yield strength parameters

There are many different ways to define the yield point of a stress-strain curve (Desai and Wu, 1976; Chen and Saleeb 1982; Desai and Siriwardane 1984; Chen and Baladi 1985; Hovind 1990; Wong 1992; and Wong et al, 1995). The secant modulus method (Hovind 1990; Wong 1992; and Wong et al, 1995) ensures that linear stress-strain modeling below yield remains within a predetermined tolerance and provides continuity at yield between non-yield and yield conditions.

Wong (1992) defined the yield stress in terms of the ultimate deviator stress by the equation

$$\sigma_y' = (1 - A) \sigma_{ult}' \qquad (3)$$

where σ_y' is effective yield strength, and σ_{ult}' is the ultimate strength, and the proportionality constant, A, is a function of the hyperbolic stress-strain parameters and a tolerance or limiting strain (Wong, 1992; Wong et al, 1995).

By establishing the hyperbolic parameters for the till and choosing an appropriate limiting strain, the yield stress and the value of A can be calculated. As the value for limiting strain increases, so does the value of the yield strength. When the value of the limiting strain is in the range of 0.2 to 0.4 % the yield strength is about half of the ultimate strength (Wong, 1992; Wong et al, 1995).

Yield strength parameters were determined from five normally consolidated triaxial strength tests and nine normally consolidated direct simple shear tests. Hyperbolic stress-strain parameters and yield strength parameters and values of A are shown in Table 1. Yield strength parameters for the normally consolidated specimens were calculated using the secant modulus method (Wong's, 1992; Wong et al, 1995). Limiting strain of 0.2% (recommended by Wong's, 1992; Wong et al, 1995) was selected to evaluate the value of A as defined by equation (3). The two-point estimation method from Duncan et al (1978) was used to determine the yield strength for all overconsolidated strength tests.

As is typical (Holtz and Kovacs, 1981), the yield stress increased linearly with increased confining stress in all of the normally consolidated tests (Table 1). The overconsolidated tests showed a consistent increase in yield strength but not in as linear a fashion. In particular, test number TXS-oc1 gives inconsistent values when compared to the other confining stresses. Also, in TXS-oc1, the yield strength is higher than in

Table 1. Hyperbolic parameters and yield strength parameters for drained triaxial and direct simple shear strength tests as determined from the secant modulus method.

Test Number	Effective Confining Stress, σ_3' (kPa)	Deviator Stress at Failure, σ_{df}' (kPa)	Initial Tangent Modulus E_i	Ultimate Deviator Stress, σ_{dult}' (kPa)	A	Effective Stress at Yield, σ_y' (kPa)
TXS-nc1	48	121	6600	138	.58	57
TXS-nc2	103	247	9000	298	.64	106
TXS-nc3	172	415	22300	463	.58	193
TXS-nc4	276	675	34300	774	.59	315
TXS-nc5	345	870	51200	966	.57	416
DSSS-nc1	56	45	2300	49	.59	20
DSSS-nc2	107	92	30300	97	.38	60
DSSS-nc3	171	156	14000	161	.50	80
DSSS-nc4	284	258	8800	321	.66	109
DSSS-nc5	348	317	18700	347	.57	149
TXS-oc1	48	454	33400	1660	.69	277
TXS-oc2	103	546	24500	1310	.70	213
TXS-oc3	172	856	72900	2100	.63	438
TXS-oc4	276	276	1300	77900	.62	453
TXS-oc5	345	345	1800	66500	.72	646

subsequent tests with higher confining stresses. The higher value for yield stress in this case is attributed to the high overconsolidation ratio of 32 and the high principal stress ratio in the test. Except for test DSSS-nc2, the values of A compare well with tests with the same stress conditions. A linear dependence also exists between both the ultimate stress and failure stress with respect to the confining stress.

Time-Rate Parameters

In the time-rate tests, the till specimen is loaded incrementally at stress levels between yield strength and failure of the till. Stress increments are selected as one-fifth increments between 90 % of yield strength and failure. The stress levels are held constant for 24 hours; nonlinear viscous deformation is measured. For analysis of the time-rate parameters, it is useful to formulate the constitutive equation in terms of the stress invariant (T_e, II_{Te}) and deformation-rate invariant (D, II_D), because the invariants satisfy the law of material frame indifference. In terms of the second principal stress invariants, equation (2) can be simplified as a power function:

$$\sqrt{II_{T_e}} = b \left[\sqrt{II_D}\right]^m \qquad (4)$$

or linearized as

$$\log\left(\sqrt{II_{T_e}}\right) = \log(b) + m \cdot \log\left(\sqrt{II_D}\right) \qquad (5)$$

where

$$b = 2\mu_o D_o^{\frac{(n-1)}{n}} \qquad (6)$$

and

$$m = \frac{1}{n} \qquad (7)$$

Equation (5) is a two-parameter power function from which the parameters b and m can be determined. The n parameter, and therefore m, quantify the non-linear strain-rate dependence of viscosity for a non-Newtonian fluid; the b parameter is an arithmetic combination of a reference viscosity (μ_0), a reference strain rate (D_0). The second principal invariants, II_{Te} and II_D, can be calculated for triaxial tests and simple shear tests, and then the values of b and m can be determined by regression, and thereby be used to determine the time-rate parameters n, μ_o, and D_o.

Note here that μ_o and D_o cannot be determined separately. For any given value of n, however, they can be derived from the parameter, b. b and n (and therefore m) can thus be regarded as the fundamental flow parameters for nonlinear flow behavior. D_o and μ_o serve simply to calibrate the parameter against a Newtonian analogue. This can be seen by examining the relationship between b and a viscosity term, F, defined in the fundamental one-dimensional form of the Iverson equation for vertically uniform shear stress (Jenson, 1993) as

$$\frac{du}{dz} = F \left[|\tau| - S_o - S_z \right]^{\frac{1}{m}} \qquad (8)$$

where

$$F = \left[\frac{1}{2 D_o} \right]^{\frac{1-m}{m}} \mu_o^{-\frac{1}{m}} \qquad (9)$$

From equations (6) and (9) it can be shown that

$$F = 2 b^{-\frac{1}{m}} \qquad (10)$$

For any given value of m, the viscous behavior of the constitutive equation (8) is fully determined by F, which in turn is determined uniquely by b. Thus, even if the Newtonian reference parameters are not available, the rheological behavior of a material can be characterized fully so long as b and m are known (Jenson, 1993).

In Iverson's theoretical development, these reference parameters are evaluated at a

Newtonian reference depth, which is the depth at which the viscosity of the till is the same as if the depth were modeled as a Newtonian fluid exhibiting the same velocity at the upper boundary. For observable mass movement phenomena such as landslides, it can be calculated from field data (Iverson 1984; Wong 1992; Wong et al 1995). For behavior that is not observable in the field, however, Newtonian reference parameters can be evaluated by calculating the velocity profile using the test parameters, b and m and equations (8) and (9), then deriving the Newtonian reference parameters from a graphic analysis of the calculated velocity profile (Jenson, 1993).

Calculation of Time-Rate Parameters b and n

Using methods developed by Hovind (1990) and applied to mass movement by Wong (1992), the time-rate parameters for the nonlinear viscoplastic model were determined using triaxial time-rate tests. In the test programs, elapsed time, axial load, axial deformation and volumetric change of each specimen were recorded. From these data, the stress invariant, II_T, and the strain-rate invariant, II_D, can be determined.

At yield conditions established during the strengths tests, the II_T becomes II_{To}. The excess stress intensity, II_{Te}, which is the stress intensity in excess of the condition established for II_{To}, can then be found by

$$\sqrt{II_{T_e}} = \sqrt{II_T} - \sqrt{II_{T_o}} \qquad (11)$$

With the known values of II_{Te} and II_D then, the time-rate parameters can be derived from regression analysis applied to equation (4).

Time-Rate Test Results

Six triaxial time-rate tests were conducted on normally consolidated specimens and two triaxial time-rate tests on the overconsolidated till (Table 2). Time-rate parameters were evaluated based on a yield strength for 0.2% limiting strain. TXT-nc5 was run twice to check repeatability of trials. Three time-rate tests were conducted at K_o conditions using the direct simple shear device (Table 2). Typical stress-strain curves for each the triaxial and direct simple shear time-rate tests (Figure 1) show a continuous deformation at a constant stress level. The decrease in deviatoric stress for the triaxial test is attributed to an increase in cross-sectional area with large axial strain with a constant axial load.

The parameters b and m were determined using three alternative techniques: (1) simultaneous evaluation of data from all six measured time increments at a given confining stress (Figure 2); (2) sequential evaluation of data from each of the six time increments for each test; and (Figure 3); and (3) evaluation of the long-term trend of m vs. time and of b vs. time (Figures 4 and 5).

In each test the samples were initially loaded at 90% of yield and at five load increments between yield and failure as determined from strength tests. All samples for the time-rate tests were consolidated and strained to at least 10% strain. Test were labeled using the following convention: TX = triaxial test; DSS = direct simple shear;

SUBGLACIAL TILL EVALUATION

Test Number	Description	m	log b	n	b (N-s/m^2)
TXT-nc1	Triaxial time-rate test, σ_3' = 48 kPa	.43	6.75	2.33	5.69E+6
TXT-nc2	Triaxial time-rate test, σ_3' = 103 kPa	.26	6.24	3.90	1.73E+6
TXT-nc3	Triaxial time-rate test, σ_3' = 172 kPa	.23	6.41	4.39	2.55E+6
TXT-nc4	Triaxial time-rate test, σ_3' = 276 kPa	.65	8.68	1.55	4.75E+8
TXT-nc5a	Triaxial time-rate test, σ_3' = 345 kPa First run.	.80	9.70	1.26	5.00E+9
TXT-nc5b	Triaxial time-rate test, σ_3' = 345 kPa Second run.	.75	9.59	1.34	3.88E+9
TXT-oc1	Triaxial time-rate test, σ_3' = 172 kPa	.45	2.95	2.20	9.00E+2
TXT-oc2	Triaxial time-rate test, σ_3' = 345 kPa	.28	3.21	3.52	1.62E+3
DSST-nc1	Simple shear time-rate test, σ_3' = 172 kPa.	.18	6.80	5.65	6.26E+6
DSST-nc2	Simple shear time-rate test, σ_3' = 276 kPa.	.22	7.58	4.58	3.83E+7

S=shear strength test; T=time-rate parameter test; nc=normally consolidated; oc=overconsolidate; and an ordinal number of the test for a given lateral confining stress.

Time-Rate Parameters from All Time Increments for a Single Confining Stress

A regression analysis yields the parameters m and log b using the data from all time increments for a single confining stress and the linear relationship of equation (5). The regression analysis defines the terms in the power function of equation (10) and subsequently the parameters b and n of equations (11) and (12) (Table 2), where the parameters b and n characterize the nonlinear viscosity of a material.

Student t-tests were conducted to compare the means of the experimental data and regressed values. Two-tail (overlapping) probabilities of the time-rate parameters based on the student t-tests were approximately 1.0, indicating that the mean of the regression data is the same as that of the experimental data and that all the regression values are within an average confidence limit of a little better than 95%. This implies that the time-rate parameters based on the regression method represent the experimental data. Regression results of the nonlinear flow on the excess stress intensity, $\Pi_{Te}^{1/2}$, versus the strain-rate intensity, $\Pi_D^{1/2}$, for each time-rate test are shown in Figure 2. The value of n is between 1.26 and 5.65 at the established limiting strain of 0.2%, suggesting a shear-thinning behavior (n > 1). Shear-thinning behavior (the reduction of viscosity with increased shear strain rate) is regarded generally as characteristic of long-term mass movement in fine-grained particulate materials (Mitchell, 1976; Iverson, 1984).

For each time-rate test, the sample was subjected to at least five levels of constant stress between yield and failure. At each stress level, measurements were made at all time increments. Figure 3 shows the excess stress intensity, $\Pi_{Te}^{1/2}$, plotted against the strain-rate intensity, $\Pi_D^{1/2}$, for two time measurements as test results were analyzed for each of the six time intervals. Equation (5) can be used to find the parameters m and log b through regression analysis. The parameters n and b can then be derived (Table 3). The value of n is between 0.64 and 6.50 for the normally consolidated triaxial (TXT-nc) tests, and from 1.41 to 3.08 for the normally consolidated direct simple shear (DSST-nc) tests (Table 3). The variability

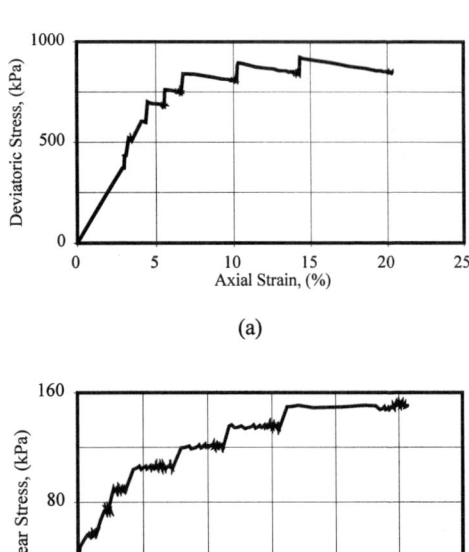

Figure 1. Typical stress-strain curve for a) triaxial time-rate test; and b) simple shear time-rate test.

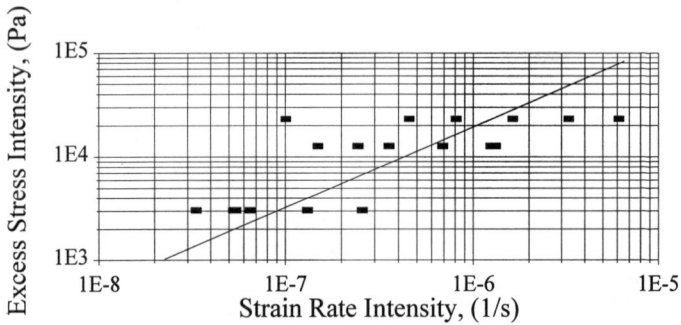

Figure 2. Excess stress intensity, $\Pi_{Te}^{1/2}$, versus strain-rate intensity, $\Pi_D^{1/2}$, for a normally consolidated time-rate triaxial test. Data for all time measurements (1, 2, 4, 8, 16, and 24 hour after increase of stress level) plotted and regressed at once.

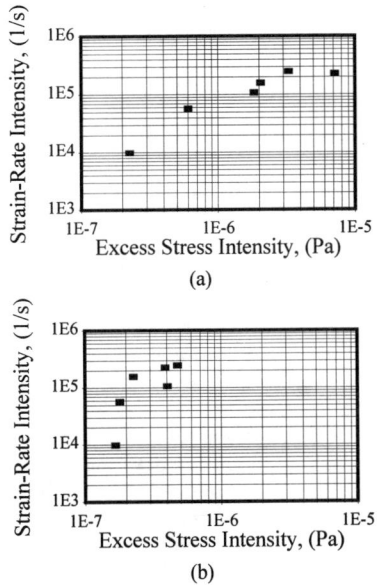

Figure 3. Excess stress intensity, $\Pi_{Te}^{1/2}$, versus strain-rate intensity, $\Pi_D^{1/2}$, for a normally consolidated time-rate triaxial test. Individual time measurements: a) 1 hour; and b) 24 hr.

from n < 1, indicating shear-thickening behavior (the increase of viscosity with increased shear strain rate), to n > 1, indicating shear-thinning behavior, in the TXT-nc tests indicates a change in the deformation style of the till. This result, which is inconsistent with the shear-thinning behavior of long-term mass movement (Mitchell, 1976; Iverson, 1984), can be attributed to the decrease of the deviatoric stress undergoing large strains in the triaxial device (mentioned earlier), and thus leading to difficulty in assessing final stress and strain conditions for longer time intervals.

Values of n for the DSST-nc tests, from 1.32 to 3.08 (Table 3) are the most reliable of the values obtained by the various tests. The improved ability to keep the stress level constant over the sample through the entire test gives more consistent and reliable results. Shear-thinning behavior is consistent among all three direct simple shear tests and all time intervals. The overconsolidated triaxial time-rate tests (TXT-oc) yield values of n between 0.49 and 4.17, again suggesting contradictory rheological behavior. In the TXT-oc however, the inconsistent values of n are associated with only one of the tests. TXT-oc5 yields values of b that are within the acceptable range.

Sequential evaluation of data from each of the time increments for each test

Mean values taken for all confining stresses and for given time increments show that the value of n decreases from n = 3.84 at 1 hour to n = 1.68 at 24 hours. As time increases, therefore, the soil deformation behavior appears to approach some limiting state. This characteristic is indicated also in the reduction of strain-rate intensity with an increase in time measurements (Figure 3). Values of b vary from 2.80×10^5 to 7.36×10^{14} in the TXT-nc, and from 8.42×10^9 to 7.40×10^{14} in the DSST-nc tests (Table 3). The values for b are consistently lower in the TXT-oc tests, varying from 1.34×10^3 to 8.31×10^7. Although the value of b varies over several orders of magnitude, within

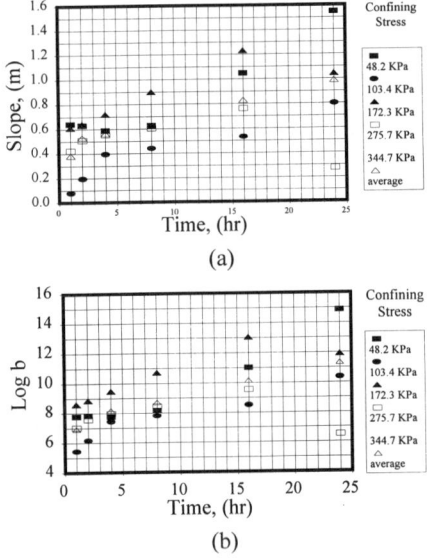

Figure 4. Long-term trend of time-rate parameters from normally consolidated triaxial (TX) tests: a) trend for power law parameter n; and b) trend for parameter b.

each time interval, the variation is small. As with the values for n, mean values of b for all confining stresses and for given time increase as the time increments increase in the TXT-nc. The same is true when the test DSST-nc1 is omitted. As noted before, there was a volume change measurement device problem with this test.

Time-Rate Parameters from Observed Long-Term Trends

The previous analysis showed that the mean values of m and b with respect to time increased for the normally consolidated cases. The general trend of the regressed values for the slope, m, and the intercept, log b, against time increment measurements for the triaxial tests and the direct simple shear tests, respectively, is a hyperbolic shape (Figures 4 and 5). These results suggest that m and b approaches some long-term value (m_{hyp} and b_{hyp}) given in equations 12 and 13.

$$m = \frac{t}{C_m + m_{hyp} \, t} \tag{12}$$

$$\log(b) = \frac{t}{C_b + \log(b)_{hyp} \, t} \tag{13}$$

The ultimate values of log b and m for each of the data sets were determined using hyperbolic fit. Triaxial test results were evaluated two different ways: first for all data; and again excluding the 24 hour measurements because of increased scatter of these measurements. The direct simple shear test data were analyzed in two ways: by individual test; and with all of the test data combined (Table 4).

Comparison of Results and Analysis of Reliability

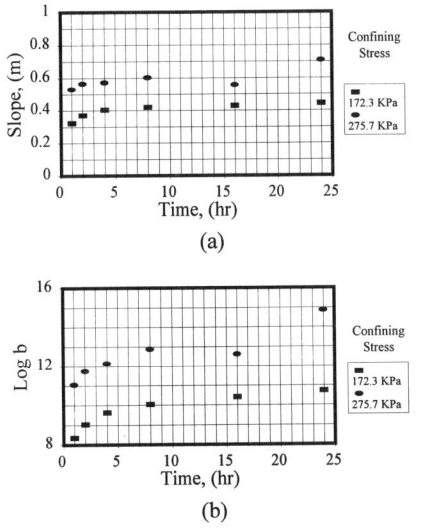

Figure 5. Long-term trend of time-rate parameters over time for direct simple shear (DSS) tests: a) trend for power law parameter n; and b) trend for parameter b.

The mean values of n based on the experimental data are greater than one indicating shear-thinning behavior. The three techniques employed for evaluating visco-plastic parameters from triaxial and simple shear tests yielded consistent results, although the third method--deriving parameters from the hyperbolic trend in the plots of m and log b over time--provide a reasonable assessment of long-term behavior in response to constant stress. Long-duration (24-hour) experiments show that the values of the time-rate parameters, b and n, converge to constants. The trend is particularly apparent in the simple shear tests (Figure 5). Since the short-term values of the parameters are transient, short-duration experiments are likely to overstate the long-term value of n.

For the Batestown soil n and b approach 1.31 and 5.58×10^{10} with standard deviations of 0.26 and 4.40×10^9, respectively for the triaxial data (Table 4). The long-term value of n and b derived from the direct simple shear tests are centered at 1.85 and 2.88×10^{12} with standard deviations of 0.18 and 2.10×10^{11} (Table 4). The n value from the triaxial tests indicate shear thickening behavior where as the n value form the direct simple shear tests indicate shear thinning behavior. This discrepancy may arise from the direct simple shear test more accurately modeling subglacial till stress conditions and variations during sample preparation.

Conclusions

A laboratory method has been established to evaluate time-rate parameters for stress-dependent mass movement. These time-rate parameters can characterize the rheological behavior of a material fully when no Newtonian reference parameters are available. The laboratory method utilizes a nonlinear viscoplastic constitutive model, combined testing

Table 3. Viscoplastic parameters evaluated at each time interval. Units of b are N-s/m^2.

Test Number	σ^{ct} (kPa)	1 hr	2 hr	4 hr	8 hr	16 hr	24 hr
TXT-nc1	48						
b		5.87x10^7	6.68x10^7	5.78x10^7	1.47x10^8	1.03x10^{11}	7.36x10^{14}
n		1.56	1.59	1.69	1.59	.95	.64
TXT-nc2	103						
b		1.05x10^7	7.36x10^8	9.58x10^7	2.82x10^8	3.64x10^8	3.26x10^6
n		2.35	1.96	1.81	1.65	1.31	3.64
TXT-nc3	172						
b		2.80x10^5	1.46x10^6	2.78x10^7	6.43x10^7	3.24x10^8	2.32x10^{10}
n		12.4	5.11	2.52	2.27	1.88	1.25
TXT-nc4	276						
b		6.71x10^5	6.23x10^8	1.89x10^8	2.76x10^8	4.12x10^8	9.23x10^{12}
n		6.50	1.49	1.79	1.79	1.77	.79
TXT-nc5a	345						
b		3.98x10^8	7.36x10^8	3.06x10^9	5.31x10^{10}	1.12x10^{13}	8.33x10^{11}
n		1.64	1.56	1.39	1.11	.81	.96
TXT-nc5b	345						
b		5.84x10^9	6.02x10^{11}	1.64x10^{14}	1.51x10^{17}	2.20x10^{21}	9.78x10^{23}
n		1.17	.84	.64	.50	.38	.33
TXT-oc1	172						
b		6.10x10^3	2.95x10^3	2.68x10^3	3.02x10^3	4.19x10^6	8.31x10^7
n		.97	1.29	1.47	.73	.60	.49
TXT-oc2	345						
b		3.16x10^4	1.74x10^4	1.60x10^3	1.34x10^3	2.01x10^4	2.26x10^4
n		1.14	1.39	3.55	4.17	1.72	1.77
DSST-nc1	172						
b		2.28x10^8	1.08x10^9	4.34x10^9	1.11x10^{10}	2.56x10^{10}	5.35x10^{10}
n		3.08	2.70	2.47	2.38	2.33	2.25
DSST-nc2	276						
b		1.14x10^{11}	5.84x10^{11}	1.44x10^{12}	7.78x10^{12}	4.49x10^{12}	7.4x10^{14}
n		1.88	1.76	1.74	1.66	1.79	1.41
Mean values (nc data)							
$b(TXT)$		1.05x10^9	1.11x10^{11}	2.73x10^{13}	2.52x10^{16}	3.67x10^{20}	1.63x10^{23}
$n(TXT)$		4.27	2.09	1.64	1.49	1.18	1.27
$b(DSST)$		5.71x10^{10}	2.92x10^{11}	7.22x10^{11}	3.90x10^{12}	2.26x10^{12}	3.70x10^{14}
$n(DSST)$		2.48	2.23	2.11	2.02	2.06	1.83

techniques, and procedures to predict the displacement rate behavior of persistent mass movement. The nonlinear viscoplastic constitutive model (Iverson, 1985, 1986) is a theoretical general model of mass movement.

The new procedures for evaluation of the time-rate parameters included evaluation of individual time increments within a confining stress and evaluation of time-rate parameters of an entire group of tests, through a hyperbolic fit method, as their dependency on time was determined. The hyperbolic regression method over all of the data shows an ultimate, constant state value to be reached by each of the parameters.

Table 4. Time-rate parameters as evaluated for long-term trend of all test data.

Test	n_{hyp}	std. dev. (n_{hyp})	b_{hyp} (N-s/m^2)	std. dev. (b_{hyp})
TXT-nc (all data)	1.31	0.26	5.62×10^{10}	4.40×10^9
TXT-nc (24 hr excluded)	1.17	0.16	1.58×10^{10}	2.55×10^9
DSST-nc (all data)	1.85	0.18	2.88×10^{12}	2.10×10^{11}
DSST-nc1	2.23		7.08×10^{10}	
DSST-nc2	1.47		4.27×10^{14}	

These procedures can reasonably determine the yield parameters and the time-rate parameters of mass movement.

The following conclusions are drawn from this study:

1) Time-rate parameters can be established through the two-test methodology test program. By using triaxial and simple shear strength and time-rate tests, the yield parameters as well as the time-rate parameters can be determined and used in the Iverson model to model mass movement. The time-rate parameters, b and n, can characterize the rheological behavior of a soil fully, and the influence of the mass movement of soil on the ice sheet dynamics even when no Newtonian reference parameters are available.

2) The measurement and evaluation of time-rate parameters is dependent on the duration of the test. Short-term deformation behavior of the soil is transient. Long-term deformation behavior appears to be constant, and that long-term time-rate parameters are approached.

3) The variation of time-rate parameters can be modeled as a hyperbolic function of time approaching a long-term steady state value. It is reasonable to assume steady-state viscous flow during the long time periods of glacial movements. Long-term parameters characterize shear thinning behavior in the soil, consistent with expectations.

The authors recognize that true long-term behavior could require time increments much greater than 24 hr. The work presented in this paper was an initial attempt to determine rheological parameters for the sub-glacial till. Asymptotic behavior of the parameters was observed during the analysis of the data. Much longer experiments would be needed to make a better characterization of the long-term rheological behavior.

Bibliography

Casagrande, A., (1936) "The determination of the preconsolidation load and its practical significance." Discussion D-34, *Proc. 1st Int. Conf. Soil Mech. Foundation Engrg.*, Cambridge, **III**, p. 60-64.

Chen, W.F., and Baladi, G.Y., (1985) "Soil plasticity-theory and implementation," *Developments in Geotechnical Engineering*, **38**, Elsevier.

Chen, W.F., and Saleeb, A.F., (1982) *Constitutive Equations for Engineering Materials. Vol 1: Elasticity and Modeling.* Wiley.

Clayton, L., Teller, J.T., and Attig, J.W., (1985) "Surging of the southwestern part of the Laurentide ice sheet." *Boreas*, **14**, p. 235-242.

Desai, C.S., and Siriwardane, H.J., (1984) *Constitutive Laws for Engineering Materials with Emphasis on Geologic Materials*. Prentice-Hall, Inc., Englewood Cliffs, N.J.

Desai, C.S., and Wu, T.H., (1976) "A general function for stress-strain curves." *Proc. Int. Conf. Numerical Methods Geomech.*, ASCE, **1**, p. 306-317.

Duncan, J.M., Byrne, P., Wong, K.S., and Maybery, P., (1978) "Strength stress/strain bulk modulus parameters for finite element analyses of stresses and moments in soil masses." *Report* - UCB/GT/78-02. University of California, Berkeley.

Harrison, W., (1958). "Marginal zones of vanished glaciers reconstructed from preconsolidation pressure values of overridden silts." *J. Geo.*, **66**, p. 72-95.

Hovind, C.L., (1990) "Determination of time-dependent parameters by triaxial tests," *MS thesis*, Wash. St. Univ.

Iverson, R.M., (1984) "Unsteady Nonuniform Landslide Motion: Theory and Measurement." *PhD dissertation*, Stanford Univ.

Iverson, R.M., (1985) "A constitutive equation for mass movement behavior." *J. Geology*, **93**, p. 143-160.

Iverson, R.M., (1986) "Unsteady, nonuniform landslide motion: 1. Theoretical dynamics and the steady datum state." *J. Geology*, **94**, p. 1-15.

Jenson, J.W., (1993) "A nonlinear numerical model of the Lake Michigan lobe, Laurentide ice sheet." *PhD thesis*, Ore. St. Univ.

Jenson, J.W., Clark, P.U., MacAyeal, D.R., Ho, C., and Vela, J.C., (1995) "Numerical modeling of advective transport of saturated deforming sediment beneath the Lake Michigan Lobe, Laurentide Ice Sheet," *Geomorphology*, **14**, p. 157-166.

Jenson, J.W., MacAyeal, D.R., Clark, P.U., Ho, C.L., and Vela, J.C, (1996) "Numerical modeling of subglacial sediment deformation: implications for the behavior of the Lake Michigan Lobe, Laurentide Ice Sheet," *J. Geophys. Res.*, in press.

Mitchell, J.K., (1976) Fundamentals of Soil Behavior, Wiley, New York.

Sauer, E.K., and Christiansen, E.A., (1991) "Preconsolidation pressures in the Battleford formation, southern Saskatchewan, Canada," *Can J. Earth Sciences*, **28**, p. 1613-1623.

Vela, J.C., Ho, C.L., Jenson, J.W., and Clark, P.U., (1996) "Rheology of till deposited by the Lake Michigan Lobe of the Laurentide Ice Sheet," *J. Geo.*, submitted.

Wong, W.W.-H., (1992). "Evaluation of time-rate slope movement based on laboratory tests," *PhD dissertation*, Wash. St. Univ.

Wong, Ho, C.L., Iverson, R.M., and Hovind, C.L., (1995) "Viscoplastic soil parameters from triaxial tests," *Reviews Engrg. Geo.*, **X**, p. 39-50.

STRAIN RATE AND STRUCTURING EFFECTS ON THE COMPRESSIBILITY OF A YOUNG CLAY

Serge Leroueil[1], Didier Perret[2] and Jacques Locat[3]

Abstract

An artificially sedimented clay has been subjected to a variety of oedometer tests in order to evaluate the influence of time and strain rate on the process of structuration and the viscous response of the soil as a function of the degree of structuration. The results indicate a structure development with time when strain rate is sufficiently small and an effect of strain rate on the soil response, independent of the degree of structuration. However, the structuring phenomena involved are not clearly identified.

Introduction

Based on results from different types of oedometer tests (Conventional 24-hrs tests, Multiple stage loading tests with reloading at the end of primary consolidation, Constant rate of strain (CRS) tests and long term creep tests) performed on clays from Canada and Sweden, Leroueil et al. (1985) showed that during one dimensional compression, the behavior is controlled by a unique effective stress-strain-strain rate (σ'_v - ε_v - $\dot{\varepsilon}_v$) relationship. At a given strain or void ratio, vertical effective stress increases with strain rate. Leroueil et al. (1985) also showed that effective stress-strain curves obtained at different strain rates can be normalized with respect to the preconsolidation pressure measured at the corresponding strain rate. That is, the ratio of the effective stresses measured at two different strain rates is constant, whatever the strain. Typically the effective stress, and in particular the preconsolidation pressure, varies by 10 to 15% per logarithm cycle of strain rate.

As already noted by Leroueil et al. (1985), this relationship represents quite well the behavior of natural clays in the normally consolidated range when strain is increasing. In the overconsolidated range or during relaxation tests, it is necessary to apply the viscous model to the plastic component of strain only and to add an elastic component.

---

[1] Prof., Dept. of Civil Engrg., Université Laval, Ste-Foy, Québec, Canada
[2] Former student, Dept. of Geology and Engrg. Geology, Université Laval, Ste-Foy, Québec, Canada
[3] Prof., Dept. of Geology and Engrg. Geology, Université Laval, Ste-Foy, Québec, Canada

Constant rate of strain tests performed at strain rates between 1.43×10^{-5} s^{-1} and 1.07×10^{-7} s^{-1} give the (σ'_v - ε_v - $\dot{\varepsilon}_v$) relationship for Batiscan clay (Fig.1). The CRS test performed at the strain rate of 1.69×10^{-8} s^{-1} does not fit with the other results and will be discussed later on.

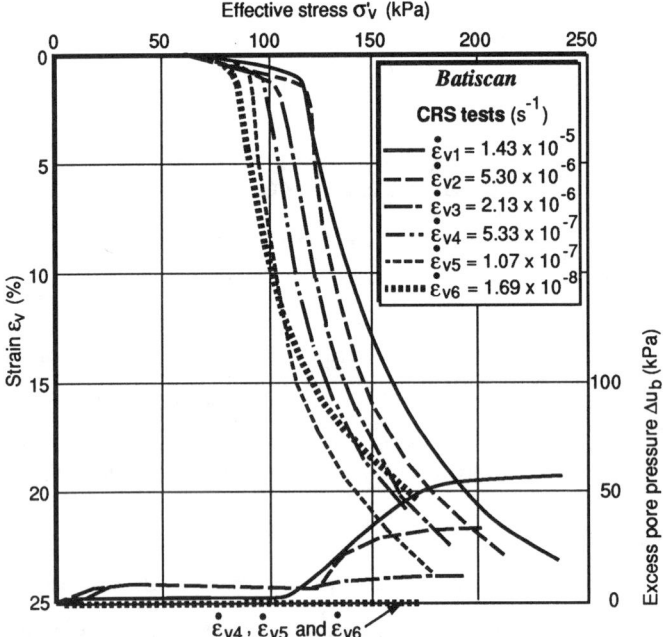

Figure 1 - Typical CRS oedometer tests on Batiscan clay (from Leroueil et al., 1985)

The effective stress-strain-strain rate model has several implications which have been explained elsewhere (Leroueil et al. 1985; Leroueil 1988 and 1996). Only those which are of direct interest for this paper are mentioned here.

A first implication of this model is that, when the strain rate is changed during a test, the compression curve jumps from one constant rate of strain curve to another. This was shown by Graham et al. (1983) and by Leroueil et al. (1985) on the Batiscan clay (Fig. 2).

A second implication comes from the fact that the strain rate at the end of the loading periods in conventional 24-hrs oedometer tests is close to 10^{-7} s^{-1} (Leroueil 1988). CRS tests performed at strain rates different from 10^{-7} s^{-1} thus give compression curves which are different from those obtained in conventional 24-hrs tests. As shown by Leroueil (1996) for natural clays from different countries (Canada,

Finland, Italy, Japan, Sweden, United Kingdom), the ratio between the preconsolidation pressures measured in CRS tests performed at a strain rate of 1 to 4×10^{-6} s^{-1} and in conventional 24-hrs tests is typically 1.25.

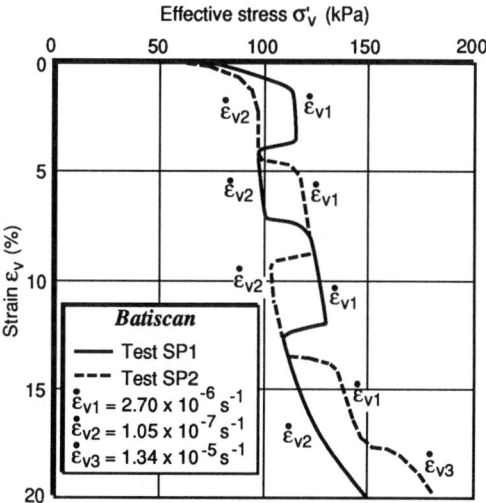

Figure 2 -   CRS oedometer tests with strain rate changes on Batiscan clay (from Leroueil et al. 1985)

A third implication concerns the increase in preconsolidation pressure due to secondary consolidation. When a soil is subjected to a constant effective stress for a long period of time, its void ratio and strain rate progressively decrease. If the soil is reloaded, there is an increase in strain rate, and the compression curve moves to the curve of constant strain rate corresponding to the new strain rate. The compression curve then shows a preconsolidation pressure associated with the void ratio of the soil and this new strain rate. The development of this "quasi-preconsolidation pressure" has been clearly explained by Bjerrum (1967 and 1972).

Oedometer tests performed by Leonards and Altschaeffl (1964) on an artificially sedimented clay ($I_p = 31\%$; < 2 µm = 40 %) however give results which are at variance with the viscous model previously described. As shown in Fig. 3, after 90 days of consolidation under a constant effective stress, and then unloading for another three months, the tested soil exhibited a preconsolidation pressure much higher than that due to its new void ratio only. Such a behavior is associated with the development of bonds between particles and aggregates, hereunder called structuration.

Perret et al. (1995) also observed structuration of inorganic clays from Saguenay Fjord during secondary consolidation. To study this aspect, and more generally diagenesis of recent sediments, the same authors performed a series of one-dimensional compression tests on an artificially sedimented clay from Jonquière, Québec (Perret 1995; Perret et al. 1996). The tests were performed in large diameter (200 mm) cells equipped with bender elements to follow the evolution of maximum shear modulus G_0. As G_0 is directly related to the preconsolidation pressure (σ'_p) of the clay, its measurement made it possible to follow the evolution of σ'_p, and thus of structure, during consolidation.

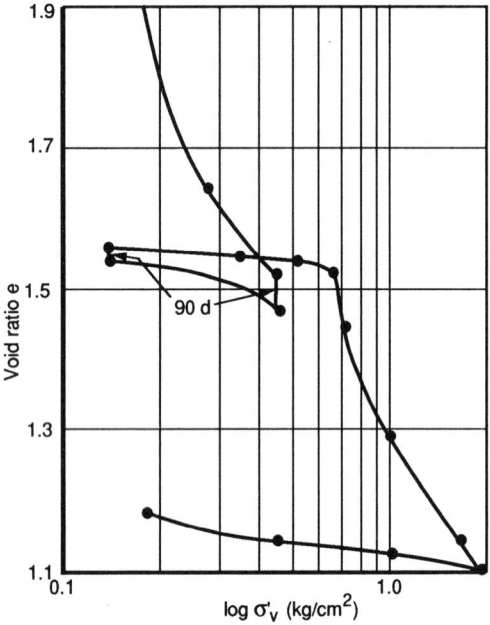

Figure 3 -   Sedimentation-compression curve of an artificially sedimented clay (after Leonards and Altschaeffl 1964)

Figure 4 presents results of a test in which the soil was consolidated for 120 days under an applied stress of 10 kPa before being loaded again. Just on the basis of the change in void ratio during secondary consolidation, the preconsolidation pressure would have increased from 10 to 11.5 kPa (Fig. 4a). However, due to structuration, the preconsolidation pressure increased to 18.5 kPa. Variation in G_0 under the sustained load of 10 kPa is shown with black dots in Fig. 4b. If G_0 was only associated with void ratio, it would follow the $\alpha\alpha$ line. In fact, G_0 first goes on the left side of $\alpha\alpha$, indicating a destructuration of the clay just after loading, and then goes on the right side of $\alpha\alpha$, as structure develops.

Such test results obtained by Perret (1995) and Perret et al. (1996) show that viscous and structuring phenomena are simultaneously present during both primary and secondary consolidation. However, several questions remained:

- Are structuring phenomena controlled by time or strain rate?
- Is strain rate effect the same in structured and non-structured materials?
- Do structuring effects develop in all clayey materials?

To answer the first two questions, the clay specimens used by Perret (1995) were loaded into the normally consolidated range, unloaded, dismantled and trimmed into sub-specimens for performing special oedometer tests. The main objective of this paper is to present and analyse the results of these tests. Data available in literature are then examined for answering the third question.

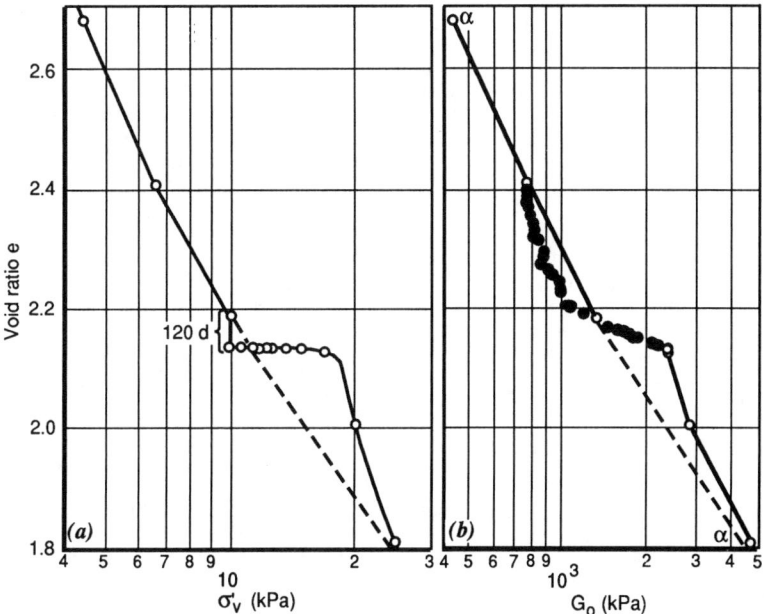

Figure 4 - Compression curve (a) and variation of the maximum shear modulus G_0 with void ratio (b) for the artificially sedimented Jonquière clay (from Perret 1995)

General characteristics of Jonquière clay

The material used in this study is a postglacial marine clay deposited about 8,000 years ago in the Jonquière area, 300 km North of Quebec City, which has been remoulded and then artificially sedimented under laboratory conditions. The physical properties of the natural clay were a liquid limit of 51%, a plastic limit of 22%, and a clay fraction (< 2 μm) of about 60%. The Atterberg limits were obtained at the natural pore water salinity of the clay, *i.e.* about 0.1‰. The organic matter content was between 0.2 and 0.4% by weight of dry soil, thus very small. A mineralogical analysis revealed approximately 15% by weight of clay minerals, the most common being chlorite and illite, although vermiculite and traces of an interstratified mineral were also detected. Non-clay minerals with particles smaller than 2 μm are generally rock flour made predominantly from quartz, amphibole and pyroxene (Locat et al. 1984; Locat 1995).

Specimen preparation

Specimens were prepared in a two-stage sedimentation-compression protocol to simulate the natural process of marine sediment formation. Details of the procedure are given by Perret (1995), and will only be briefly described here. After a thorough remoulding of intact samples, the clay paste was mixed with de-aired distilled water in order to obtain a creamy slurry at a water content of 115%. Then, to simulate recent sediments at the bottom of Saguenay Fjord, de-aired saline water was added to the slurry until reaching a particle concentration of 120 g/l and a water salinity of 30‰. This increased the plasticity index of the clay from 29 to 32%. One liter of this suspension was poured into a 20 cm diameter by 50 cm high one-dimensional consolidation cell filled with saline water. This process was repeated ten times at 12 hour intervals to create a homogeneous soil column with minimal grain size segregation.

The soil column was then compressed by step-loading starting at a vertical stress of $\approx 10^{-2}$ kPa with a load-increment ratio of about 50%. Each stage was maintained until the strain rate reached a value on the order of 10^{-7} s^{-1}, which is the strain rate typically observed at the end of the loading periods in conventional 24-hrs oedometer tests. Eight specimens were prepared that way and loaded until they reached an applied stress somewhere between 1 and 50 kPa which was maintained for 120 days. Then, the specimens were loaded step by step, until they reached the normally consolidated range; their thickness was then on the order of 4 cm. After unloading and dismantling of the specimens, four of them (1C4, 2C2, 4C3, and 5C4) were trimmed into sub-specimens, in preparation for the oedometer testing program relevant to present study.

As previously indicated, the long term creep tests performed for studying the diagenesis of recent sediments, and in particular the respective influence of viscous and structuring phenomena on the strength increase with time, are presented by Perret (1995) and Perret et al. (1996).

Testing program and procedure

Both conventional and constant rate of strain (CRS) oedometer tests were performed on artificially sedimented Jonquière clay. The conventional tests were carried out with a load-increment ratio of 50% on 50.4 mm diameter by 19.0 mm high specimens, each load being applied for 24 hours. The CRS tests were carried out on 75 mm diameter by 20 mm high specimens, at strain rates varying from 1.0×10^{-7} s^{-1} to 1.27×10^{-5} s^{-1}. Such strain rates resulted in test durations typically between 5 hours for the most rapid ones and about one month for the slowest ones. These CRS

tests were of two types: 1) CRS tests with strain rate maintained constant during the entire
test; 2) special CRS tests in which strain rate was changed at various strains. A back-pressure of 100 kPa was used, and pore pressure was measured at the base of the specimen during the CRS tests.

Table 1 presents the test program performed on the sub-specimens taken from the specimens 1C4, 2C2, 4C3, and 5C4. As the results were essentially the same in the four series, only the series issued from specimens 2C2 and 4C3 will be presented here.

For all oedometer tests, rings were made of stainless steel, and a silicon oil film was applied inside the ring to reduce friction. Synthetic filters of low compressibility and high durability were installed at the top and bottom of the specimen in order to minimize the compliance of the system and bacterial activity. Saline water at the same salinity as the pore water was used in the drainage leads and in the cell chamber in order to avoid osmotic-controlled flows. All consolidation tests were conducted under isothermal conditions at 21±1°C. Also, the reference for strain was taken under an effective stress of 5 kPa for all tests.

Specimen	Test	Strain rate (s^{-1})	e_0
1C4	CRS	1.27×10^{-5}	1.159
	CRS	1.00×10^{-7}	1.165
	Conventional	-	1.161
2C2	CRS	1.27×10^{-5}	1.768
	CRS	1.00×10^{-7}	1.768
	Special CRS	1.27×10^{-5}; 2.54×10^{-6}; 1.00×10^{-7}	1.800
	Conventional	-	1.804
4C3	CRS	1.27×10^{-5}	1.906
	Special CRS	1.27×10^{-5}; 1.00×10^{-7}	1.915
	Conventional	-	1.917
5C4	CRS	1.27×10^{-5}	1.288
	Special CRS (1)	1.27×10^{-5}; 2.54×10^{-6}; 1.00×10^{-7}	1.274
	Special CRS (2)	1.27×10^{-5}; 2.54×10^{-6}; 1.00×10^{-7}	1.285
	Conventional	-	1.277

Table 1 - One-dimensional test program

Data analysis

Figure 5 shows several oedometer test results obtained on sub-specimens of 2C2. There are a conventional 24-hrs oedometer test and two CRS tests, one performed at a strain rate of 1.27×10^{-5} s^{-1} and the other performed at a strain rate of 10^{-7} s^{-1}. A comparison of the curves obtained in the CRS test performed at the most rapid strain rate and in the conventional 24-hrs test shows at the same strain, a ratio of effective stresses typically equal to 1.3. This is in agreement with the previously mentioned (σ'_v - ε_v - $\dot{\varepsilon}_v$) model, considering that the strain rate in conventional 24-hrs oedometer tests is typically on the order of 10^{-7} s^{-1}, *i.e.* two orders of magnitude slower than the strain rate in the rapid CRS test.

The effective stress-strain curve deduced from the CRS test performed at the strain rate of 1.0×10^{-7} s^{-1} is very different from that obtained in the conventional 24-hrs oedometer test, yet corresponding to essentially the same strain rate. It is even above the effective stress-strain curve obtained at the much higher strain rate of 1.27×10^{-5} s^{-1}. This is obviously at variance with the (σ'_v - ε_v - $\dot{\varepsilon}_v$) model, and can be explained by the effects of structuring phenomena developing with time.

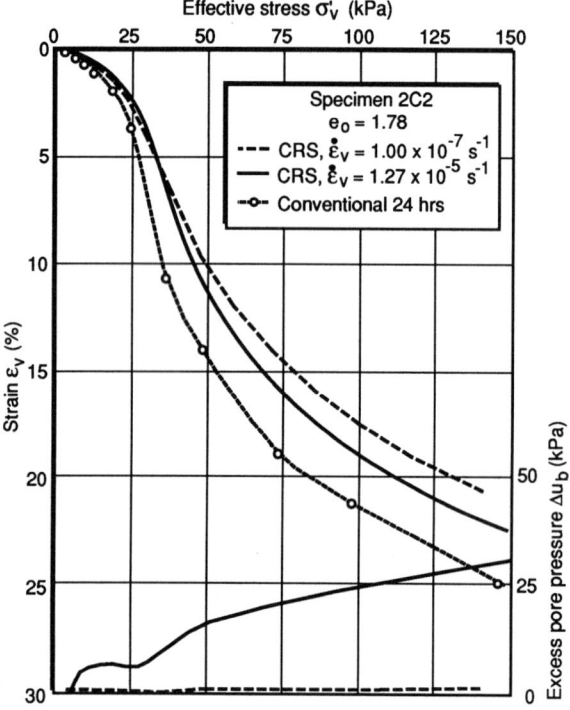

Figure 5 -   CRS and conventional 24-hrs oedometer tests on the resedimented Jonquière clay

The effective stress-strain curves obtained at the rapid 1.25×10^{-5} s^{-1} strain rate and in the conventional test are in good agreement with the stress-strain-strain rate model proposed by Leroueil et al. (1985) for natural clays, which can be explained as follows: in the rapid CRS test, the strain rate is so high that structuration, *i.e.*

strengthening of the contacts between particles or aggregates due to thixotropy or cementation, has no time to develop; in the conventional 24-hrs test, the structure which could have developed during the last hours of the previous loading stage is destroyed by the following one in a manner similar to what has been observed by Perret et al. (1995) in the early stages of the loading under 10 kPa (see Fig. 4b). As a consequence, effects of the structuring phenomena are undiscernible, and only the strain rate effect appears when comparing the two types of test results. On the other hand, the slow CRS test, lasting almost one month, allows structure to develop in the young clay, in such a way that accumulated strain under a given effective stress becomes smaller than that obtained at 1.27×10^{-5} s^{-1}.

These observations are to some extent similar to those made by Sills and Thomas (1984). Studying deposition of a silty clay (I_p = 32%) in laboratory, these authors observed that the higher the rate of deposition, the more compacted is the resulting settled mud. This can also be attributed to the progressive development of structure when the soil is slowly deposited.

Another series of tests, performed on sub-specimens of 4C3, is shown in Fig. 6. As in Fig. 5, the CRS test performed at the strain rate of 1.27×10^{-5} s^{-1} and the conventional 24-hrs test are in accordance with the (σ'_v - ε_v - $\dot{\varepsilon}_v$) model. The third test is a CRS test in which the strain rate was changed from 1.0×10^{-7} s^{-1} to 1.27×10^{-5} s^{-1} several times during the test.

The test began at a strain rate of 1.0×10^{-7} s^{-1}, and a progressive structuration is evidenced by the fact that the effective stress-strain curve is slightly above the curve obtained at the strain rate of 1.27×10^{-5} s^{-1}. At a strain of 5.1%, strain rate was increased to 1.27×10^{-5} s^{-1}. An increase in effective stress can be observed, and then, with the accumulation of strain, the effective stress-strain curve becomes progressively closer to the curve obtained in the CRS test continuously performed at the same strain rate. At the strain of 10.3%, the strain rate was reduced to 1.0×10^{-7} s^{-1}. This reduction is reflected by a rapid reduction in effective stress. Then, the effective stress progressively increases with a stiffness which is about twice that observed in the other tests at the same strain, again indicating an important structuration of the soil. At 13.8%, the effective stress exceeds that measured in the continuous CRS test at 1.27×10^{-5} s^{-1}. When the strain rate is again increased to 1.27×10^{-5} s^{-1}, there is a rapid increase in effective stress followed by an abrupt increase in strain at an essentially constant effective stress. Then, the effective stress-strain curve becomes progressively closer to the compression curve obtained in the CRS test at 1.27×10^{-5} s^{-1}, indicating a progressive destructuration of the clay at high strain rates.

In this series of tests, there is also a progressive structuration with time when the clay is continuously loaded at a low strain rate. However, the CRS tests performed with changes in strain rate give an additional opportunity to examine in detail the viscous aspect of soil behavior. Indeed, at strains where strain rate was changed (5.1%, 10.3%, and 13.8% for the test shown in Fig. 6), *i.e.* at given fabric and structure, it is possible to isolate the effect of strain rate. Figure 7 shows the ratio of the effective stresses obtained at these particular strains, at strain rates of 1.27×10^{-5} s^{-1} and 1.0×10^{-7} s^{-1}. Data from other tests, and for the same change in strain rate, are also shown on the figure. When strain rate is increasing, the effective stress ratio is the same as that observed for natural aged Eastern Canada clays, indicating that the viscous

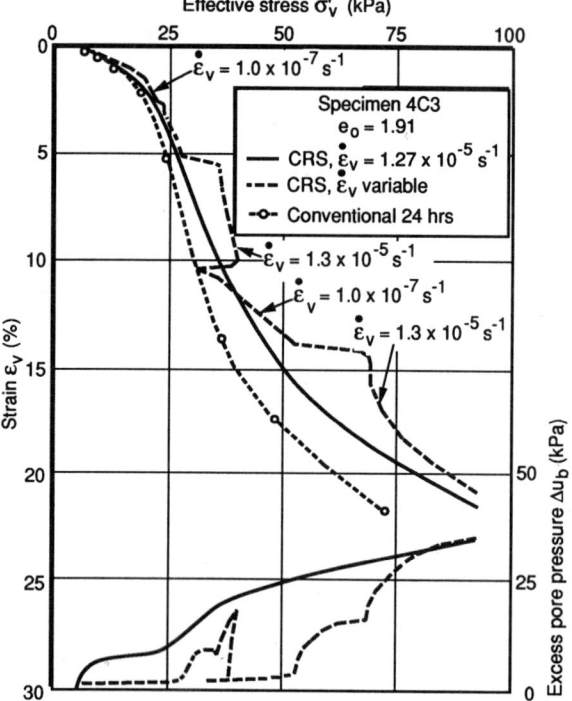

Figure 6 - CRS and conventional 24-hrs oedometer tests on the resedimented Jonquière clay

effect is the same whether the clay is young or aged. Even if there are only few data for a decreasing strain rate, they tend to indicate a slightly smaller effective stress ratio. This could be due to the fact that when the strain rate is reduced, the time necessary to take few reliable measurements could be long enough for the clay to become slightly structured, thus reducing the real effect of strain rate.

It is important to point out that structuration is not simply related to strain rate, but rather to the combined effect of time and strain rate. For example, in Fig.6, at a strain of about 10.3%, both the conventional 24-hrs test and the CRS test with variable strain rate have about the same effective stress at about the same strain rate (10^{-7} s^{-1}) because structure which could have existed in the specimens has been destroyed by previous rapid loading and straining in both tests. At a strain of 13.8 %, the specimen subjected to the conventional 24-hrs oedometer test is, for the same reason, not or only slightly structured, and its effective stress is 36 kPa, whereas the specimen that was loaded in the CRS test at the slow strain rate of 10^{-7} s^{-1} for 4 days (strain increase from 10.3 to 13.8%), had time to become structured and sustains a much higher effective stress of 53 kPa.

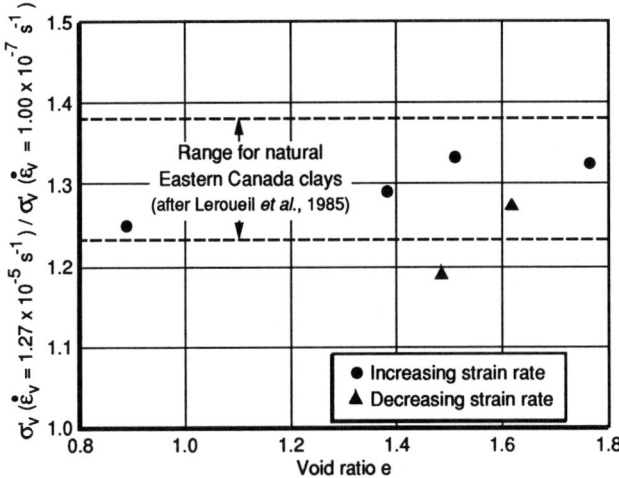

Figure 7 - Influence of change in strain rate on change in vertical effective stress

Discussion

Test results on young Jonquière clay show that, at a given fabric and structure, there is a strain rate effect very similar to that observed in aged clays. They also show strong structuring phenomena which, in slow or long duration tests, can compensate and even overshadow the effects of strain rate. A review of the literature on the compression of young clays can show if the observed phenomena are general or not.

Burghignoli (1979) presented constant rate of loading oedometer tests performed on remoulded Fiumicino clay. This clay has a clay fraction of 45%, a plasticity index of 40%, and was loaded under constant rates of loading varying from 5.6 to 557.0 kPa/hr. The results, analysed in terms of strain rate by Kabbaj (1985), are in agreement with the (σ'_v - ε_v -$\dot{\varepsilon}_v$) model. The step-loading oedometer tests performed by Imai and Tang (1992) on interconnected specimens of Yokohama remoulded clay (I_p = 72 %), also confirm the validity of the effective stress-strain-strain rate model.

On the other hand, the CRS test results obtained by Smith and Wahls (1969) on a commercial kaolinite having a plasticity index of 34% deviate from the behavior observed by Burghignoli (1979) and Imai and Tang (1992). Smith and Wahls (1969) did not observe any viscous effect for strain rates varying between 4×10^{-5} and 4×10^{-7} s^{-1}. Wissa et al. (1971) and Gorman et al. (1978) also do not indicate significant strain rate effects. However, their tests considered relatively small changes in strain rate (factor of about 5) for which the effect of strain rate is less than 10% and often smaller than the effect of the natural variability of the material tested.

Because it has been observed on a large variety of natural clays, and because it has been systematically observed, at a given fabric and structure, on young Jonquière clay, it is thought that viscous effects are general in clayey soils, whatever the degree of structuration. It is also thought that behavior such as that observed by Smith and Wahls (1969) or in the present study, when CRS tests performed at low strain rates show smaller compressibility than CRS tests performed at higher strain rates, are due to structuring phenomena. However, according to Burghignoli's data, these phenomena do not seem to be involved, or at least not with the same intensity, in all young clays.

When aged natural clays are considered, the effect of structuring phenomena is not clear either, but generally seems less important. Performing long term oedometer tests on natural clay specimens from Saguenay Fjord, Perret et al. (1995) observed the development of a significant strengthening in a specimen with a low organic matter content (1.1%) while a specimen of the same geological origin, but with an organic matter content of 4.6%, did not show any significant structuration. Organic matter thus seems to be one factor influencing the development of structure in soils. Kabbaj (1985) and Leroueil et al. (1985) performed oedometer tests, and in particular CRS tests, at different strain rates, including very small ones, on different non-organic clays from the Champlain Sea Basin. As shown in Fig. 1, the soil behavior observed on Batiscan clay at the strain rate of 1.7×10^{-8} s^{-1} is at variance with the effective stress-strain-strain rate relationship, and indicates some progressive structuration of the clay. A similar behavior was observed on St-Alban clay (I_p = 18%), at a strain rate of 6.7×10^{-9} s^{-1}. On the other hand, slow CRS tests performed at a strain rate of 6.7×10^{-9} s^{-1} on Berthierville clay from two different depths (I_p = 23 and 16%) do not show any indication of structuration. It thus appears that, even in non-organic clays, structuring effects can vary from clay to clay.

Several phenomena could be responsible for the observed behavior (see Bennet et al., 1991, for more details). Thixotropy, considered by Mitchell (1976) as related to time-dependent particle reorientations and water structure changes, is certainly a factor which is influenced by clay mineralogy. From experience gained by Perret et al. (1995) with clays from Saguenay Fjord, organic matter content also seems to be an influencing factor. Dissolution and precipitation of silica, considered by Mitchell and Solymar (1984) to explain structuration of sands previously disturbed by blasting, dynamic compaction and vibrocompaction, is also worth mentioning, especially in the context of the Jonquière clay which contains a high percentage of rock flour. Other factors could be bacterial activity and chemical reactions, such as oxydation, which, in particular, could be different in laboratory and in situ. All these conditions vary from clay to clay, but also with the testing environment and the age of the material. Therefore, at the present time, the manifestation of structuring processes is difficult, if not impossible to predict. However, the possibility of having structuration in clayey soils does exist, may have important practical implications, and thus has to be considered.

Conclusions

A variety of oedometer tests were performed on young Jonquière clay. The conclusions can be summarized as follows:

- There is an important structuration of clay with time when strain rate is small (10^{-7} s^{-1}).

- When the clay is rapidly loaded, as in CRS tests with a strain rate on the order of 10^{-5} s^{-1}, structure has no time to develop; also, in step-loading tests, as in conventional 24-hrs tests, the structure which could develop during one step is destroyed by the following one, so that the effects of the structuring phenomena are undiscernible. As a result, only the effect of strain rate appears when the two types of test are compared.

- Changes in strain rate are accompanied by a change in effective stress which seems independent of the degree of structuration.

Several structuring factors are known. However, their presence varies from clay to clay, and their effect seems to be influenced by the environment and the age of the clay. The manifestation of these structuring factors thus seems difficult to predict. However, the possibility of their occurrence must be recognized.

References

Bennett, R.H., Bryant, W.R., and Hulbert, M.H. (1991). "Microstructure of fine-grained sediments, from mud to shale" *Springler-Verlag*, 582p.

Bjerrum, L. (1967). "Engineering geology of Norwegian normally consolidated marine clays as related to the settlements of buildings". *Géotechnique*, 17(2),83-118.

Bjerrum, L. (1972). "Embankments on soft ground". *ASCE Specialty Conf. on Performance of Earth and Earth -supported Structures*, West Lafayette, 2, 1-54.

Burghignoli, A. (1979). "An experimental study of the structural viscosity of soft clays by means of continuous consolidation tests." *Proc. 7th ECSMF*, Brighton, 2, 23-28.

Gorman, C.T., Hopkins, T.C., Deen, R.C., and Drnevich, V.P. (1978). "Constant-rate-of-strain and controlled-gradient consolidation testing." *Geotechnical Testing J.*, 1(1), 3-15.

Graham, J., Crooks, J. H. A., and Bell, A. L. (1983). "Time effects on the stress-strain behavior of natural soft clays." *Géotechnique*, 33(3), 327-340.

Imai, G., and Tang, Y. X. (1992). "A constitutive equation of one-dimensional consolidation derived from inter-connected tests." *Soils and Foundations*, 32 (2), 83-96.

Kabbaj, M. (1985). "Aspects rhéologiques des argiles naturelles en consolidation." Ph.D. thesis, Laval University, Quebec City, Québec, Canada.

Leonards, G. A., and Altschaeffl, A. G. (1964). "Compressibility of clays." *J. Soil Mech. Found. Div.*, Proc. ASCE, 90 (SM5), 133-155.

Leroueil, S. (1988). " Recent developments in consolidation of natural clays." *Can. Geotech. J.*, 25(1), 85-107.

Leroueil, S. (1996). "Compressibility of clays: Fundamental and practical aspects." *J. Geotech. Eng.*, ASCE, 122(7), 534-543.

Leroueil, S., Kabbaj, M., Tavenas, F., and Bouchard, R. (1985). "Stress-strain-strain rate relation for the compressibility of sensitive natural clays." *Géotechnique*, 35(2), 159-180.

Locat, J. (1995). "On the development of microstructure in collapsible soils. Lessons from the study of recent sediments and artificial cementation". In Genesis and Properties of Collapsible Soils, Derbyshire et al., (eds). *Kluwer Academic Publishers*, 93-128.

Locat, J., Lefebvre, G., and Ballivy, G. (1984). "Mineralogy, chemistry, and physical properties interrelationships of some sensitive clays from Eastern Canada." *Can. Geot. J.*, 21, 530-540.

Mitchell, J.K. (1976). "Fundamentals of soil behavior". Wiley, New York.

Mitchell, J.K., and Solymar, Z.V. (1984). "Time-dependent strength gain in freshly deposited or densified sand." *J. Geotech. Eng.*, ASCE, 110(11), 1559-1576.

Perret, D. (1995). "Diagénèse mécanique précoce des sédiments fins du Fjord du Saguenay". Ph.D. Thesis, *Université Laval*, Ste-Foy, Québec, Canada.

Perret, D., Locat, J., and Leroueil, S. (1995). "Strength development with burial in fine-grained sediments from the Saguenay Fjord, Québec". *Can. Geotech. J.*, 32(2), 247-262.

Perret, D., Locat, J., and Leroueil, S. (1996). "Structuring effects during compression of a resedimented clay." *In preparation*.

Sills, G.C., and Thomas, R.C. (1984). "Settlement and consolidation in the laboratory of steadily deposited sediment." Proc. IUTAM/IUGG Symp., Univ. of Newcastle-upon-Tyne, Sept. 1983, Seabed Mechanics, Ed. Bruce Denness, Graham & Trotman, Ltd., London.

Smith, R. E., and Wahls, H. E. (1969). "Consolidation under constant rates of strain." *J. Soil Mech. Found. Div.*, ASCE, 95 (SM2), 519-539.

Wissa, A. E. Z., Christian, J.T., Davis, E. H., and Heiberg, S. (1971). "Consolidation at constant rate of strain." *J. Soil Mech. Found. Div.*, ASCE, 97(SM10), 1393-1413.

Effects of Stress Ratio on Behavior of Quasi-Preconsolidated Compacted Clay under Plane Strain Compression

Hoe I. Ling[1] Member, ASCE, and Fumio Tatsuoka[2]

Abstract

This paper presents the results of a series of plane strain compression tests conducted on compacted volcanic ash clay specimens. The soil specimens were prepared by dynamic compaction and consolidated isotropically or anisotropically. The effects of consolidation stress ratio on the strength and deformation characteristics were investigated under both drained and undrained shears. It was found that compacted clay behaved differently after having been consolidated anisotropically when compared to isotropically consolidated conditions. A small consolidation stress ratio, defined as the ratio of minor to major principal stresses, led to a higher strength. A large stiffness measured in anisotropically consolidated specimens was related to quasi-preconsolidated soil behavior. Smaller excess pore water pressure and volumetric strain were obtained for anisotropically consolidated specimens when compared to the corresponding isotropically consolidated specimens. The test results also revealed that conventional critical state family of soil models are not very appropriate for simulating the behavior of this compacted clay.

Introduction

The strength and deformation characteristics of cohesive soil, particularly those of soft clays, have been studied extensively in the past few decades (e.g., Jamiolkowski et al., 1985; Ladd, 1991). The design parameter used in total stress analysis, such as the undrained shear strength, has been one of the main subjects of study. The realization of the recompression technique (Bjerrum, 1973) and SHANSEP approach (Ladd and Foott, 1974) led to the acceptance of consolidated drained or undrained tests, conducted using a triaxial (ASTM D 4767) or direct shear device (ASTM D

---

[1] Visiting Assistant Professor, Department of Civil and Environmental Engineering, University of Delaware, Newark, DE 19716. (E-mail: *Ling@ce.udel.edu*)

[2] Professor, Department of Civil Engineering, University of Tokyo, Hongo, Bunkyo-ku, Tokyo 113, Japan.

anisotropically consolidated or K_o-consolidated specimen in a specially designed triaxial apparatus. Effects of overconsolidation, inherent and induced anisotropy on the stress-strain-strength behavior of clay have been studied to a certain extent. These experimental works have enabled development and verification of constitutive models with different degrees of sophistication (e.g., Schofield and Wroth, 1968; Sekiguchi and Ohta, 1977; Kaliakin and Dafalias, 1990; Whittle, 1993).

There have been few studies to compare strengths from triaxial and direct shear tests with that determined under plane strain conditions (e.g., Vaid and Campanella, 1974), and therefore, the relevance of applying axisymmetric strength parameters to design soil structures under plane strain conditions needs to be further investigated. While normally consolidated or slightly overconsolidated clays have been the target of laboratory study, the behavior of compacted and stiff clays is not as well understood as that of the soft clays.

This paper describes the strength and deformation behavior of a compacted clay, known as Kanto Loam, under plane strain compression. The soil properties were studied in conjunction with the development of a new technique to reinforce soil structures constructed of low-quality on-site soils. The effects of consolidation stress ratio and drainage conditions on quasi-preconsolidated clay specimens are investigated. A description of laboratory testing procedures is given, followed by the test results and discussions.

Specimen Preparation and Saturation

Kanto Loam is a volcanic ash clay available in the Metropolitan Tokyo area, Japan. It is a silty clay having a high natural water content (Fig. 1). The soil used in this study was obtained from a site where a full scale geosynthetic-reinforced soil retaining wall was constructed (Tatsuoka et al., 1992).

The soil specimens were prepared in the laboratory according to the dynamic compaction method (ASTM D 698). A cubical mold having inner dimensions of 12.0 cm × 9.72 cm × 6.25 cm (height × length in σ_3-direction × width in σ_2-direction) was used. The soil was compacted in six layers in this mold simulating a Standard Proctor test compaction energy. At the end of compaction, the specimen was removed from the mold and a piece of filter paper was placed on its top and bottom ends. Consolidation drainage was improved using strips of filter paper placed along the sides of the specimen in the σ_3-plane.

The specimen was then transferred to the testing apparatus and enclosed in a rubber membrane. A layer of Dow vacuum silicone grease, 0.05 mm thick, was smeared on the confining plates to minimize friction with the membrane. The top and bottom ends of the specimen were lubricated using a rubber membrane disc smeared with Dow silicone grease so that the end friction was minimized (Tatsuoka et al., 1984). The pressure cell was then assembled and filled with water before saturation was initiated. A schematic sketch of experimental setup is shown in Fig. 2. It has to be noted that a pair of specially designed wedges and sliding wheels were installed in this device to control the confining plates after isotropic consolidation such that a full confinement along the σ_2-direction could be ensured during shearing.

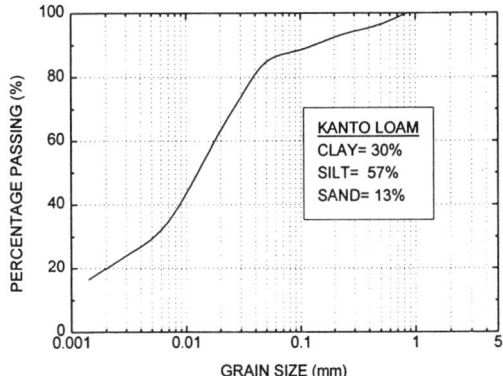

Figure 1. Grain Size Distribution Curve of Kanto Loam Soil

Saturation of soil specimen was conducted using a dry setting method (ASTM D 4767; Ampadu and Tatsuoka, 1993). First, a suction (negative confining pressure) was applied to the soil specimen and cell water simultaneously. An initial suction was applied to the specimen slightly higher than that of the confining cell water, say by 5 kPa. The suction was increased incrementally until it reached about 90 kPa in the specimen or 85 kPa in the cell water. An effective confining stress of 5 kPa was maintained in the specimen. This suction was maintained for another three hours to facilitate deairing (Fig. 2). The suction in the specimen and cell water was then reduced slowly. Water was drawn into the specimen during release of suction.

The back pressure was then applied to the specimen and cell water. Again, a constant effective confining stress was maintained in the soil specimen. This back pressurizing procedure was continued until it measured 100 kPa in the specimen. It was then left for about three hours before degree of saturation was checked. This procedure was found to be effective in saturating the soil specimen. All specimens had a B parameter (Skempton, 1954) measured greater than 0.96. Note that during saturation, the specimen was subjected to isotropic consolidation with $\sigma_2 = \sigma_3$.

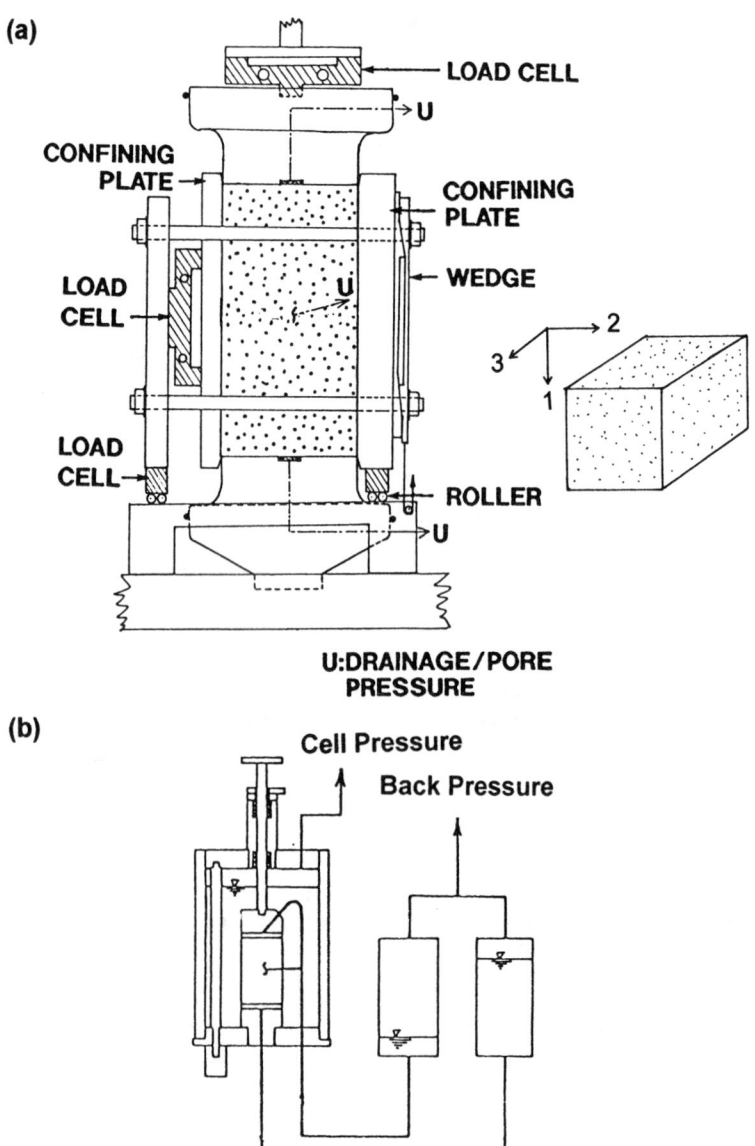

Figure 2. Testing Setup: (a) Plane Strain Apparatus, (b) Specimen Saturation

Testing Conditions

A series of drained and undrained tests were conducted under plane strain conditions. The specimens were consolidated either isotropically or anisotropically prior to shearing. Table 1 gives the details of this testing program with the water content, dry density and initial void ratio of the specimens, measured prior to consolidation. The specimens were consolidated under several different stress ratios, $K = \sigma'_{3c}/\sigma'_{1c}$. For anisotropically consolidated specimens $K = 0.15$ and 0.3 were selected as typical values to simulate possible loading conditions in an embankment. The axial load and load acting on the confining plates were measured using load cells. The confining pressure and volume change in the specimen were measured using high and low capacity differential pressure transducers. The electro-pneumatic (E/P) transducer, analog-digital (A/D) and digital-analog (D/A) converters were used to control the air pressure, loading motor and data acquisition.

Table 1. Testing Conditions

Test	Conso. Stress Ratio K	Drainage Condition	Minor Effective Stress σ'_{3c} (kPa)	Bulk Density γ_t (kN/m^3)	Dry Density γ_d (kN/m^3)	Water Content w (%)	Void Ratio e
1	0.15	Undrained	50	13.3	6.6	97	3.14
2	0.3	Undrained	25	13.2	6.5	102	3.28
3	0.3	Undrained	50	14.3	7.7	85	2.61
4	1.0	Undrained	50	13.7	6.7	105	3.18
5	0.15	Drained	30	13.0	6.6	97	3.24
6	0.15	Drained	50	13.4	6.7	98	3.14
7	0.3	Drained	25	13.5	7.3	85	2.82
8	0.3	Drained	50	14.3	7.5	91	2.72
9	0.3	Drained	100	13.8	7.4	86	2.78
10	1.0	Drained	50	13.6	6.5	111	3.32
11	1.0	Drained	100	13.9	7.4	88	2.79

During isotropic consolidation, air pressure was applied to the cell water through the E/P transducer in a small increment of 5 kPa. The specimen was allowed to consolidate for 30 minutes in each increment. After having attained the targeted effective confining stress, the specimen was consolidated further for 8 hours before shearing was initiated. The isotropic consolidation up to an effective confining stress of 50 kPa typically took 24 hours.

During anisotropic consolidation, a constant axial strain of 0.01 %/minute was used. The E/P transducer was instructed to apply the required air pressure to the cell water. A specified stress ratio, K, was used to control the stress path following an increase in the axial load. One unloading and reloading cycle was included in some of the tests. The slow loading rate resulted in negligible excess pore water pressure during consolidation. Consolidation was allowed for an additional 8 hours after reaching the targeted effective stresses. The time taken for anisotropic consolidation depended on the value of K; it extended over 2 days when K= 0.15.

Shearing was conducted at an axial strain rate of 0.01 %/minute in both undrained and drained tests to eliminate possible rate effects. The drained tests were performed by leaving the drainage valves at the top and bottom open; whereas in the undrained tests, all the valves were closed. A typical shear phase took about 24 hours.

Stress-Strain-Strength Relationships

In contrast to conventional triaxial tests, in which equal magnitudes of minor and intermediate principal stresses (i.e., $\sigma'_2 = \sigma'_3$) are applied, plane strain compression tests mobilize σ'_2 to a value between σ'_1 and σ'_3. For convenience, two different expressions were used to represent the deviatoric stress:

$$t = \sigma'_1 - \sigma'_3 \tag{1}$$

$$q = \sqrt{(\sigma'_1 - \sigma'_2)^2 + (\sigma'_2 - \sigma'_3)^2 + (\sigma'_3 - \sigma'_1)^2} \tag{2}$$

where t is the difference between the major and minor principal stresses as is used in conventional triaxial testing, and q is the second stress invariant which includes the intermediate principal stress and better defines the stress states in a soil specimen under three dimensional stress conditions.

Figs. 3 and 4 show the stress-strain relationships of the drained and undrained tests, respectively. For K= 0.3 and 0.15, the deviatoric stress increments, $q-q_o$ and $t-t_o$, are presented; where t_o and q_o are the initial values of deviatoric stress. It is seen that these two expressions give rise to different values of mobilized stress at an identical axial strain level. t is always greater than q. The difference between t and q is small in isotropically consolidated specimens, but they differ considerably over the entire strain levels for anisotropically consolidated specimens. This difference is apparently due to the intermediate principal stress, which is mobilized to a larger extent in an anisotropically consolidated specimen than that consolidated isotropically.

For the test with K= 0.15, the peak values of q and t were attained at very small axial strain (see insert in Fig. 3c). It is of interest to see that positive volumetric strain increased after this peak stress. This is in contradiction to the isotropically consolidated drained test results of most normally consolidated soils in which the specimen dilates (increase in volume) after attaining the peak stress. Thus, the peak stress shown in this particular test indicated failure of the soil fabric structure formed during anisotropic consolidation so that its stress-dilatancy relationships were different from the isotropically consolidated soil specimens.

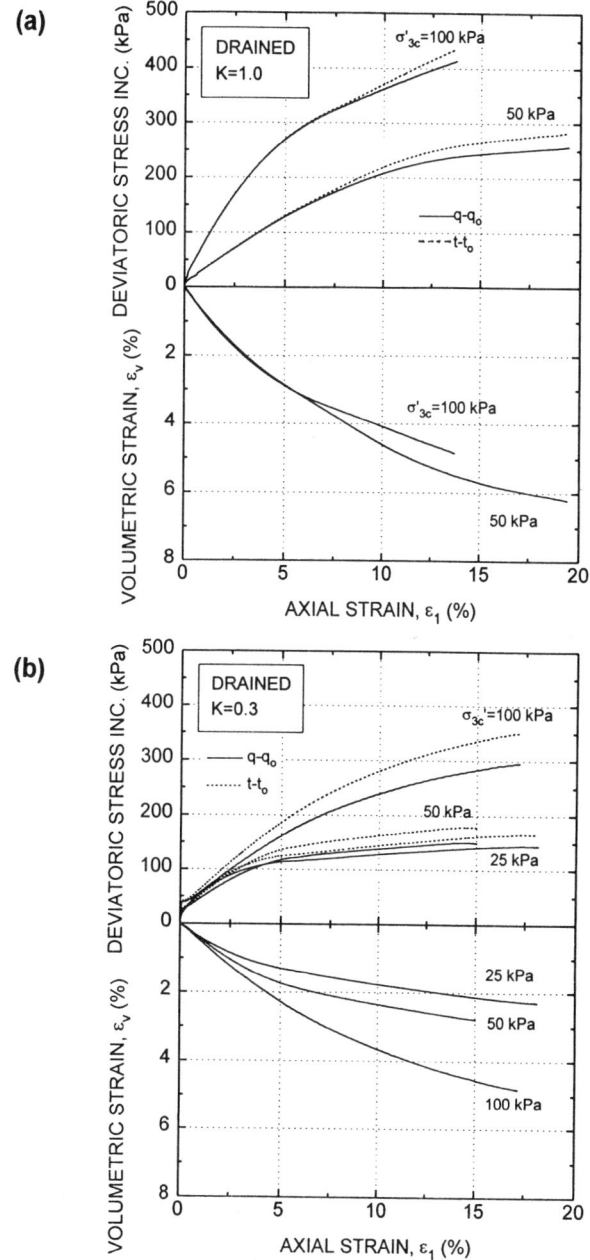

Figure 3. Stress-Strain Relationships for Drained Tests: (a) K= 1.0, (b) K= 0.3

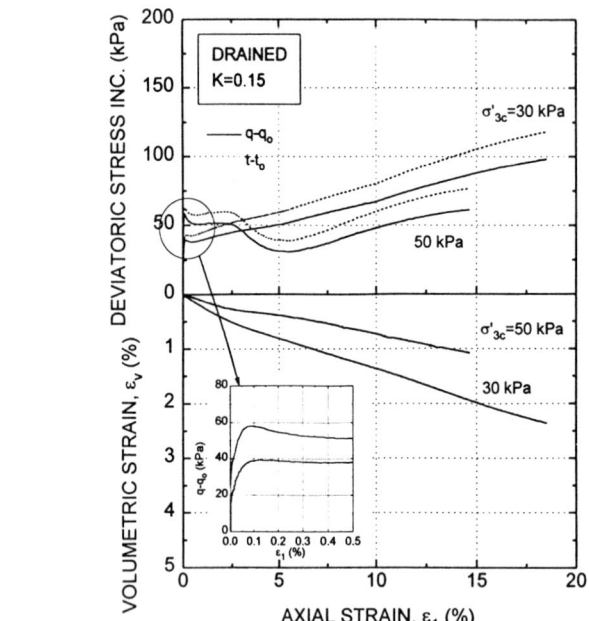

Figure 3 (Cont). Stress-Strain Relationships for Drained Tests: (c) K= 0.15

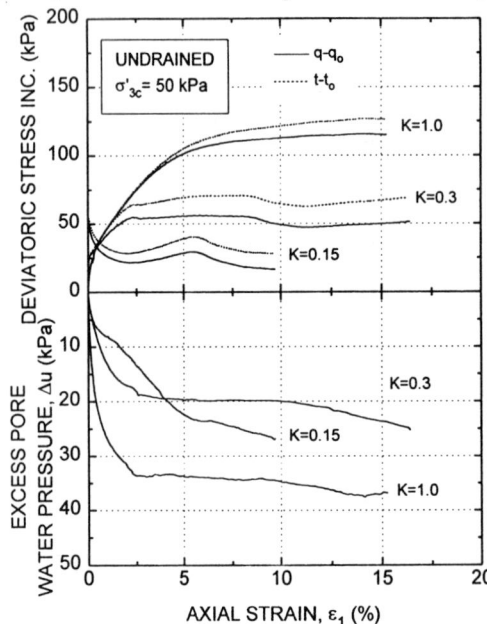

Figure 4. Stress-Strain Relationships for Undrained Tests, σ'_{3c}= 0.5 kPa

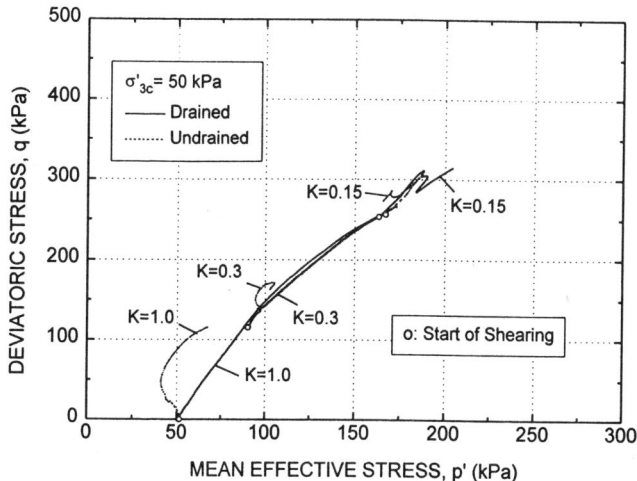

Figure 5. Effective Stress Paths for Drained and Undrained Tests, σ'_{3c}= 50 kPa

Mean effective stress is expressed using s' and p'. s' does not include the intermediate principal stress σ_2':

$$s' = \tfrac{1}{2}(\sigma_1' + \sigma_3') \tag{3}$$

$$p' = \tfrac{1}{3}(\sigma_1' + \sigma_2' + \sigma_3') \tag{4}$$

The typical effective stress paths of drained and undrained shearing tests are shown in Fig. 5. The drained stress paths at different values of K seem to fall within a similar curve, except the initial portion of that having K= 0.15. The effective stress paths deviated slightly from linearity as shearing proceeded due to mobilization of the intermediate principal stress. In undrained tests, the initial portion of the effective stress paths is apparently linear, but it deviated from linearity as shearing proceeded. The stress path of undrained tests is quite different from that used to formulate conventional critical state model (Schofield and Wroth, 1968).

Figs. 3(a) to (c) also show the volumetric strain of soil specimens during the drained shearing. Except at K= 0.15, a higher initial confining stress resulted in a larger positive volumetric strain (contraction). However, the specimens having a smaller consolidation stress ratio recorded a smaller volumetric strain at subsequent shearing.

There is a trend that a lower value of consolidation stress ratio induced a smaller excess pore water pressure when compared to the isotropically consolidated specimen (Fig. 4). The overconsolidated behavior of compacted clay can be inferred from the pore pressure coefficient proposed by Skempton (1954) or Henkel (1960), where Henkel (1960) considers the intermediate principal stress at failure. Negative values of pore pressure coefficients at failure obtained for these tests corresponded to the range of values of compacted clays.

Fig. 6 shows the stress states at failure in the effective stress space. s'-t and p'-q envelopes represent Coulomb failure envelope and critical sate line, respectively. The Coulomb failure envelope fits the strength under 2-dimensional stress conditions satisfactorily for specimens consolidated with different stress ratios and drainage conditions. The effective cohesion and internal friction angle of Kanto Loam were determined as c'= 10 kPa and ϕ'= 36°. The critical state line of this compacted clay does not resemble that proposed by Schofield and Wroth (1968), but exhibiting a strength intercept. The best fit curve should be considered nonlinear instead of a straight line.

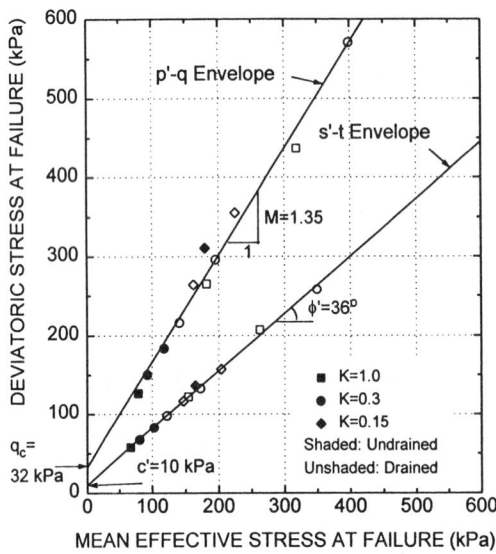

Figure 6. Failure Envelopes Under 2- and 3-Dimensional Stress Conditions

Stiffness

The stiffness of soils can be conveniently expressed using the Young's modulus or shear modulus. In the plane strain compression test, assuming that the material is linear elastic and isotropic (Hooke's law), its secant Young's modulus and shear modulus are determined from the following expressions:

$$E_{\sec} = \frac{t-t_o}{\varepsilon_1}(1-v'^2_{\sec}) \qquad (5)$$

$$G_{\sec} = \frac{t-t_o}{2(\varepsilon_1 - \varepsilon_3)} \qquad (6)$$

where t_o is the deviatoric stress at the start of shearing, and $v'_{\sec} = \sigma'_2/\sigma'_3$ is the secant Poisson's ratio determined from plane strain compression test. In this paper, the 'plane strain' Young's modulus, E_{psc}, was used in lieu of the Young's modulus:

$$(E_{\sec})_{psc} = \frac{t-t_o}{\varepsilon_1} \qquad (7)$$

Fig. 7 shows the relationships between K and the Young's and shear moduli determined at very small value of axial strain ($\varepsilon_1 = 10^{-4}$). Under an identical confining stress, $\sigma'_3 = 50$ kPa, the elastic moduli are larger in the undrained tests when compared to those of the drained tests. Consolidation stress ratio affected the stiffness of compacted clay. In both drained and undrained tests, a higher stiffness was obtained as the consolidation stress ratio becomes smaller. For instance, the elastic moduli at K= 0.15 is about 10 times larger than that of K= 1.0.

Fig. 8 shows the relationships between shear modulus, G, and initial mean effective stress, p'_o, for all the tests. The effect of consolidation stress ratio is seen at different stress levels. The results show that the shear modulus may be normalized by p'_o for this compacted clay. That is, G/p'_o is equal to a constant at a particular value of K. For example, $G/p'_o = 250$ for K= 0.3 under drained conditions.

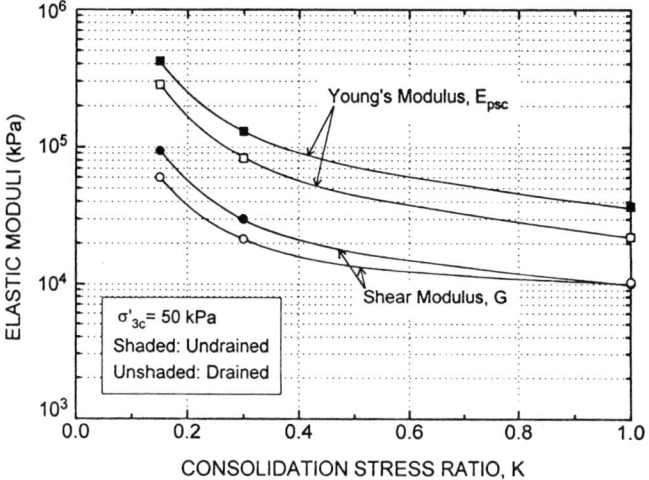

Figure 7. Relationships Between Elastic Moduli and Stress Ratio, $\sigma'_{3c} = 50$ kPa

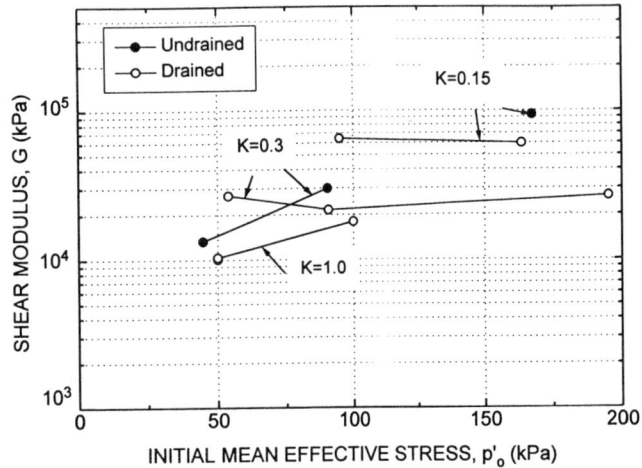

Figure 8. Relationship Between Shear Modulus and Initial Mean Effective Stress

The large stiffness at the initial stage of shearing, in particular for anisotropically consolidated specimens, was due to delayed consolidation (Leonards and Ramiah, 1960). Figs. 9(a) and (b) show the relationships at small strains for the specimens consolidated to $\sigma'_{3c} = 50$ kPa. Nearly simultaneous yielding may be seen in the corresponding $q-q_o$ vs ε_1 and $q-q_o$ vs ε_v (or Δu) relationships.

It is seen in Fig. 10 that when subjected to a constant mean effective stress, p'_o, the specimen behaved as if it has been mechanically preconsolidated by a mean effective stress p'_c. Although not presented herein, the results of testing indicated that the magnitude of p'_c depended on K. A specific soil fabric structure was formed during delayed consolidation resulting in a high initial stiffness. This soil fabric was able to sustain axial load without exhibiting lateral strain and excess pore water pressure. Upon breaking down of this fabric structure, the shearing curve resembled that of normal consolidation.

Conclusions

Based on a series of plane strain compression tests conducted on compacted Kanto Loam soil, the following conclusions were drawn with respect to its stress-strain-strength characteristics:

(a)
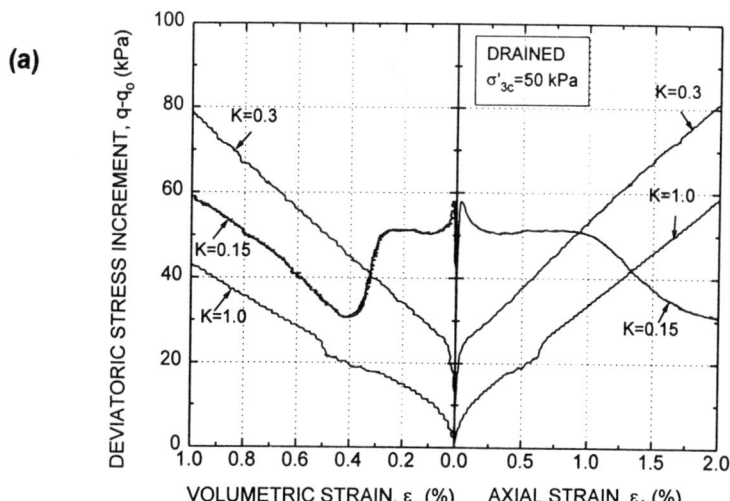

(b)
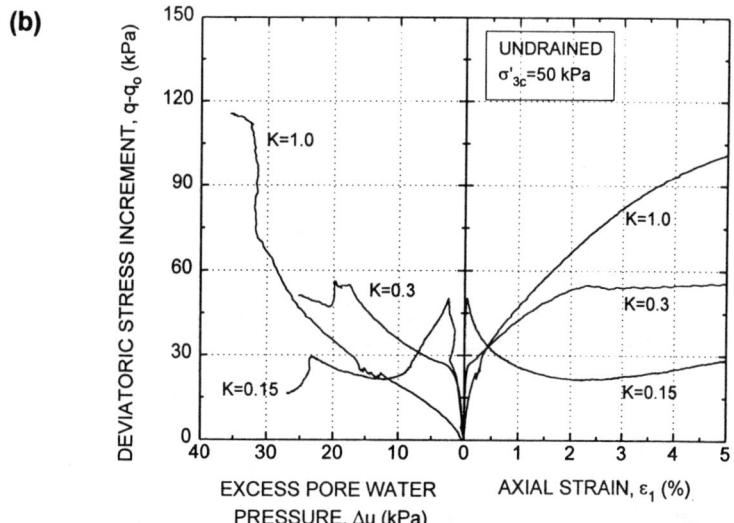

Figure 9. Yielding in (a) Drained and (b) Undrained Tests, σ'_{3c} = 50 kPa

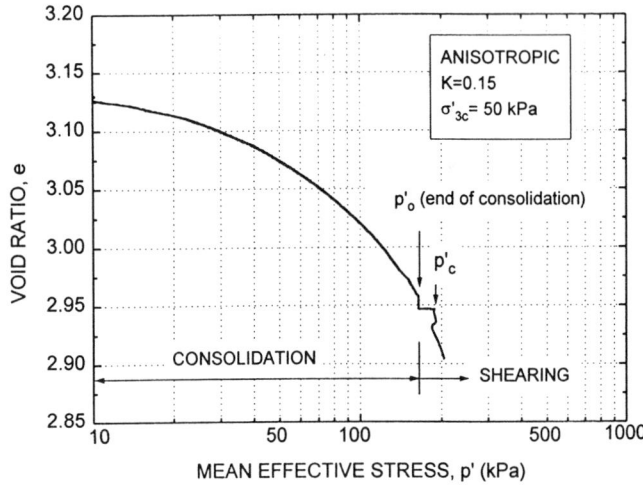

Figure 10. Change in Void Ratio During Anisotropic Consolidation and Shearing

1. A quasi-preconsolidated behavior was observed for compacted Kanto Loam soil. Negligible excess pore water pressure and lateral strain were measured during initial stage of shearing. Quasi-preconsolidation gave rise to a higher stiffness depending on the consolidation stress ratio.
2. Specimens consolidated at a stress ratio $K= 0.15$ showed peak strength at very small strain levels ($\approx 0.05\%$). In the post-peak softening regime, the specimen volume continued to decrease in the drained test while the mean effective stress continued to decrease in the undrained test.
3. The consolidation stress ratio affected the behavior of soil. In an anisotropically consolidated soil specimen, the stiffness was larger than that of the isotropically consolidated specimens subjected to a similar effective confining stress. The shear modulus may be normalized by the initial mean effective stress.
4. The effective strength parameters for the Coulomb failure envelope were found to be independent of the drainage conditions and consolidation stress ratio.

This experimental investigation showed that compacted clay exhibits behavior different from typical soft clays. It revealed the importance of fabric formed during anisotropic consolidation and limitations of conventional critical state models in expressing the strength and deformation characteristics of compacted clay. It is, thus, hoped that the results presented herein will stimulate application of nonconventional plasticity models for simulating performance of earth structures constructed of compacted clays.

References

Ampadu, S. and Tatsuoka, F. (1993) "Effect of setting method on the behavior of clays in triaxial compression from saturation to undrained shear." *Soils and Foundations*, Tokyo, 33(2), 14-34.

ASTM D 698. Laboratory Compaction Characteristics of Soil Using Standard Effort. *Annual Book of ASTM Standards*, Vol. 04.08.

ASTM D 3080. Standard Test Method for Direct Shear Test of Soils Under Consolidated Drained Conditions. *Annual Book of ASTM Standards*, Vol. 04.08.

ASTM D 4767. Standard Test Method for Consolidated-Undrained Triaxial Compression Test on Cohesive Soils. *Annual Book of ASTM Standards*, Vol. 04.08.

Bjerrum, L. (1973). "Problems of soil mechanics and construction on soft clays." *Proceedings of 8th International Conference on SMFE*, Moscow, 111-159.

Jamiolkowski, M., Ladd, C.C., Germaine, J.T., and Lancellotta, R. (1985). "New developments in field and laboratory testing of soils." *Proceedings of 11th International Conference on SMFE*, San Francisco, 57-153.

Henkel, D.J. (1960). "The shear strength of saturated remolded clays." *Proceedings of Research Conference on Shear Strength of Cohesive Soils*, Boulder, 533-554.

Kaliakin, V.N. and Dafalias, Y.F. (1990). "Theoretical aspects of elastoplastic-viscoplastic bounding surface model for cohesive soils." *Soils and Foundations*, 30(3), 11-24.

Ladd, C.C. and Foott, R. (1974). "New design procedure for stability of soft clays." *Journal of Geotechnical Engineering*, ASCE, 100(7), 763-786.

Ladd, C.C. (1991). "Stability evaluation during staged construction." *Journal of Geotechnical Engineering*, ASCE, 117(4), 540-615.

Leonards, G.A. and Ramiah, B.K. (1960). "Time effects in the consolidation of clays." *Papers on Soils 1959 Meetings, ASTM STP 254*, ASTM, 116-130.

Schofield, A.N. and Wroth, P. (1968). *Critical State Soil Mechanics*, McGraw-Hill, London, 310 pp.

Sekiguchi, H. and Ohta, H. (1977). "Induced anisotropy and time dependency in clay." *Proceedings of 9th International Conference on SMFE*, Tokyo, 229-237.

Skempton, A.W. (1954). "The pore pressure coefficients A and B." *Geotechnique*, 4, 143-147.

Tatsuoka, F., Molenkamp, F., Torii, T., and Hino, T. (1984). "Behavior of lubricated layers of platens in element tests." *Soils and Foundations*, 24(1), 113-128.

Tatsuoka, F., Murata, O, and Tateyama, M. (1992). "Permanent geosynthetic-reinforced soil retaining walls used for railway embankments in Japan." *Geosynthetic-Reinforced Soil Retaining Walls*, Wu (ed.), Balkema, 101-130.

Vaid, Y.P. and Campanella, R.G. (1974). "Triaxial and plane strain behavior of natural clay." *Journal of Geotechnical Engineering*, ASCE, 100(3), 207-224.

Whittle, A.J. (1993). *"*Evaluation of a constitutive model for overconsolidated clays.*"* *Geotechnique*, 43(2), 289-314.

Rate and Creep Effect on the Stiffness of Soils

Diego C. F. Lo Presti[1], Michele Jamiolkowski[2], Oronzo Pallara[3] and Antonio Cavallaro[4]

Abstract

The shear modulus of two undisturbed Italian clays was measured in the laboratory by means of a Resonant Column/Torsional shear apparatus for a strain range of 0.0001 % up to 1 %. Three different kinds of tests were performed on hollow cylindrical specimens reconsolidated to the supposed in situ geostatic stress: a) static monotonic loading tests at constant stress rate, b) cyclic loading tests at constant strain rate, c) Resonant Column tests. Moreover, the small strain shear modulus G_o, determined at strain level less than 0.001 % was measured during the drained creep following reconsolidation. The increase of G_o with time during the drained creep was therefore assessed and compared with available correlations from literature. The data obtained from the above mentioned research enabled one to examine the influence of the strain rate, loading conditions (cyclic vs. monotonic) and strain level on the deformation moduli of the tested clays. It was possible to assess that the normalised curves (G/G_o vs. γ) are dependent on loading conditions and strain rate.

Introduction

Shear modulus (G) and damping ratio (D) of soils are basic input parameters used to compute the equivalent-linear seismic response of soil deposits and the interaction between soil and structures. A great deal of experimental data are available in the literature concerning the dependence of these two important parameters on several factors such as the shear strain level (γ) consolidation stresses $(\sigma'_{vc}, \sigma'_{hc})$, void ratio (e), overconsolidation ratio (OCR), etc. However there is;

---

[1] Research Associate, Department of structural engineering, Politecnico di Torino, Corso Duca degli Abruzzi 24, 10129 Torino, Italy
[2] Professor, ditto
[3] Post-grauduate student, ditto
[4] Graduate student, University of Catania

limited information about the influence of loading rate, or strain rate, on G and D. Moreover, the majority of the, available experimental studies consider the dependence of G and D on the loading frequency (f) rather than on the strain rate, even though many authors have pointed out that the strain rate is a much more significant parameter (see Drnevich and Ashmawy 1995 among others).

In laboratory cyclic tests, loading frequency and loading rate are linked to each other by the following well known relationship:

$$\dot{\gamma} = 4 \cdot \gamma_{SA} \cdot f \, [\%/s] \quad (1)$$

where: $\dot{\gamma}$ = shear strain rate; γ_{SA} = single amplitude shear strain [%]; f = frequency [Hz];

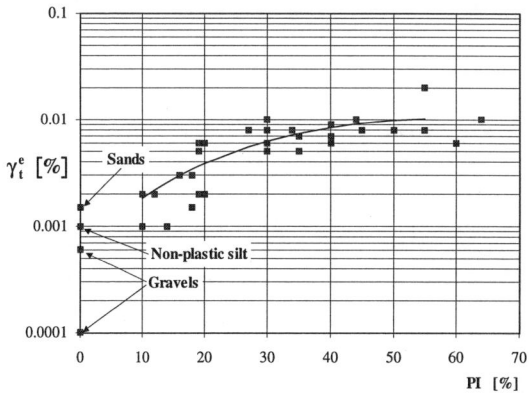

Figure 1 Elastic treshold shear strain from resonant column tests (Lo Presti 1989).

The above relationship holds for strain controlled tests performed using triangular loading. In the case of sinusoidal loading or stress controlled tests the value computed using eq. (1) represents the average strain rate during each cycle. It is therefore obvious from eq. (1) that in a constant frequency test the strain rate can increase with the strain level by several orders of magnitude. An increase of the stiffness for an increase of the strain rate is generally recognised (Dobry and Vucetic 1987, Tatsuoka and Shibuya 1992); therefore, the stress-strain relationship, in a typical quasi-static test performed at constant frequency, is distorted, as observed first by Isenhower and Stokoe (1981). In particular, the effect of an increase in the strain rate is that of an increase in the elastic limit which is also called the elastic threshold shear strain (γ_t^e) (Vucetic 1994), that is, the strain level below which the stress-strain relationship is linear. It is generally recognised that the rate dependence of soil stiffness is due to the viscosity of the soil skeleton (Dobry and Vucetic 1987). A larger influence of the strain rate for soils with larger plasticity index (PI) is therefore expected. The increase of the elastic limit with PI shown in Figure 1 (Lo

Presti 1989) supports this idea. Similar data have also been shown by other authors (see Vucetic 1994). The elastic limiting line (ELL) concept proposed by Tatsuoka and Shibuya (1992) is a more general framework to explain the influence of the strain rate on the soil stiffness. The ELL concept is supported by the experimental observation that the strain rate effect on the soil stiffness increases for increasing strain level as shown in Figure 2. This Figure shows the coefficient of strain rate vs. the axial strain as obtained from undrained triaxial compression tests on different intact and reconstituted clay specimens. The coefficient of strain rate is defined as:

$$\alpha(\gamma) = \frac{\Delta G(\gamma)}{\Delta(\log \dot{\gamma}) \cdot G(\gamma, \dot{\gamma}_{REF})} \quad (2)$$

i.e. the increase in shear modulus for one logarithmic cycle of strain rate normalised with respect to the shear modulus obtained with a reference strain rate.

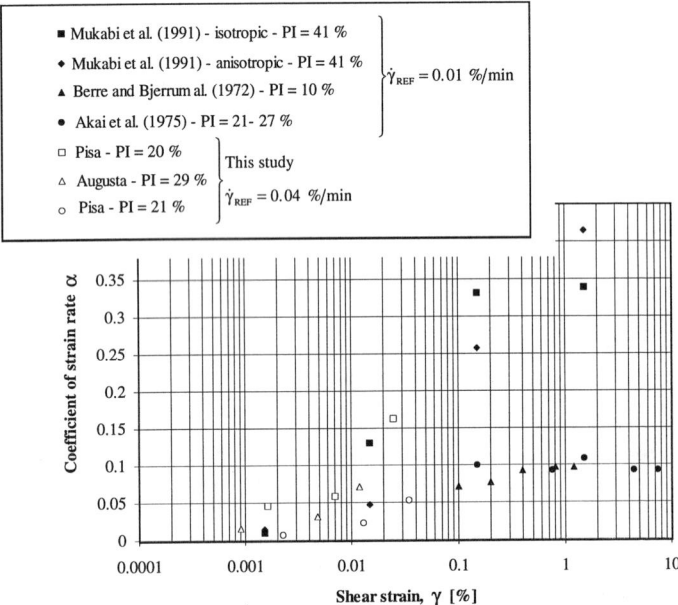

Figure 2 Coefficient of strain rate vs. shear strain.

The values of $\alpha(\gamma)$ in Figure 2 increase with an increase of the strain level and PI. Unfortunately, only data by Mukabi et al. (1991) cover strain interval from 0.001 % to 1 %, whereas the other data consider shear strains larger than 0.05 %. In order to experimentally determine $\alpha(\gamma)$ values at small and intermediate strains and to assess the degree of distortion of stress-strain curves obtained using the current practice (constant frequency tests), it was decided to perform cyclic loading torsional shear tests (CLTSTs) and conventional resonant column tests (RCTs) on two different Italian clays. Monotonic loading torsional shear tests (MLTSTs) were also performed in order to compare the soil response obtained in MLTSTs with that from CLTSTs and in particular to verify the validity of the second Masing Rule (Masing 1926).

Testing apparatus and experimental procedures

The Torsional Shear/Resonant Column apparatus used for this research is the same as that described by Lo Presti et al. (1993). The main characteristics of the apparatus are summarised below.

The apparatus is equipped with an electric motor which has 8 coils and four magnets (Figure 3). This motor was provided by SBEL (Arizona). It is capable of a maximum torque of about 1 Nm and can accurately operate under stress control conditions. The motor is driven by means of an arbitrary function generator and a power amplifier. Both arbitrary function generator and data acquisition system are computer controlled.

A) Coils and magnets;
B) Proximity transducers for rotation measurements;
B1) Target;
C) System used to move the position of the proximity transducer;
D) LVDT for axial displacement measurements.

Figure 3 General view of the apparatus.

The apparatus can house hollow clay specimens with internal diameter of 30 mm, external diameter of 50 mm and height of 100 mm. The specimens can be anisotropically consolidated with consolidation stress ratio $K_c = \sigma'_{hc}/\sigma'_{vc} \leq 1$. A bath of silicon oil is provided all around the specimen to avoid air migration through the membranes.

An accelerometer is used to determine resonance during RCTs. The shear strain during MLTSTs or CLTSTs is determined by means of a pair of proximity transducers with a resolution of 0.1 μm (Figure 3). The relative distance between the proximity transducers and their targets can be set by means of the system shown in Figure 3 which enable one to operate from outside the cell. This system enable one to move the transducers along the horizontal direction. A glass window is available for visual inspection inside the pressure cell. The applied torque is computed by means of a torque-voltage calibration curve (Lo Presti et al. 1993).

Pisa and Augusta clay specimens were tested with this apparatus. The Pisa clay samples were retrieved from depths of between 12 and 17 meters at the site of the Leaning Tower by means of Laval sampler (La Rochelle et al. 1991). The tested samples belong to the deposit locally called "Upper Pancone clay" which is a soft, lightly overconsolidated (OCR = 1.5 to 2.0), quaternary, marine clay with low to medium PI. Detailed information on the Pisa clay deposit is given by Berardi et al. (1991), Costanzo et al. (1994) and Jamiolkowski et al. (1994). The Augusta site is located on the east coast of Sicily. A civil building (228 flats), which was damaged by the Sicilian earthquake of December 13th 1990 and is now under repair, is located in that area. The Augusta clay is a medium stiff, overconsolidated (OCR = 2.0 to 6.0), quaternary marine clay with low to medium PI. Samples were retrieved with a Shelby sampler of 86 mm diameter. Detailed information on the Augusta clay deposit is given by Maugeri et al. (1994).

The Pisa clay specimens were reconsolidated to the in situ vertical effective stress (σ'_{vo}) with $K_c = 1$ or $K_c = K_o = 0.65$ (i.e. the best estimate of the coefficient of earth pressure at rest reported by Berardi et al. 1991). In the case of anisotropic consolidation, the specimen was first isotropically reconsolidated to $\sigma'_{vc} = \sigma'_{hc} = K_o \cdot \sigma'_{vo}$ and then the vertical effective stress was raised to $\sigma'_{vc} = \sigma'_{vo}$. The coefficient of earth pressure at rest for the Augusta clay ranges from 0.9 to 1.5. In this case the specimens were isotropically reconsolidated to $\sigma'_m = \sigma'_{vo}(1 + 2 \cdot K_o)/3$.

The degree of saturation of the system was assessed during the isotropic consolidation stage by checking the Skempton's B parameter. High values of B (larger than 0.95) were achieved in the case of Pisa clay specimens by means of conventional back pressurization procedure. For Augusta clay it was not possible to obtain B values greater than 0.9. The final consolidation stage was maintained for 24 hours and then the specimens was subjected to MLTST with closed (U) or opened (D) drainage, as specified in Table 1. At the end of this stage the drainage was opened for at least 24 hours allowing the dissipation of the pore pressure generated during shear. The drainage was again closed and the specimens experienced CLTST and/or RCT. In some tests two stages of CLTSTs with different strain rates were applied. A 24 hour period of drainage was allowed after each shearing stage. The permanent shear deformation after MLTST was so large for only one Pisa clay specimen that it was decided to duplicate the specimen to perform cyclic tests. Table 1 summarises the tests performed.

Table 1 Test Conditions

Site	Test number	σ'_{vc}	K_c	e	PI	MLTST	CLTST	RCT
Pisa	1	112	0.65	1.606	55	D		U
Pisa	2	112	1	1.572	47	D		U
Pisa	3	138	1	1.023	21		U	U
Pisa	4	114	1	0.901	22	U		
Pisa	5	104	1	0.896	22		two stages (U)	
Augusta	6	182	1	0.684	29	U	two stages (U)	U

MLTSTs were performed at a constant stress rate (0.5±0.035 kPa/min). CLTSTs were performed with triangular loading under stress control. However, considering that the cyclic degradation of shear modulus was not severe in this case, the average strain rate during cyclic loading was maintained constant throughout the test by reducing the frequency as the strain level increased. Sinusoidal loading was used for RCTs.

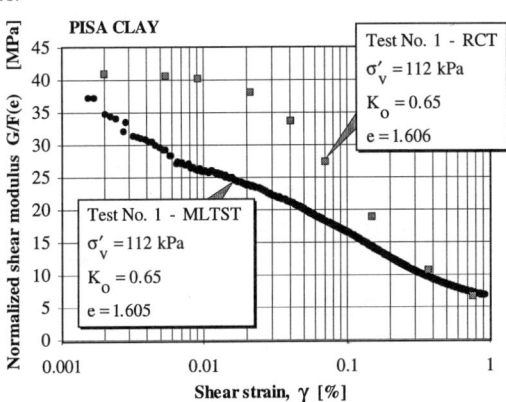

Figure 4 Shear modulus of anisotropically consolidated upper Pisa clay

Experimental results

Typical G-γ curves obtained from drained MLTSTs and undrained RCTs on Pisa clay specimens (test No. 1, see Table 1) are shown in Figure 4. The shear moduli plotted in Figure 4 are normalised by means of a void ratio function

$F(e) = e^{-1.3}$ (Lo Presti 1989, Jamiolkowski et al 1994) in order to take even very small variations of the void ratio into account. The authors believe that the differences between these two curves are mainly due to strain rate effects. The data shown in Figure 4 indicate that the strain rate effect is relatively unimportant at very small strains, whereas this rate effect increases with an increase in the strain levels. Figure 5 shows the strain rates applied during the MLTST and those observed during RCT. In the same Figure, the strain rates that the specimens usually experience during conventional quasi-static tests performed with f = 0.1 or 1 Hz, are also shown. The strain rate values, observed during RCTs, are typical of vibrations induced by traffic but are much higher then those occurring under seismic or wave loading conditions, especially for $\gamma > 0.01\%$ (Shibuya et al. 1995). During undrained RCT a very small pore pressure build-up was observed for a shear strain larger than 0.1 %. Considering that MLTST was performed under drained conditions, this pore pressure build-up could explain why the differences between MLTST and RCT curves tend to disappear for $\gamma > 0.1\%$. However, the curves compared in Figure 4 have been obtained under different test conditions: drainage, loading history and strain rate, it is therefore not possible to draw definitive conclusions from this comparison. In order to assess the influence on shear modulus of each of the above indicated factor undrained MLTST, CLTST and RCT were performed.

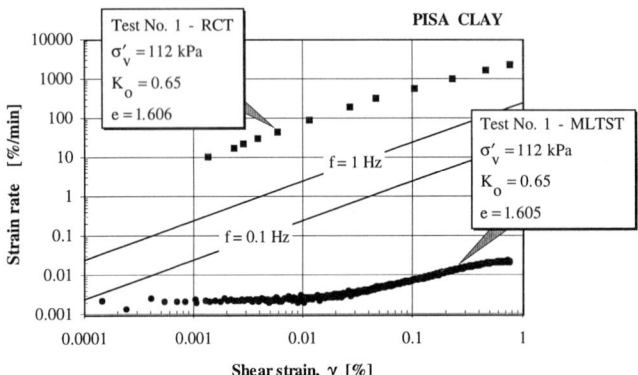

Figure 5   Strain rate during MLTST, cyclic quasi-static and RCT tests.

Shear modulus degradation during untrained CLTSTs

Figures 6a and 6b show the stress-strain relationships determined during undrained CLTST on Pisa and Augusta clays respectively. In particular the Figures show 30 loading cycles applied under stress control with triangular loading. It is possible to observe that for these clays and for the considered strain interval the cyclic degradation of shear modulus (Dobry and Vucetic 1987) is negligible. The above consideration implies that the 30 loading cycles shown in the Figures 6a and

6b were performed at constant average strain rate. The value of this strain rate can be computed using equation (1).

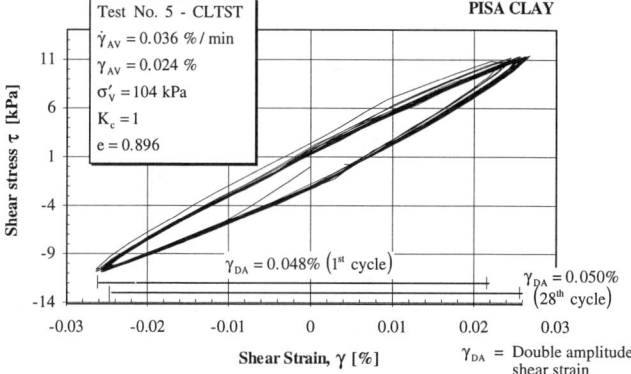

Figure 6a Stress - strain loops for Pisa clay.

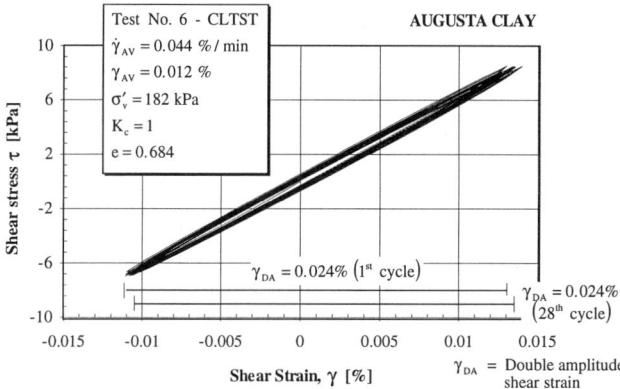

Figure 6b Stress - strain loops for Augusta clay.

Pore pressure build up

The volumetric threshold shear strain (γ_t^v) (Vucetic 1994, Jamiolkowski et al. 1994) indicates the strain level above which the build up of permanent volumetric strains in drained tests or the pore pressure build up in undrained tests occur. In order to evaluate γ_t^v The above phenomena, obviously, have to be observed in monotonic or cyclic tests that do not involve a permanent change of the mean total stress. The accumulated excess pore pressure, measured during undrained tests, is plotted in Figure 7 vs. the shear strain. It is possible to observe that:

- the values of γ_t^v increase with an increase of the strain rate.

- the values of the accumulated Δu, for a given strain level increase with an increase of the strain rate.
- greater values of γ_t^v occur during monotonic tests in comparison to cyclic tests, regardless of the strain rate.
- the pore pressure build-up, for a given strain level is greater in the case of cyclic tests than for monotonic tests, regardless of the strain rate.

Unfortunately, the majority of the data plotted in Figure 7 are from tests on Augusta clay where a not very high value of the B parameter was achieved. It is therefore not possible to draw definitive conclusions on the effective values of γ_t^v and Δu.

Coefficient of strain rate

The $G/F(e)-\gamma$ curves of Pisa clay obtained from test No. 4 during the MLTST stage (curve A) and that obtained from test No, 5 during the CLTST (curve B) are compared in Figure 8. Specimens for tests No. 4 and 5 were obtained from the same Laval sample and were reconsolidated to very similar isotropic consolidation stresses (see Table 1). The average strain rate values experienced by

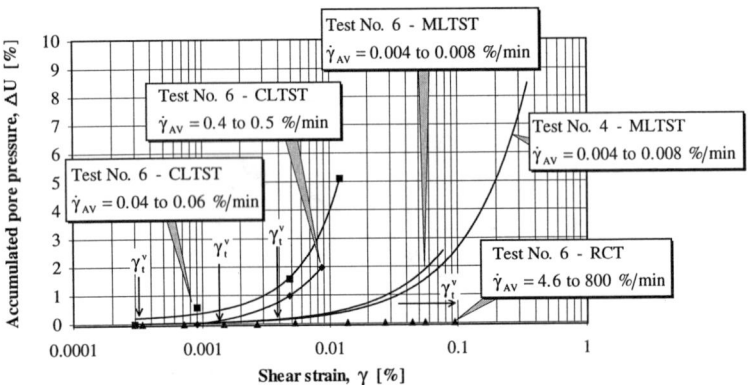

Figure 7 Pore pressure build up during cyclic and monotonic tests on Pisa and Augusta clays

the specimens during different stages are also indicated in Figure 8. It is possible to observe that:
- the, G/F(e)-γ curve from MLTST (curve A) and that from CLTST (curve B) are different from each other. The strain rate during CLTST was 5 to 10 times larger than those experienced by the specimen during MLTST. Considering that the coefficient of strain rate, for strain levels of 0.001 % to 0.05 % is relatively small (about 5 %, see Figure 2), the differences between the two curves are mainly due to the different loading conditions (monotonic and cyclic).
- the strain rate of the MLTST increases with an increase of the strain level, while that of CLTST is almost constant. Moreover, according to the data shown in Figure 7, higher values of the accumulated pore pressure, build-up are expected in

the case of CLTST for $\gamma > \gamma_t^v$. The differences between the G/F(e)-γ curves from MLTST and CLTST should therefore decrease at larger strains.
- the differences observed at small strains between the shear modulus obtained from MLTST and that from CLTST indicate that, even at very small strains (i.e. 0.001% in Figure 8), the soil response is not perfectly elastic. It is reasonable to consider that the elastic limit in this case is less than 0.001 %, even though, for practical purpose, it is possible to assume $\gamma_t^e \cong 0.001$ % as suggested in the literature (Hardin 1978).
- the stress-strain relationship obtained from the MLTST represents the so called backbone curve. The Masing's second rule, implicitly, assume that , for a given strain level, the secant stiffness obtained from the backbone curve has to be the same as that obtained from the unload reload cycle. The above discussed differences between the G/F(e)-γ curves from MLTST and CLTST, therefore, devalue the use of the Masing's second rule.

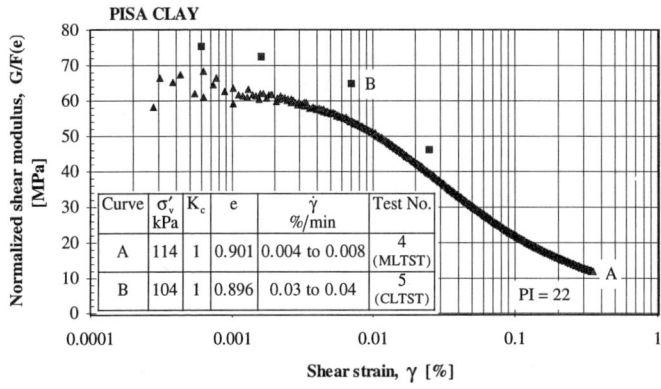

Figure 8   Shear modulus of Pisa clay from MLTST and CLTST tests

The shear modulus of Pisa clay obtained from CLTST and RCT (Test No. 3 of Table 1) is plotted vs. the log. of shear strain in Figure 9. It is possible to observe that the differences between the G-γ curves from CLTST and RCT increase with an increase of the shear strain. Moreover the RCT shows a constant shear modulus for a large strain interval, while for the CLTST the shear modulus continuously decreases. Similar observations can be done for the results of test No. 6 on Augusta clay. The G/G_o-γ curves obtained from two CLTSTs with different strain rates and the RCT performed on Augusta clay are plotted in Figure 10. G_o is the small strain or maximum shear modulus determined at $\gamma \leq 0.001$ %. Data of Figures 11 and 12 enable one to compute the coefficient of strain rate of Pisa and Augusta clays in the following way.

$$\alpha(\gamma) = \frac{G(\gamma,\dot{\gamma}_1) - G(\gamma,\dot{\gamma}_2)}{(\log \dot{\gamma}_1 / \dot{\gamma}_2) \cdot G(\gamma, \dot{\gamma} \approx 0.04\% / \min)} \qquad (2b)$$

The α values obtained from test No. 3 on Pisa clay and test No. 6 on Augusta clay are plotted on Figure 2. These values increase with strain level and their trend is in good agreement with the data obtained by other researchers from undrained triaxial compression test (Akai et al. 1975, Berre and Bjerrum 1972, Mukabi et al. 1991). However a larger database is needed in order to provide a mathematical function of $\alpha(\gamma)$ for a given value of PI.

Drained creep

Under constant consolidation stresses, G_o increases with time when subjected to drained creep. The increase of G_o with time during secondary compression is often quantified by means of the following parameter (Anderson and Stokoe 1978):

$$N_G = \frac{\Delta G}{G_o(t=t_p)} \cdot \frac{1}{\Delta(\log t)} \qquad (3)$$

which represents the increase of shear modulus per log cycle of time of secondary compression, normalised with respect to the small strain shear modulus at the end of primary consolidation or some other reference modulus [i.e. G_o (t = 1000 min)]. Different authors have proposed empirical correlations between N_G and simple soil properties such as D_{50}, PI and the secondary compression index C_α. Lee et al. (1995) have reviewed the available correlations of N_G and have proposed an empirical formula which take both PI and C_α into account. Mesri and Choi (1983) proposed the following correlation:

$$N_G = 10^{C_\alpha/C_c} - 1 \qquad (4)$$

where: C_c is the compression index.

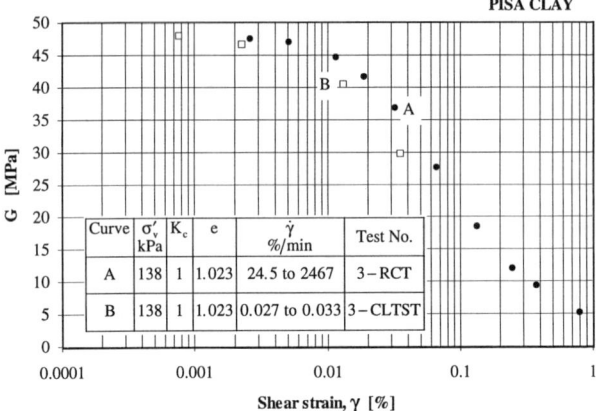

Figure 9   Shear modulus of Pisa clay from CLTST and RCT tests

The database concerning N_G, available at the authors' laboratory, is reported in Table 2.

Values of N_G reported in Table 2 cannot be successfully correlated with D_{50} or PI. The correlation proposed by Lee et al. (1995) consider that N_G increases with PI and decreases with C_α which is not consistent with the data of Table 2. The equation by Mesri and Choi (1983) seems more rational and capable of considering both cohesive and cohesionless soils.

The values of N_G reported in Table 2 were plotted in Figure 11 vs. the secondary compression index (C_α). The correlation between N_G and C_α seems quite consistent. However a larger number of experimental data is required in order to assess the validity of this correlation for cohesionless and cohesive soils.

Conclusions

The test results shown in this paper enable one to draw the following conclusions:
- The small strain stiffness is scarcely affected by the loading conditions and strain rate. Therefore it is possible to assume that this stiffness represents, from a practical point of view, the elastic stiffness or initial tangent modulus of the soils which only depends on the soil state. In reality, even at very small strains (i.e. 0.001 %), the soil response is not perfectly elastic and a certain difference between the shear modulus from MLTST and CLTST is observed.

Table 2 Values of N_G determined at the authors' laboratory for sands and clays

Soil	D_{50} [mm]	N_G [%]	PI [%]	C_α	Notes
Quiou sand	0.71	5.3	-	0.002 to 0.003	Carbonate
Kenya sand	0.13	12	-	0.004	Carbonate
Ticino sand	0.54	1.2	-	0.00015	Silicate
Hokksund sand	0.45	1.1	-	0.00015	Silicate
Messina sand and gravel	2.10	2.2 to 3.5	-	0.0007 to 0.0008	Silicate
Messina sand and gravel	4.00	2.2 to 3.5	-	0.0007 to 0.0008	Silicate
Glauconite sand	0.22	3.9	-	0.0025	50% quartz 50% glauconite
Fucino silty clay	-	7 to 11	60	0.022 to 0.035	
Taranto clay	-	9 to 13	14 to 15	0.01 to 0.015	
Pisa clay	-	13 to 19	23 to 38	0.013 to 0.05	
Augusta clay	-	8	35	0.01 to 0.02	

- For increasing strain levels an increasing effect of the strain rate on shear modulus is experimentally observed. The main practical implication of this finding is that the G/G_0 - γ curves are rate dependent.

- The coefficient of strain rate (α) expresses the stiffness dependence on the strain rate. The values of α determined from this study were compared with other data available in literature. It was possible to assess that α is a function of strain level and PI.

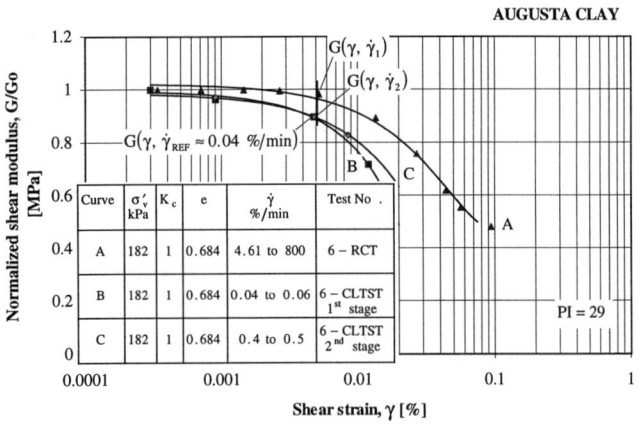

Figure 10 Shear modulus of Augusta clay from CLTST and RCT tests.

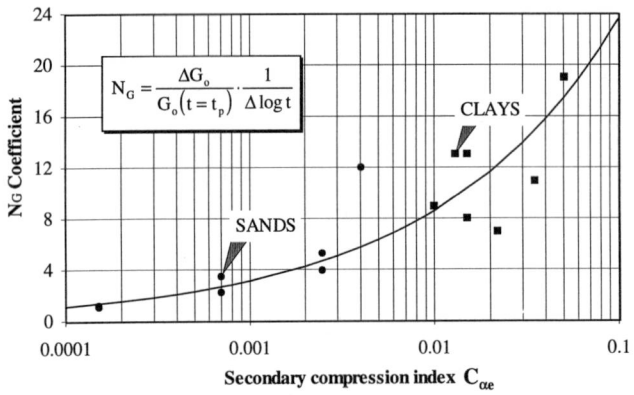

Figure 11 N_G coefficient vs. secondary compression index.

- The G-γ curves obtained from monotonic and cyclic loading torsional shear tests exhibit large differences, except at small strains. For the tests performed in this study the differences were due to the different loading conditions and different strain rates, even though the effect of the loading conditions resulted to be predominant. The different response of a soil element to monotonic and cyclic loading devalues the second Masing's Rule.

- The volumetric threshold shear strain γ_t^v seemed to be dependent on strain rate, as well as on loading conditions (monotonic or cyclic. In particular larger values of γ_t^v were observed in MLTST than in CLTST tests. Moreover, for a given strain level, larger values of the accumulated pore pressure were observed during CLTST.
- The available empirical correlations between N_G and other simple soil characteristics (D_{50} or PI) do not fit very well with the experimental data of cohesionless and cohesive soils available at the authors' laboratory. The N_G coefficient experimentally determined at the authors' laboratory had a stronger correlation with the secondary compression index (Cα).

Acknowledgements

The Authors would like to thank Prof. Maugeri of the University of Catania, who provided the Augusta samples, Mr. R. Maniscalco, the laboratory technician for the help given in the apparatus setting up, the undergraduate students Mr. A. Squadrito and G. Tommasi for the help given in performing the test and Mrs. M. Jones who reviewed the literary form of the text.

References

Anderson D.G., and Stokoe K.H. 1978, Shear Modulus: A Time Dependent Soil Property. Dynamic Geotechnical Testing. ASTM STP 654, pp. 66-90.
Akai K., Adachi T. and Ando N. 1975, Existence of a Unique Stress-Strain-Time Relation of Clays. Soils and Foundations, Vol 15, No 1, pp. 1-16
Berardi, G., Caroti, L., Giunta, G., Jamiolkowski, M., Lancellotta, R., 1991, Mechanical Properties of Upper Pisa Clay. Proceeding of the X ECSMFE Firenze 1991, Associazione Geotecnica Italiana, Vol. 1, pp 7 - 10
Berre T. and Bjerrum L 1972, Shear Strength of Normally Consolidated Clays. Proceedings ICSMFE, Moscow, paper No. 1/6, pp. 39-49.
Costanzo D., Jamiolkowski M., Lancellotta R. and Pepe M.C. 1994, Leaning Tower of Pisa - Description of the Behaviour. Invited Lecture - Proceedings Settlement '94, Texas A&M University
Dobry R. and Vucetic M. 1987, DynamicProperties and Response of Soft Clay deposits. State of the art Report Proceedings of the Int. Symposium on Geotechnical Engineering of Soft Soils, Mexico city, Vol. 2, pp. 51-87.
Drnevich V. P. and Ashmawy A. K. 1995, Discussion on paper tided "Deformation Characteristics of Soils and Soft Rocks under monotonic and Cyclic Loads and Their Relationships by Tatsuoka et al. Proceedings 3rd International Conference on Recent Advances in Geotechnical Earthquake Engineering and Soil Dynamics, S. Louis 1995, in print.
Hardin B.O. 1978, The Nature of Stree Strain Behaviour for Soils. Proc. ASCE Specialty Conference on Earthquake Engineering and Soil Dynamics, Pasadena, CA, pp. 3-90.

Isenhower, W.M. and Stokoe, K.H. II 1981, Strain Rate Dependent Shear Modulus of San Francisco Bay Mud. Proceedings International Conference on Recent Advances in Geotechnical Earthquake Engineering and Soil Dynamics, St Louis, Missouri.

Jamiolkowski M., Lancellotta R. and Lo Presti D.C.F. 1994, Remarks on the Stiffness at Small Strains of Six Italian Clays. Theme lecture Session Ia, Proceedings IS Hokkaido, Volume 2, pp. 817-836.

La Rochelle, P., Sarrailh, J., Tavenas, F. and Leroueil, S. 1981, Causes of Sampling Disturbance and Design of a New Sampler for Sensitive Soils. Canadian Geotechnical Journal, Vol. 18, pp. 52-66.

Lee Hsien, H., Tsai Tzung, R., and Chen Dung, Y. 1995, Prediction of Secondary Increment Rate of Dynamic Shear Modulus. Chinese Journal of Geotechnical Engineering, Vol. 17, No. 3, pp. 10 - 18, (in Chinese).

Lo Presti, D.C.F., 1989, Proprietà Dinamiche dei Terreni, Proceedings XIV Conferenza Geotecnica di Torino, Department of Structural Engineering, Politecnico di Torino. (in Italian)

Lo Presti, D.C.F., Pallara, O., Lancellotta, R., Armandi, M. and Maniscalco, R. 1993, Monotonic and Cyclic loading Behaviour of Two Sands at Small Strains, Geotechnical Testing Journal, Vol. 16, No. 4, pp. 409-424.

Mukabi J.N., Tatsuoka F. and Hirose K., 1991, Effect of Strain Rate on Small Strain Stiffness of Kaolin In CU Triaxial Compression, Proc. 26th Japan National Conf. on SMFE, Nagano, pp 659-662

Masing G. 1926, Eigenspannungen und verfestigung beim messing. Proc. 2nd International Congress of Applied Mechanics, Zurich, Swisse. (in German)

Maugeri M., Castelli F. and Motta E. 1994, Pile Foundation Performance of an Earthquake Damaged Building, Proc. of the Italian-French Symposium on Strengthening and Repair of Structures in Seismic Areas, Nice, France

Mesri G. and Choi Y.K. 1983, Dynamic Properties of Soft Clays for a Wide Strain Range, Discussion, Soils and Fondations, Vol. 23, No. 1, pp. 125-127.

Shibuya S., Mitachi T., Fukuda F. and Degoshi T. 1995, Strain Rate Effect on Shear Modulus and Damping of Normally Consolidated Clay. Geotechnical Testing Journal, Vol. 18, No. 3, pp. 365-375.

Tatsuoka, F. and Shibuya, S. 1992, Deformation Characteristics of Soils and Rocks from Field and Laboratory Tests, Keynote Lecture, IX Asian Conference on SMFE Bangkok, 1991, vol. 2, pp. 101-190.

Vucetic, M. 1994, Cyclic Threshold Shear Strain in Soils. Journal of Geotechnical Engineering, ASCE, Vol. 120, No. 12, pp. 2208-2228.

Prediction of time-dependent behaviour of remolded soft marine clay
in axi-symmetric undrained conditions

Satoshi Murakami[1], Kazuya Yasuhara[2], and Kaoru Bessho[3]

Abstract

This paper describes characteristics of normally consolidated clay subjected to undrained creep loading and a method for predicting time-dependent behaviours in axi-symmetric undrained conditions. For these purposes, the triaxial undrained creep tests were carried out. On the basis of the experimental results, characteristics of both the effective stress - shear strain rate and the effective stress - pore water pressure ratio rate relations have been investigated in detail. A model for the prediction of the undrained creep behaviour has been proposed based on characteristics and the validity of the proposed model is confirmed by comparison with experimental results. Moreover, an applicabilies of the model to the other time-dependent aspects of normally consolidated clay being subjected to stepped loading and constant strain rate were verified by comparing experimental results.

Introduction

It is known that stress-strain relationship of clays is dependent on time (Singh and Mitchell 1968 ; Arulanandan 1971 ; Kurihara 1972 ; Murayama et al. 1974 ; Vaid and Campanella 1977 ; Hyde and Brawn 1976 ; Borja and Kavazanjian 1985 ; Kuhn and Mitchell 1993). Most of their models expressing the time-dependent behaviours are either rheological or elasto visco-plastic in which the rheological viscous behaviour is incorporated into the elasto plastic behaviour. Thus, the model is based on the hypothesis that the shear strain can be decomposed into elastic, plastic, and viscous parts. In the models, the viscous part of the shear strain or the shear strain rate is generally expressed by as a function of the preconsolidation stress, the shear stress, and the elapsed time. The pore water pressure during undrained loading is predicted by considering the volumetric strain during drained loading. Undrained creep behaviour of normally consolidated clay, which is the typical of time-dependent behaviours can be considered from another point of view. If the clay is regarded as a frictional

[1]Res. Asst., Dept. of Urban and Civ. Engrg., Ibaraki Univ. 4-12-1, Nakanarusawa, Hitachi, 316, Japan.
[2]Prof., Dept. of Urban and Civ. Engrg., Ibaraki Univ. 4-12-1, Nakanarusawa, Hitachi, 316, Japan.
[3]Metropolitan Expressway Public Corp.

material with cohesion, it is thought that the shear strain develops because of increasing pore water pressure which is equivalent to the reduction of the mean effective stress. It is thought that from the moment when the creep stress is applied the pore water pressure continues to increase until the clay reaches the most stable effective stress state. Therefore, the time-dependent behaviour of clay subjected to undrained creep loading depends on the change of pore water pressure rate with elapsed time. In this respect, the shear strain increment during undrained creep loading may not be viscous but plastic. It is suggested that undrained shear characteristics particularly depend on the present effective stress state from the experimental fact that the undrained shear properties of normally consolidated clay after undrained cyclic loading are similar to those of an apparently overconsolidated clay in the same effective stress state induced by cyclic laoding (Matsui et al. 1980, Yasuhara et al. 1992).

For above-stated purposes of this study, the triaxial undrained creep tests were carried out. On the basis of the experimental results, the characteristics of both the effective stress state - shear strain rate relationships and the effective stress state - pore water pressure ratio rate relationships have been investigated in detail. A model for the prediction of the undrained creep behaviour has been proposed on the basis of the characteristics and the validity of the proposed model is confirmed by comparison with experimental results. Moreover, an application of the model to the other time-dependent behaviours is verified by comparing with experimental results.

Definitions

The stress and strain parameters used in this paper are as follows:

$$\tau_{oct} = \sqrt{\frac{1}{3}(\sigma_{ij} - \sigma_m \delta_{ij})(\sigma_{ij} - \sigma_m \delta_{ij})} \ , \ \sigma'_m = \sigma_m - u \ , \ \sigma_m = \frac{1}{3}\sigma_{ij}\delta_{ij} \quad (1\text{-a})$$

$$\gamma_{oct} = \sqrt{\frac{1}{3}\varepsilon_{ij}\varepsilon_{ij}} \ , \ \varepsilon_v = \varepsilon_{ij}\delta_{ij} \quad (1\text{-b})$$

where τ_{oct} is the octahedral shear stress, σ'_m is the mean effective stress, σ_m is the total mean principal stress, u is the pore water pressure, γ_{oct} is the octahedral shear strain, ε_v is the volumetric strain, σ_{ij} is the effective stress tensor (i,j=1,2,3), ε_{ij} is the strain tensor, and δ_{ij} is Kronecker delta. In this paper, τ_{oct} and γ_{oct} are simply called shear stress and shear strain.

In the case of axi-symmetric undrained triaxial conditions, each stress and strain parameter is expressed by:

$$\tau_{oct} = \frac{\sqrt{2}}{3}(\sigma_a - \sigma_r) \ , \ \sigma'_m = \sigma_m - u \ , \ \sigma_m = \frac{1}{3}(\sigma_a + 2\sigma_r) \quad (2\text{-a})$$

$$\gamma_{oct} = \frac{\sqrt{2}}{2}\varepsilon_a \quad (2\text{-b})$$

where σ_a is the axial stress, σ_r is the radial stress, and ε_a is the axial strain.

Let us assume that the strain rate tensor $\dot{\varepsilon}_{ij}$ consists of recoverable and irrecoverable strains given by:

$$\dot{\varepsilon}_{ij} = \dot{\varepsilon}^e_{ij} + \dot{\varepsilon}^p_{ij} \quad (3)$$

where the superscripts e and p denote the recoverable (elastic) and irrecoverable (plastic) parts, respectively. Following Eq. (3), the shear strain rate $\dot{\gamma}_{oct}$ can be written as:

$$\dot{\gamma}_{oct} = \dot{\gamma}^e_{oct} + \dot{\gamma}^p_{oct} \tag{4}$$

Since the elastic part of the shear strain rate can be expressed by Hooke's law, we have:

$$\dot{\gamma}^e_{oct} = \dot{\tau}_{oct}/G \tag{5}$$

where G is the shear modulus. Therefore, it is important to estimate the plastic part of the shear strain.

It is generally considered that the rate of excess pore water pressure in undrained conditions consists of a consolidation component due to the isotropic stress and a dilatancy component due to the shear stress (Lo 1969). Since the rate of pore water pressure change due to the isotropic stress is proportional to the rate of the total mean principal stress change, it is recoverable and independent of the current effective stress state and the stress and strain histories. On the other hand, the rate of pore water pressure change due to the shear stress is irrecoverable. It depends on the current effective stress state and the stress and strain histories. That is:

$$\dot{u} = \dot{u}_i + \dot{u}_s \;,\; \dot{u}_i = \dot{\sigma}_m \;,\; \dot{u}_s = \dot{u}_s(\tau_{oct}, \sigma'_m, s_1, s_2, \cdots, s_n) \tag{6}$$

where the subscripts i and s denote the pore pressure rates due to isotropic stress and shear stress, respectively. s_k (k=1,2,···,n) is a state variable or a state function which expresses stress or strain history, or consolidation process, preconsolidation pressure, preconsolidation period, and loading rate.

Experimental procedure

Specimens were made from a remolded soft marine clay. A clay slurry was thoroughly mixed with an initial water content of about 200% and poured into a large consolidation vessel to achieve homogeneous clay specimens in every test. Then, a preconsolidation pressure of 49kPa was applied to the clay slurry for about a week. The specimen with 5cm diameter and 10cm height was trimmed from the large clay block formed in the vessel. The index properties of the clay specimen are tabulated in Table 1.

TABLE 1. Index properties of clay

Specific gravity	2.607
Liquid limit	100.3 %
Plastic limit	46.8 %
Plasticity index	58.5

TABLE 2. Testing conditions for undrained creep loading

Test No.	Preconsolidation isotropic stress σ_{me} kPa	Sustained deviator stress $\Delta\sigma_a$ kPa	Duration of test min
CIUC1	196	78	1800
CIUC2	196	118	1800
CIUC3	196	127	2200
CIUC4	196	142	2272
CIUC5	196	144	117
CIUC6	196	153	11

All the specimens in the undrained triaxial creep tests were subjected to a prescribed isotropic stress σ_{mc} for 24 hours of preconsolidation. Then, each specimen was subjected to creep loading with a sustained deviator stress $\Delta\sigma_a$ under undrained conditions until either the axial strain reached 15% or the prescribed time had elapsed. During all of the tests a back pressure of 196kPa was applied. The pore water pressure was measured during undrained creep loading by means of uniting drainage lines from the centre of the top and bottom platens and the filter paper surrounding the specimen. The testing conditions are summarized in Table 2.

Experimental results

Effective stress - shear strain relationship
It is known that the excess pore water pressure of normally consolidated clay under undrained creep loading increases with time. The increment of excess pore water pressure is equivalent to the reduction of the mean effective principal stress if the total stress state is maintained constant. Therefore, shear strain develops during undrained creep loading as the effective mean principal stress decreases in spite of sustained shear stress. Singh and Mitchell (1968) and Kurihara (1972) showed that undrained creep behaviour of clay could be divided into three types based on the relationships between shear strain and the elapsed time as shown in Fig.1-a. Type I is the case where only elastic shear strain occurs at the moment when the shear stress is applied and no shear strain takes place after that time. Type II is the case where only elastic shear strain occurs at the moment when the shear stress is applied and thereafter the shear strain develops with elapsed time. However, the specimen does not reach creep failure because the shear strain rate decreases with the elapsed time and the shear strain converges to a constant value. Type III is the case where the shear strain proceeds with the elapsed time and the specimen reaches creep failure. In this case, in the first stage the shear strain rate decreases with elapsed time, keeps a constant value in the second stage, and increases in the last stage. The same kind of creep behaviour as the above three types was recognized in the experimental results in this study. Excess pore water pressure behaviour corresponding to each type is schematically shown in Fig.1-b. The pore water pressure rate decreases with elapsed time regardless of whether the specimen reaches failure or not. It was recognized from the experimental results in the cases of failure that the pore water pressure converged to a certain value depending on the shear stress. This condition is regarded as the critical state under undrained creep loading.

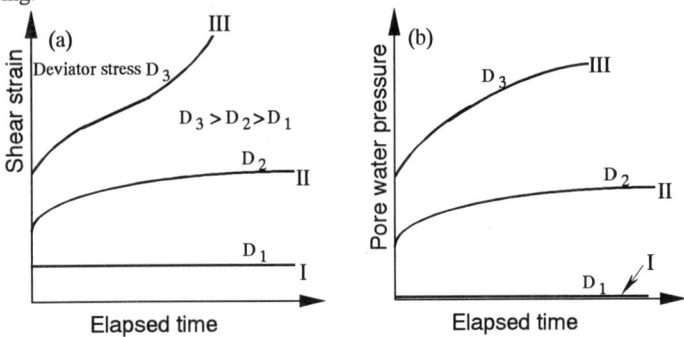

Figure 1. Typical undrained creep behaviour of clay subjected to a constant deviator stress : (a) shear strain and (b) pore water pressure

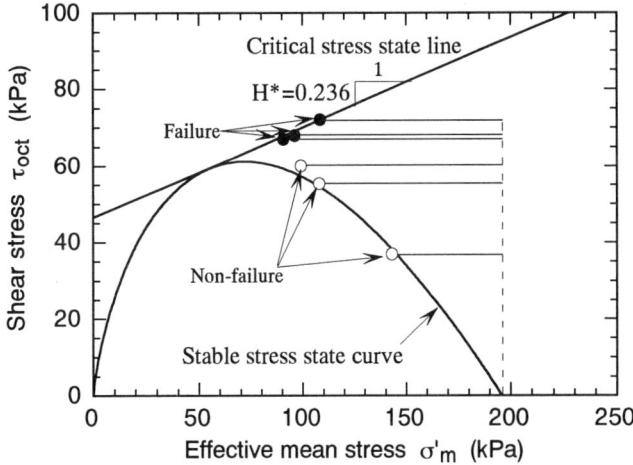

Figure 2. The critical stress state and the stable stress state under undrained creep loading

This critical state was observed in a series of undrained creep tests of three different large creep loads. The critical stress states are plotted on the τ_{oct} - σ'_m plane in Fig. 2. The final stress states in the cases where the specimens did not reach failure are shown in the same figure. In this figure, a unique straight line, which expresses the critical stress state under undrained creep loading, can be drawn in the τ_{oct} - σ'_m plane. This line can be called the critical stress state line under undrained creep loading (CSL). A formula for this line can be written as :

$$\tau_{oct} = H^* \left(\sigma'_m - c^* \sigma_{me} \right) \tag{7}$$

where H^* is the slope of the CSL and $c^* \sigma_{me}$ is the σ'_m-axis intercept of the CSL. c^* is named as the coefficient of bond stress. Thus, the CSL moves up with increasing σ_{me}.

A stress parameter η^* for estimating the plastic shear strain is introduced as :

$$\eta^* = \frac{\tau_{oct}}{\sigma'_m - c^* \sigma_{me}} \tag{8}$$

Fig. 3 shows the relationships between the stress parameter η^* and plastic shear strain which was obtained from all the experimental results regardless of whether failure occured. Here it is assumed that the difference between the measured shear strain and the elastic strain calculated by Eq. (5) is equal to the plastic shear strain and the shear modulus was decided as initial slope shear stress-shear strain curve from the undrained triaxial compression tests of normally consolidation clay. The relationships are independent of the magnitude of creep load and the elapsed time. This unique relationship is expressed by the following formula :

$$\gamma^p_{oct} = \mu_s \ln \frac{H^{*2}}{H^{*2} - \eta^{*2}} \tag{9}$$

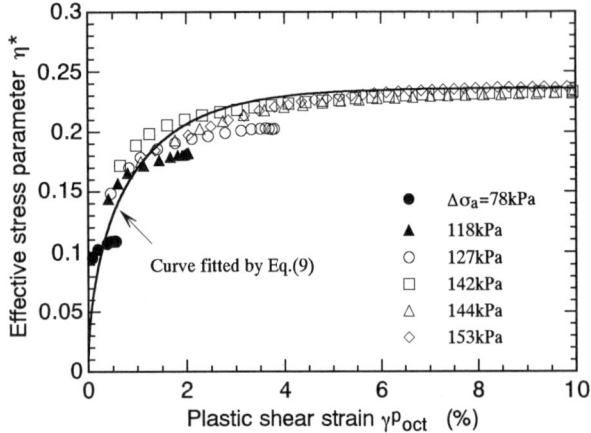

Figure 3. The effective stress parameter- plastic shear strain relationships

where μ_s is the strain hardening coefficient. Since this relationship is independent of time, the following equation can be obtained:

$$\dot{\gamma}^p_{oct} = \frac{2\mu_s \eta^*}{H^{*2} - \eta^{*2}} \dot{\eta}^* = \frac{2\mu_s \eta^{*2}}{\tau_{oct}\left(H^{*2} - \eta^{*2}\right)} \dot{\tau}_{oct} - \frac{2\mu_s \eta^{*3}}{\tau_{oct}\left(H^{*2} - \eta^{*2}\right)} \dot{\sigma}'_m \quad (10)$$

Combination of Eq. (10) with Eq. (5) gives:

$$\dot{\gamma}_{oct} = \left\{ \frac{2\mu_s \eta^{*2}}{\tau_{oct}\left(H^{*2} - \eta^{*2}\right)} + \frac{1}{G} \right\} \dot{\tau}_{oct} - \frac{2\mu_s \eta^{*3}}{\tau_{oct}\left(H^{*2} - \eta^{*2}\right)} \dot{\sigma}'_m \quad (11)$$

Stable stress state and critical stress state

Specimens subjected to a constant deviator stress less than a certain critical value may reach a stable state on which there are no variations in increments of pore water pressure and shear strain with elapsed time. The effective stress state at the stable state may correspond to that of normally consolidated clay subjected to undrained quasi-static loading (very slow monotonic-incremental loading). Such a stress state on the τ_{oct} - σ'_m plane may be expressed by the effective stress path of the normally consolidated clay which is of the same type as that given by original Cam-clay model (Roscoe et al. 1963). That is:

$$\tau_{oct} = \frac{H^*}{1 - \kappa/\lambda} \sigma'_m \ln \frac{\sigma_{me}}{\sigma'_m} \quad (12)$$

where λ is the compression index and κ is the swelling index. Judging from the fact that a clay element subjected to undrained quasi-static shear certainly reaches the

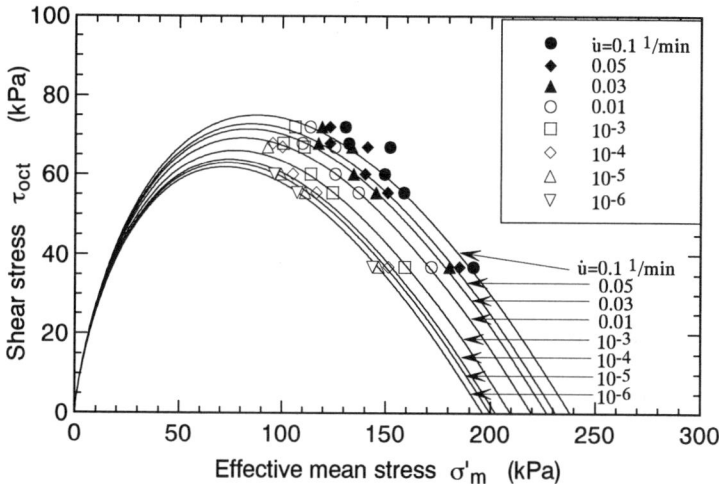

Figure 4. The rate of excess pore water pressure ratio and the contour curves on the τ_{oct} - σ'_m plane

critical state and the stable stress state surface expressed by Eq. (12) without crossing the critical stress state line on the τ_{oct} - σ'_m plane, the stable stress state surface must be in contact with the critical state line. By using λ and κ, the parameter c^* is expressed by:

$$c^* = -\frac{1}{1-\kappa/\lambda} exp\left(\frac{\kappa}{\lambda} - 2\right) \qquad (13)$$

Pore water pressure rate - effective stress state relationship

In this study, the rate of the pore water pressure ratio is defined as:

$$\dot{u} = \frac{d(u_s/\sigma_{me})}{dt} \qquad (14)$$

Different rates of pore water pressure ratio due to shear stress were found from the experimental results and the stress states at $\dot{u}$ = 0.1, 0.5, 0.3, 10^{-2}, 10^{-3}, 10^{-4}, 10^{-5}, and 10^{-6} /min are shown in Fig.4. It is recognized that the rate of the pore water pressure ratio reduces with decreasing effective mean principal stress or shear stress. A family of curves which represent contours of the rate of the pore water pressure ratio, can be described as:

$$\tau_{oct} = \frac{H^*}{1-\kappa/\lambda}\sigma'_m \ln\frac{\beta\sigma_{me}}{\sigma'_m} \qquad (15)$$

This equation corresponds to Eq. (12) when $\beta=1$. Namely, this equation expresses the stable stress state surface when $\beta=1$. The relationship between β and each rate of pore water pressure ratio is shown in Fig. 5. As the rate of pore water pressure ratio increases with increasing shear stress, the rate becomes close to infinity at the maximum shear stress. In other words, when the stress state momentarily reaches the critical stress state after the start of undrained creep test. Therefore, the β-$\dot{u}$

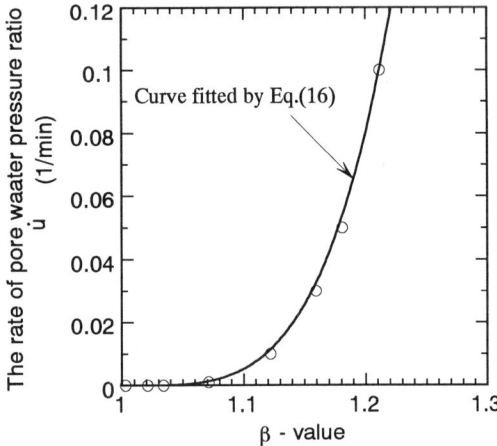

Figure 5. The relationship between β-value and the rate of excess pore water pressure ratio

relationship can be approximated by the following equation which is $\dot{u}=0$ on $\beta=1$ and $\dot{u}\to\infty$ on $\beta\to\beta_{max}$.

$$\dot{u} = \mu_u \ln \frac{(\beta_{max} - 1)^4}{(\beta_{max} - 1)^4 - (\beta - 1)^4} \tag{16}$$

where μ_u is a constant, and β_{max} is given by :

$$\beta_{max} = exp\left\{(1 - \kappa/\lambda)(1 - c^*)\right\} \tag{17}$$

Prediction of the undrained creep behaviours

The undrained creep behaviour of clay can be predicted using Eq. (9) through Eq. (17). The calculated results of the variations of the shear strain and the pore water pressure ratio with elapsed time are shown in Fig. 6. The figures show an overall agreement between the calculated and the experimental results.

Application to other time-dependent behaviours

The present model was established on the basis of the characteristics of undrained creep behaviour observed in triaxial tests. However, it is possible to extend it for predicting the time-dependent behaviours of clays with different undrained conditions. An application of this model to the other time-dependent behaviours was verified by comparing experimental results from undrained triaxial tests with a stepped increasing axial stress and a constant axial strain rate. Each test was carried out by the following procedure. All the specimens were subjected to a prescribed isotropic stress σ_{me} for a preconsolidation period. Then, one specimen was subjected to an incremental

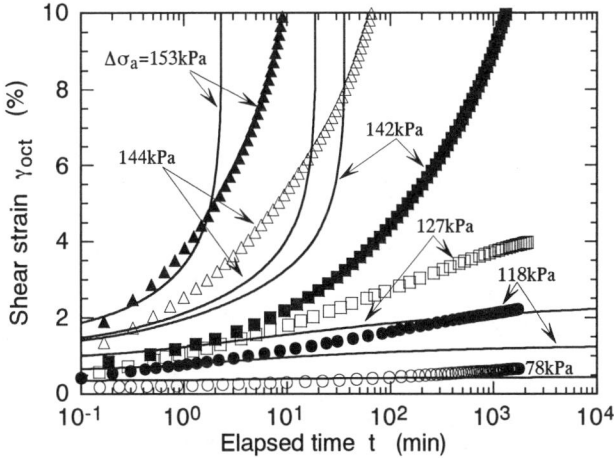

(a) Variations of the shear strain with elapsed time

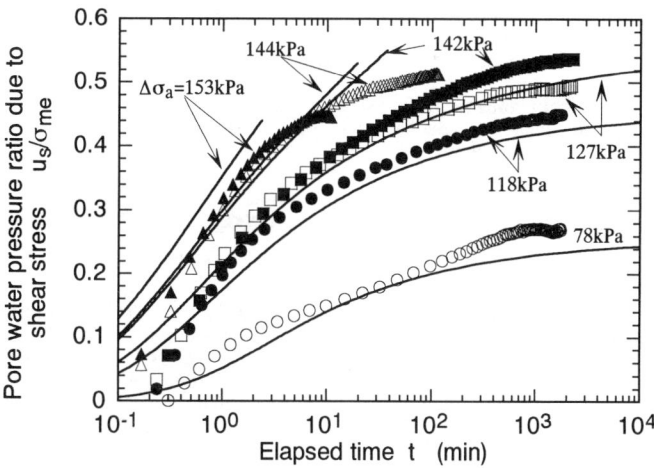

(b) Variations of the pore water pressure ratio with elapsed time

Figure 6. Comparison of the calculated results with
the experimental ones in undrained creep loading
(symbols = experimental results, lines = calculated results)

deviator stress $\Delta\sigma_a$ under undrained conditions in the case of the stepped loading, and the three specimens were subjected to a prescribed constant axial strain rate $\dot{\varepsilon}_a$ for the other case of testing. The testing conditions for both cases are summarized in Tables 3 and 4, respectively.

TABLE 3. Testing condition for undrained stepped loading

Step No.	Sustained deviator stress $\Delta\sigma_a$ kPa	Duration of the step min
1	30	100
2	59	100
3	88	100
4	118	100
5	147	120
6	176	10

σ_{me} =196kPa

TABLE 4. Testing conditions for undrained shearing at constant strain rate loading

Test No.	Preconsolidation isotropic stress σ_{me} kPa	Constant axial strain rate $\dot{\varepsilon}_a$ %/min
CIUSR1	196	0.05
CIUSR2	196	0.1
CIUSR3	196	0.5

Prediction of behaviour under stepped loading

A method for the predicting undrained creep behaviour when subjected to a stepped deviator stress is fundamentally the same as that for the undrained creep behaviour subjected to a sustained deviator stress. It was calculated when the pore water pressure increment due to shear stress was equal to zero at the moment when a new level of deviator stress was applied. The calculated and experimental results for the variations of shear strain and pore water pressure ratio with elapsed time are shown in Fig. 7. There is not shown the calculated result when the deviator stress $\Delta\sigma_a$= 176kPa because the clay reached the critical state when $\Delta\sigma_a$=147kPa. However, Fig. 7 shows an overall agreement between the calculated and the experimental results.

Prediction of behaviour under constant strain rate loading

To predict the undrained behaviour of clay subjected to constant strain rate loading, Eq. (11) is rewritten as :

$$\dot{\tau}_{oct} = \left\{ \dot{\gamma}_{oct} - \frac{2\mu_s \eta^{*3}}{\tau_{oct}\left(H^{*2} - \eta^{*2}\right)} \dot{u}_s \right\} \bigg/ \left\{ \frac{2\mu_s \eta^{*2}}{\tau_{oct}\left(H^{*2} - \eta^{*2}\right)} + \frac{1}{G} \right\} \quad (18)$$

$\dot{\gamma}_{oct}$ is a constant because $\dot{\varepsilon}_a$ is a constant and $\dot{\gamma}_{oct} = (\sqrt{2}/3)\dot{\varepsilon}_a$ in the undrained axisymmetric condition. It is possible to calculate the rate of pore water pressure by using Eq. (16) and the rate of shear stress by using Eq. (18) at a certain stress state, (σ'_m, τ_{oct}), during constant strain rate loading. Stress path and stress-strain relationships which were obtained from the experimental results and the calculated results are shown in Fig. 8 and Fig. 9. It can be seen from Fig. 8 and Fig. 9 that the calculated results were in overall agreements with the experimental ones. Fig. 10 shows the relationship between the axial strain rate and the ratio of undrained shear strength to that for $\dot{\varepsilon}_a$=1%/min from the calculated results. The relationships in Fig. 10 shows good agreement with the findings of Bjerrum (1972), Vaid et al. (1977), and Jamiolkowski (1991), indicating that the undrained shear strength increases with the axial strain rate.

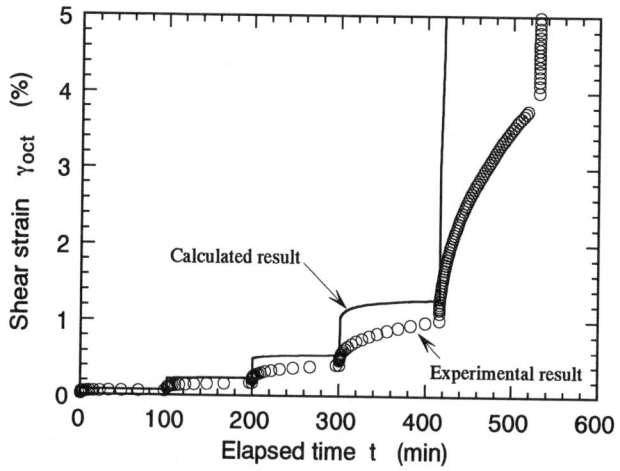

(a) Variations of the shear strain with elapsed time

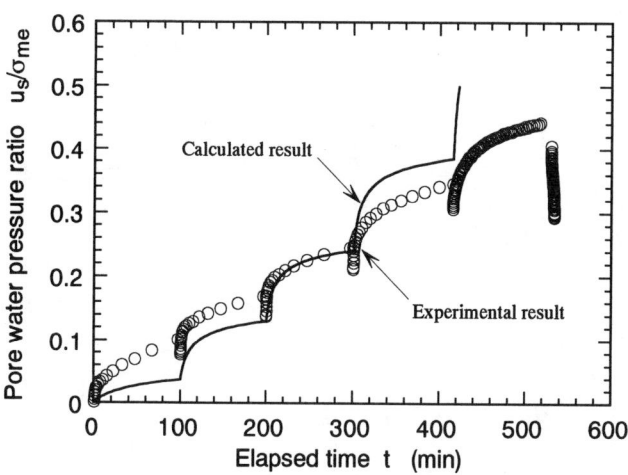

(b) Variations of the pore water pressure ratio with elapsed time

Figure 7. Comparison of the calculated results with the experimental ones in undrained stepped loading

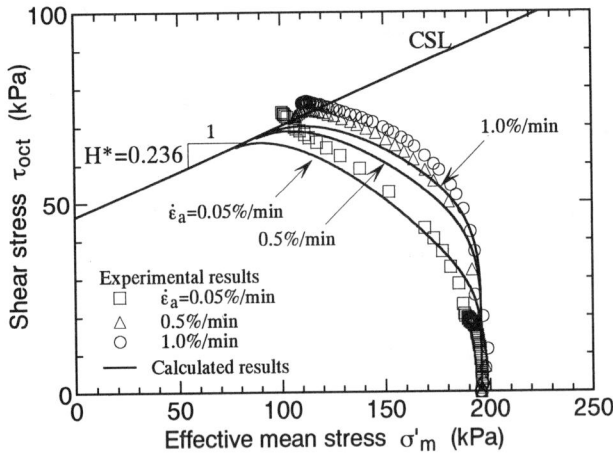

Figure 8. Comparison of calculated stress path with the experimental results of constant strain rate loading

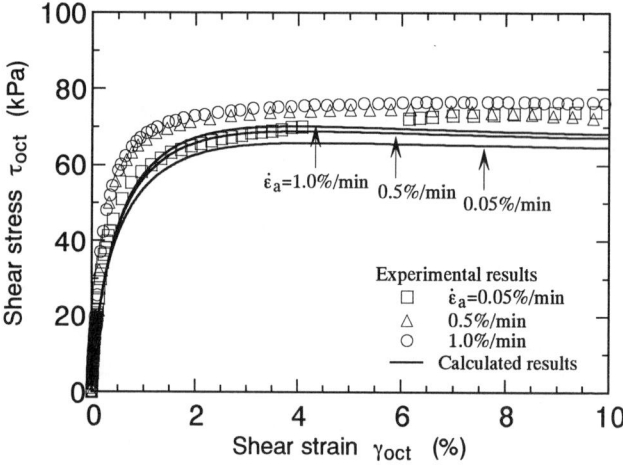

Figure 9. Comparison of calculated results with experimental results on the shear stress - shear strain relationships

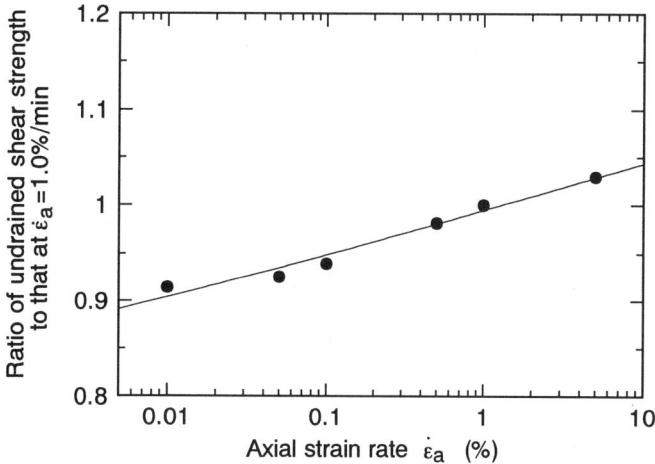

Figure 10. The relationship between the axial strain rate and the ratio of undrained shear strength to that for $\dot{\varepsilon}_a$=1%/min from the calculated results.

Conclusions

The time-dependent behaviour of normally consolidated clay was examined by means of undrained triaxial creep tests. On the basis of the experimental results, characteristics of both the effective stress state-shear strain rate relationships and the effective stress state-pore water pressure ratio rate relationships have been investigated in detail.

The following conclusions are made from the present study:
(1) A unique straight line, which expresses the critical stress state under undrained creep loading, could be drawn in the τ_{oct} - σ'_m plane.
(2) A stress parameter η^* for estimating the plastic shear strain was introduced as:

$$\eta^* = \frac{\tau_{oct}}{\sigma'_m - c^* \sigma_{me}}$$

where $c^*\sigma_{me}$ is the σ'_m-axis intercept of the critical stress state line and σ_{me} is the preconsolidation stress. The relationships between the stress parameter and the plastic shear strain were independent of the magnitude of creep load and the elapsed time.
(3) Contour curves of the rate of the pore water pressure ratio on the τ_{oct} - σ'_m plane could be described by the modified equation based on original Cam-clay model.
(4) By using equations formulated on the basis of the characteristics of clay under undrained creep loading, the creep behaviors were predicted. It was shown that the calculated results were in overall agreements with the experimental ones.
(5) Applicability of the model to the other time-dependent behaviours of normally consolidated clay subjected to a stepped loading and a constant strain rate was verified by comparing experimental results. It was shown that the calculated results agree with the experimental ones.

(6) The relationship between the axial strain rate and the ratio of undrained shear strength to that for $\dot{\varepsilon}_a=1\%/min$ from the calculated results shows the good agreement with the findings of Bjerrum (1972), Vaid et al. (1977), and Jamiolkowski (1991), who indicated that the undrained shear strength increased with the axial strain rate.

References

Arulanandan,K. Shen,C.K., and Young,R.B. (1971). "Undrained creep behaviour of a coastal organic silty clay." *Geotechnique*, 21(4), 359-375.

Bjerrum,L. (1972). "Embankment on soft ground." *Proc. ASCE Specialty Conf. on Performance of Earth and Earth-supported Structures*, 2, 1-54.

Borja,R.I. and Kavazanjian,E.Jr. (1985). "A constitutive model for the stress-strain-time behaviour of 'wet' clay." *Geotechnique*, 35(3), 283-298.

Hyde,A.F.L., and Brown,S.F. (1976). "The plastic deformation of a silty clay under creep and repeated loading." *Geotechnique*, 26(1), 173-184.

Jamiolkowski,M., Leroueil,S., and Presti,D.C.F.L. (1991). "Design parameters from theory to practice." *Proc. of Int. Conf. Geotech. Engrg. for Coastal Development*, 2, 877-917.

Kuhn,M.R. and Mitchell,J.K. (1993). "New perspective on soil creep." *J. Geotech. Engrg. Div.*, ASCE, 119(3), 507-524.

Kurihara,N. (1972). "Experimental study on creep rupture of clays." *Proc. JSCE*, 202, 59-71. (in Japanese)

Lo,K.Y. (1969). "The pore pressure-strain relationships of normally consolidated undisturbed clays. Part 1 Theoretical considerations." *Canadian Geotech. J.*, 6, pp.383-394.

Matsui,T., Ohara,H., and Ito,T. (1980). "Cyclic stress-strain history and shear characteristics of clay." *J. Geotech. Engrg. Div.*, ASCE, 106(10), 1101-1120.

Murayama,S., Sekiguchi,H., and Ueda,T. (1974). "A study of the stress-strain-time behavior of saturated clays based on a theory of nonlinear viscoelasticity." *Soils Founds*. 14(2), 19-33.

Roscoe,K.H., Schofield,A.N., and Thurairajah,A. (1963). "Yielding of clays in states wetter than critical." *Geotechnique*, 13, 211-240.

Singh,A., and Mitchell,J.K. (1968). "General stress-strain-time function for soils." *J. Soil Mech. Found. Div.*, ASCE, 94(1), 21-46.

Yasuhara,K., Hirao,K., and Hyde,A.F.L. (1992). "Effects of cyclic loading on undrained strength and compressibility of clay." *Soils Founds*. 32(1), 100-116.

Vaid,Y.P. and Campanella,R.G. (1977). "Time-dependent behaviour of undisturbed clay." *J. Geotech. Engrg. Div.*, ASCE, 103(7), 693-709.

Soil Creep and Creep Testing of Highly Weathered Tropical Soils

Peter G. Nicholson,[1] M.ASCE, Philip W. Russell[2], and Clint F. Fujii[3]

Abstract

In response to a number of recent landslides and large slope deformations occurring on the island of Oahu, Hawaii, along with a history of slope failures and slow moving slides, a laboratory study was undertaken to evaluate the creep potential and creep characteristics of a variety of tropical soils. The soils tested included high plasticity CH colluvial clays and highly weathered residual MH clays.

Since much of the concern was in the use of these materials for engineered earth structures, all testing in this study was performed on remolded specimens compacted to specified moisture and densities intended to represent field compaction conditions. All soils were prepared in the laboratory under carefully controlled conditions so that the variability between specimens of the same material would be minimized.

Each of the soils tested showed significant creep potential (defined as continued deformation under a constant application of load), as well as potential for strength loss (defined as a creep "rupture" failure under a load less than the short term tested peak strength). While several of the test specimens appeared to demonstrate attenuating creep behavior (decreasing strain with time), a number of the specimens moved into a mode of tertiary creep (increasing rate of strain with respect to time), with some reaching creep "rupture" failure.

It appears from the results of this study that a number of tropical soils may have a significant creep potential and that engineers should be cautious of the potential for creep behavior when considering construction utilizing these soils.

Introduction

Studies on soil creep and its damaging effects on structures have been conducted on soils in many of the regions around the world. The study of soil creep on tropical soils of volcanic origin, however, has been limited. This has not slowed the development in these areas where creep may be a significant problem or where soil with potential to creep is used as engineered fill. In the future, there will be a continued desire to construct projects in tropical regions with significant deposits of these often troublesome soils.

---

[1] Associate Professor, University of Hawaii, 2540 Dole St., Honolulu, HI 96822
[2,3] Research Assistant, University of Hawaii, 2540 Dole St., Honolulu, HI 96822

Tropical Hawaiian soils differ from the more commonly studied soils of temperate climates as they are generated from young oceanic basalt parent rock and weathered under conditions of high rainfall, heat, and humidity. Of particular interest in Hawai'i are:
- smectite or montmorillonite clays which are highly expansive,
- high plasticity colluvial clays, commonly found on the base and flanks of valleys between eroding silica-poor lava flows,
- residual volcanic soils typically found weathered in place in a saprolitic structure, often containing halloysite.

<u>Definition of Soil Creep</u>

To first understand the phenomenon of soil creep, an understanding of rheological processes; the science studying the flow of substances, is necessary. Rheological phenomenon in soils and rocks can be observed everywhere, with durations from a few hours and days to even a century or more. Earlier in this century, one of the first studies on the behavior of metals at elevated temperatures proved flow is an inherent property of solid metals as well as soils. Thus, a more general definition, the phenomenon of slow flow of solids, has been termed creep (Vyalov, 1986).

In geotechnical engineering, soil creep is commonly defined as deformation and movement proceeding under a state of constant stress or load. For creep stresses large enough to result ultimately in failure, deformation of a typical soil is divided into primary, secondary, and tertiary periods (Figure 1). During the primary phase after application of the load, the rate of deformation decreases. In the secondary period, continued strains accumulate at a slow, steady rate. However, soils rarely undergo an extended period of secondary creep and more often the creep rate will change with time. This change in rate may be either exhibited as attenuation of strains whereby the material reaches some kind of equilibrium showing no further deformation, or the strain rate may again begin to accelerate and eventually lead to failure or "rupture".

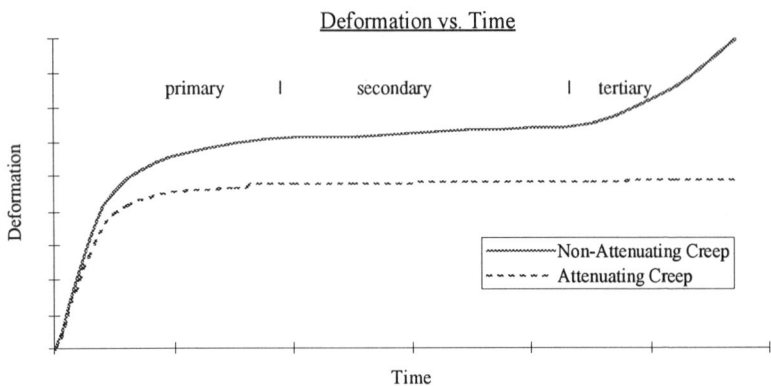

Figure 1. Typical Creep Curves

Creep is typically investigated by either one of two methodologies: examining creeping flow in the form of long term shear, or examining creeping flow in the context of deformation of the soil skeleton. Hence, the term secondary settlement is sometimes considered analogous with soil creep in examples of long term settlement of clay foundations soils. In reality, primary consolidation and creep are overlapping over a long time span. Since total settlement is the sum of the two, primary dominates at first, but may or may not be overshadowed later in the process.

Historic Examples

Many cases are known where the creep of soils below foundations has lead to long term and differential settlement (i.e., tilting of retaining walls, instability of slopes, roadway failure). Landslides, although not typically thought of as a soil creep example are in fact frequently triggered by the occurrence of soil creep. Although not dramatic in terms of surface expression (until the actual slide), slow downward movement by creep has been a very important process of mass wasting (i.e. landslides) in Hawai'i.

Peck and Wilson (1968) in a study of landsliding in Aina Haina Valley for the City and County of Honolulu, concluded that many hillsides in Honolulu are moving very slowly downward as a result of soil creep. Aina Haina and Palolo Valleys were estimated to have rates of 0.5 to 1 cm/yr. As early as 1954, signs of slope instability were recognized in Palolo Valley. Two years after the completion of the Waiomano subdivision, creep had progressed to the point that it became necessary to disconnect/reconnect utility services. During the following year, all utility lines were dug up and relocated above ground. More recently, several additional problem areas in Hawaii have displayed deformations which may be attributed to creep phenomenon.

Rate Process Theory

The theory of absolute reaction rates, commonly known as rate process theory, describes the flow of a non-Newtonian medium. It has been suggested that the flow of liquids and the creep of solids are of the same nature, both the manifestation of an oriented motion of elemental particles (molecules, atoms) induced by structural defects. Using the rate theory, Mitchell et al. (1968), found soil deformation is attributed to the "hopping" of molecules at the points of contact between mineral particles in the interparticle bonds of the soil.

Research has shown that the creep behavior of soils is in general accord with the predictions of the rate process theory. Since this theory can be applied to any process involving the time-dependent rearrangement of matter, it has the potential for providing a powerful tool for the description and prediction of soil behavior (Mitchell et al., 1968).

Using the rate process theory, a practical and simple mathematical relationship was created by Singh and Mitchell (1968) to predict the strain rate at a certain time period and, to a certain degree, time to failure for any given type of clay soil following the equation:

$$\varepsilon = A e^{\alpha D}(t_1/t)^m \quad [\text{Eq. 1}]$$

Where:

A = strain rate at time t1 and $D = 0$ (projected value)
D = deviatoric stress $(\sigma_1 - \sigma_3)$
α = value of the slope of the mid-range linear portion of a plot of the logarithmic strain vs. deviator stress, all points corresponding to the same time after load application.
t1 = unit time
m = negative slope of a logarithmic strain rate vs. logarithmic time

This relationship is independent of whether the clays are undisturbed or remolded, wet or dry, normally consolidated or overconsolidated, or tested drained or undrained. The limiting factor of this relationship is that clay soils must be subjected to loads within the range of 30% to 90% of the soil's strength, a range considered to fall within engineering interest.

The parameter "m" in Equation (1) is a key factor in defining the creep potential of a soil; the smaller the value of "m", the higher the creep potential. Under a given stress level, the more rapid the creep movements, the shorter the time to a given strain. It has been further observed that soils with m < 1 eventually fail in creep rupture under sustained loads less than their short-term strengths. Values of "m" less than unity are indicative of high creep potential, softening with increase in shear strains, lower strength after creep, and creep rupture under sustained loads. Soils with m = 1 seem to exhibit the same strength before and after creep, while soils with m > 1 exhibit cessation of creep with time (attenuation).

Changing Creep Rate and Bond Strength

It has been proposed by Kuhn and Mitchell (1993) that creep deformation is due to sliding between particles. Although this viscofrictional sliding is thought to occur at solid contacts, the sliding velocity at each contact depends on the ratio of the tangential-to-normal components of contact force. As deformation proceeds during creep, particles undergo slight rearrangements that reduce the tangential components and/or increase the normal components of the contact forces, which in turn reduces the velocity of interparticle sliding and the overall creep rate of the soil.

Several mechanisms have been proposed to account for the observed change in a soil's creep rate with time. These mechanisms can be classified into two groups: deformation-dependent mechanisms that are based on changes in the soil structure, and those based on intrinsically time-dependent processes. In deformation-dependent models, the interparticle contacts are assumed to have a range of strengths and are subjected to a range of shearing forces. The weaker contacts are the first to slide, and the resulting deformation shifts a portion of the shearing force to stronger contacts. Intrinsically time-dependent processes such as chemical cementation, the strengthening of contacting asperities under pressure (Sholtz and Engelder, 1976), and thixotropic processes also have an effect on the creep rate.

Mitchell et al. (1968) discussed the nature of interparticle bonding in soils and presented evidence that a soil's resistance to shearing deformation occurs at interparticle contacts, which are effectively solid-to-solid in nature. Even in clays, the sliding between particles takes place at solid contacts, leaving distinct scratched paths across particle surfaces and producing acoustic emissions similar to those of sands and gravels. The water present in a soil does not appear to be active in promoting the sliding between particles as wet and dry clays exhibit qualitatively similar creep behavior. Furthermore, the strength of the bonds formed within a soil, as measured by their activation energy, is much greater than the bond strength associated with the viscous flow of water (Mitchell, 1993).

Consider the interaction between solid particles separated by a film of water. Each particle is acted upon by a system of external and internal forces as well as fields of energy induced by the forces. External fields are set up by the applied load and gravitational forces. Internal fields are produced by interparticle forces arising between the soil components themselves. In general, the following inter- and intra-particle forces occur: forces of a chemical nature, molecular forces (Van der Waals), ionic-electrostatic forces, capillary-electrostatic (Coulomb) forces and magnetic forces (Vyalov, 1986).

The forces discussed above produce fields of energy and form bonds between the particles of a dispersed system. The strength of the bonds varies over a wide range (up to 20 orders of magnitude) depending on the type of prevailing bonding. It must be stressed that distinction is made between the strength of a single bond (individual contact) and the strength of a soil.

Of the six types of bonds listed in Table 1, three may have creep potential: molecular (Van der Waals) bonds, electrostatic bonds, and magnetic bonds.

Table 1 - Strength Of Bonds In Clay Soils

Type of bond	Strength of an individual contact	Strength of soil system
Chemical	Up to 102 (kg/cm^2 x 10^{-3})	107-108 (kg/cm^2 x 10^{-2})
Molecular	10^{-4} - 10^{-3} (kg/cm^2 x 10^{-3})	Up to 104 (kg/cm^2 x 10^{-3})
Ionic-electrostatic	10^{-3} - 10^{-1} (kg/cm^2 x 10^{-3})	Up to 106 (kg/cm^2 x 10)
Capillary	10^{-2} (kg/cm^2 x 10^{-3})	Up to 400 (kg/cm^2 x 10^3)
Electrostatic	10^{-5} (kg/cm^2 x 10^{-3})	103-104 (kg/cm^2 x 10^{-3})
Magnetic	10^{-10} (kg/cm^2 x 10^{-3})	102-103 (kg/cm^2 x 10^{-3})

Molecular (Van der Waals) forces significantly affect the strength characteristics of clay deposits, especially at the early stages of lithogenesis such as sedimentation (coagulation and settling of deposits) and diagenesis. Thus, in freshly settled deposits (or artificially remolded clays) the strength imparting bonds owe their origin to the lower magnitude molecular forces. The subsequent compaction of clay deposits gives rise to predominantly ionic-electrostatic forces.

The effect of electrostatic forces is felt at the early stages of lithogenesis (in young clay deposits). Since the particles are negatively charged at the sides and positively at the edges, their side-to-side contact gives rise to the electrostatic forces of repulsion; a side-to-edge contact produces forces of attraction.

The origin of magnetic forces are ferromagnetic substances (hematite, goethite, hydrometite) which occur in clayey soils in the form of a surface film varying in thickness between 0.05 and 0.5 μm. The rigid magnetic moment of the film transmits the effect of coagulation to the adjacent particles. The magnitude of magnetic forces is low, and they are a factor at the stage of sedimentation only.

Previous Creep Testing

Past research and on soil creep by Singh and Mitchell (1968), Vyalov (1986), Feda (1992), Mitchell (1993), Tavenas et al. (1978) and others have helped improve the engineering community's understanding of soil creep, relaxation, and stress-strain behavior. A problem is ensuring proper drainage conditions in terms of controlling loading rates. Some investigators have applied loads (or shear stresses) to soil samples in one large application. This application of load would certainly create "undrained" conditions in saturated clay soils, but this loading technique may not be acceptable for drained conditions where pore pressures within the saturated soil sample need to dissipate over time. To minimize the effects of unwanted pore pressure, loading rates need to be slow with loads applied in small increments.

The American Society of Testing and Materials (ASTM) lists three testing standards for creep tests of rock (ASTM D4341-84, ASTM D4405-84, ASTM D4406-84). Each test standard suggests loading samples within a thirty second period. It seems obvious that this loading technique may be too fast for soil creep tests, especially where drained conditions are being considered. The precision of such a test is questionable due to heterogeneity of the material, operator consistency, and specific laboratory testing techniques. More importantly, since there is no standard for testing soils for creep, results may vary greatly with technique. In addition, since the cost of running creep tests is very high, duplicating experiments to test for variability is generally not feasible.

From previous creep tests where temperatures have been varied, it has been shown that creep can be treated as a thermally activated process. Therefore fluctuations in temperature may cause inconsistencies in creep data. High temperatures tend to increase the strain rate while low temperatures tend to slow the strain rate down. Hence, temperature fluctuations must be kept at a minimum during creep tests.

Hawaiian Soils

Hawaiian soils are a weathering product of young volcanic rock and debris. The combination of volcanic origin and the weathering conditions of tropical heat and humidity gives the soil unique characteristics. Much of the Hawaiian soils are residual and grade downward into their parent rocks. However, some soils have been transported and have no mineralogical or geological connection with the underlying rock.

Montmorillonite CH clays are usually associated with volcanic ash deposits, the surface soils within the cone and at the toe of the cone are most likely to be expansive. The southern half of Oahu is dotted with secondary ash cones (Lum, 1982). Areas of the leeward (west) side of Oahu also contain expansive soils.

Another type of problematic soils in Hawaii are the tropical "MH" soils (which are often actually clay soils), showing a tendency to rebound and creep, in slopes steeper than 2-1/2 or 3H to 1V. The rebound and creep phenomena warrant investigation and research to determine their specific engineering characteristics.

Development of marginal lands is increasing to create communities for Hawaii's growing population. Other nations of the Pacific with similar geology and soil types are also experiencing a rapid growth in population. The research conducted at the University of Hawaii may later serve as a guide to other nations in the volcanic mid-Pacific with similar geology and soil types. Despite some inconsistencies, what we learn of the Hawaiian Islands may be applicable to other islands farther south, so long as we remain within the boundaries of the tropics in the true oceanic basin (MacDonald, 1982).

Soils Investigated

The soils tested in this study are high plasticity soils from various areas of Oahu, Hawaii, known to present foundation and slope stability problems. These soils possess a variety of undesirable characteristics such as low strength, high plasticity, high swell potential, potential creep failure, and loss of strength over time. There were three soils studied for this research project. Soil samples were obtained from the Kaneohe side of the H-3 Highway project currently under construction, from an area adjacent to the Waianae Comprehensive Health Center, and from Manoa Valley near a known area of landslides (Figure 2).

Kaneohe Silty Clay

The first soil tested was a silty-clay from Kaneohe obtained from the H-3 Highway project. This soil is classified as MH according to the Unified Soil Classification System (USCS), and represents one of the "troublesome" high plasticity soils found in Hawaii. This soil is similar in nature to many other tropical soils found in other regions around the world. These MH clays are known to lose significant strength upon remolding and typically have moderate swell potential. The Kaneohe silty clay soil has a relatively high in-situ moisture content and low density. Manipulation of the soil cover in these wet areas can affect equilibrium and lead to an unstable condition. A number of landslides have occurred in and around the construction areas and, subsequently, there is subsurface monitoring consisting of piezometers and inclinometers in certain areas of the site.

The Kaneohe soil sample contains halloysite clay. Previous research performed at the University of Hawaii has demonstrated that if halloysites are dehydrated, a permanent physio-chemical change occurs which may permanently alter many of its engineering properties, including moisture-density relationships, plasticity indices, and strength. Careful handling and procedures are necessary to ensure that laboratory results represent actual field conditions of halloysite soils. To minimize the effects of dehydration, careful drying of the soil from its moist condition must be observed to prevent a physio-chemical change.

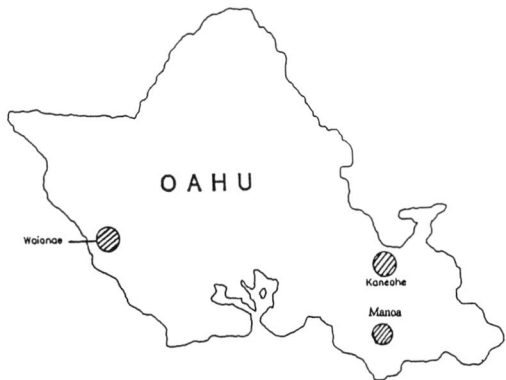

Figure 2 - Location Where Soil Samples Were Obtained

Waianae Clay

Another soil tested was a dark grey clay obtained near the Waianae Comprehensive Health Center on the southwestern portion of Oahu. Waianae Clay is much richer in kaolinite rather than the montmorillonite rich clays of the younger volcanics of the eastern parts of the island. This material is often found in a naturally dry and stiff state, typical of a region of low rainfall and dry weather. Index properties and compaction tests reveal a soil of high plasticity and high dry density. The leeward side of the Waianae Range where these soils exist exhibits components of mass wasting such as sliding of rock fragments, rain wash, and soil creep.

In the vicinity of the Waianae Health Center, large cracks and a misaligned parking lot showed evidence of significant movement of the underlying clay soil. All of the clay soil located beneath the Health Center building was excavated (no visual damage was observed inside the Center), while the clay soil remained beneath other engineered portions of the project such as roadways and parking areas.

Manoa Valley Clay

The Manoa Valley soil is a CH soil according to the Unified Soil Classification System. The soil is dark gray in color and contains a high percentage of smectite. The soil often has a very high natural water content and swell potential. This highly plastic soil is typical of the soils (locally known as "adobe") found along the sloping sides and bases of the valleys on the southern slope of the Koolau Range where rainfall is relatively high. The Manoa soil has a very high swell potential and low shear strength. There have been many problems resulting from the movement and deformation of the "adobe" soils, many of which have been attributed to "creep".

Additional soil samples from areas of eastern Oahu where damages due to subsurface movements have occurred, particularly Hawaii Kai, Palolo, and Aina Haina, are of interest for further creep studies. Conclusions by Peck and Wilson (1968) were that soil creep may have been the cause of the deformation, cracking, and in some cases, the destruction of homes and roadways, although no experimental creep testing was undertaken.

Identification of Soil Properties

Soil properties such as specific gravity, grain size distribution, Atterberg Limits, compaction, and triaxial strength were identified for each of the soils. Testing procedures were performed in general accordance with ASTM standards where applicable. A summary of soil properties determined is given in Table 2.

Triaxial CU and CD Strength Tests

Static triaxial tests were performed on each soil type to provide a baseline strength evaluation of the compacted samples. All triaxial compression tests were performed on remolded specimens isotropically consolidated to 100 kPa.

Strain-controlled triaxial CU and CD compressive strength tests were performed on remolded specimens of each of the soils. Each of the soils were first sieved through the #4 sieve and brought to a moisture content near optimum. Since the Waianae soil was initially dry, water was added to bring the soil to near optimum moisture content. Each of the soils was then cured for 24 hours before compaction into 63.5 mm (2.5 inch) diameter, 152 mm (6 inch) high tubes. Compaction was done in five layers to achieve a predetermined degree of relative compaction as determined by the Modified Proctor Compaction test. The compacted soils were then extracted and set up in the triaxial apparatus. Due to the heterogeneous nature of all soils, three newly compacted samples of each soil were tested for their undrained and drained strengths. An average value was used for each type of drainage condition and used as baseline values for the creep tests.

Kaneohe and Waianae soils were compacted to 90% of the maximum dry density and the Manoa soil samples were compacted to 85%. The compacted soils were then extracted and set up in the triaxial apparatus. The loads applied to the compacted soil samples for creep testing were determined by taking various percentages of the average shear strength (100% maximum load determined from triaxial tests) and converting these stresses into percent of maximum loads. These loads were applied axially to each soil specimen.

Consolidated-Undrained (CU) tests were run on saturated cohesive soil samples to represent conditions where loading is too rapid for pore pressure dissipation within the soil mass or where actual conditions restrict drainage. Positive pore pressure built up within the soil may weaken the soil structure. Shear stresses are usually carried by the soil structure (skeleton) since liquid water and air have negligible shear strengths. In the laboratory, the soil sample is fully saturated and the loading rate is kept sufficiently slow, so that pore pressures measured anywhere in the specimen are identical.
Undrained tests were performed as strain-controlled tests at a loading rate of 0.5%/min. Although the loading rates in the field may be faster than in the lab, the rates are still considered comparable.

Consolidated-Drained (CD) tests were prepared and consolidated identically to the CU tests, then sheared with drains closed. Loadings of the CD tests were applied at a much slower rate to allow pore pressures to dissipate from the specimens. For the CD strength tests, no excess pore pressure was allowed to develop in the soil sample. By their nature, CD tests can take a long time. The time required to fail the specimen may range from one day to several days (Bishop and Henkel, 1962).

Table 2 - Summary of Tests Performed

Tests Performed	Hawaiian Soils Tested		
	Kaneohe Silty Clay	Waianae Clay	Manoa Clay
Wet Sieve, % Passing #200	89.6	93.1	
Specific Gravity	2.89	2.91	
Atterberg Limits:	MH	MH	CH
Liquid Limit	77.6	81.5	141
Plastic Limit	50.1	43	41
Plasticity Index	27.5	38.5	100
Modified Proctor Compaction:			
Maximum Dry Density (kN/m^3)	14.5 (92.5 pcf)	15.4 (98.1 pcf)	14.6 (93.2 pcf)
Optimum Moisture Content (%)	32	22.9	29
Triaxial Strength Tests:			
Consolidated Undrained (CU) Test			
Test #1 Max. Dev. Stress (kPa)	152	83	141
Test #2 Max. Dev. Stress (kPa)	162	88	138
Test #3 Max. Dev. Stress (kPa)	137	77	143
Test #4 Max. Dev. Stress (kPa)	NA	86	NA
Average CU Strength (kPa)	150	84	141
Consolidated Undrained (CD) Test			
Test #1 Max. Dev. Stress (kPa)	247	115	NA
Test #2 Max. Dev. Stress (kPa)	265	124	NA
Test #3 Max. Dev. Stress (kPa)	250	120	NA
Average CD Strength (kPa)	254	119.7	NA
CU Creep Tests:			
90% Max. Dev. Stress	Creep Failure	Creep Failure	Creep Failure
80% Max. Dev. Stress	Still Deforming*	Creep Failure	Creep Failure
70% Max. Dev. Stress	NA	Still Deforming*	Still Deforming*
60% Max. Dev. Stress	Still Deforming*	NA	NA
CD Creep Tests:			
90% Max. Dev. Stress	Attenuated	Attenuated	NA
80% Max. Dev. Stress	Attenuated	NA	NA
70% Max. Dev. Stress	Attenuated	NA	NA

Note - * test terminated prior to failure

The loading rates of both the Consolidated-Undrained (CU) and the Consolidated-Drained (CD) triaxial compression strength tests are given as strain rates. Strain rates were determined using the data from the consolidation tests run before each strength test. ASTM has no testing standard for CD tests on soils, but there is a standard for the CU test. The testing operator has to assume when the soil sample will fail in order to determine an appropriate strain rate. There are other references to help determine the time to failure and the strain rate for these strength tests. Further discussion is presented in the following section on creep tests.

The rates of loading required for determining the effective stress strength may not be appropriate for the short-term or undrained loading situation. The stress-deformation and strength response of a clay soil is rate dependent; the faster the loading, the stronger it becomes (Holtz and Kovacs, 1981). In the short-term case, however, the field loading may be quite rapid, and so for correct modeling, the rates of loading in the laboratory should be comparable.

All strength and creep tests performed at the University of Hawaii were double-drained tests where applicable. To aid in the saturation of cohesive soils, vacuum-backpressure saturating and filter strips were used. This method has proven effective and involves very little manual assistance. The minimum differential vacuum time for acceptable saturation results was about 4 days. Controlling the amount of filter paper used is also important. Nine 0.95 cm by 16.5 cm long filter strips were placed on the sides of each specimen. By carefully limiting the amount of filter paper on the soil sample, the strength of the filter paper will minimally affect the strength of the soil sample and help reduce the time required for proper saturation.

The results of triaxial CU and CD strength tests for each of the soils tested are listed in Table 2. The variability in strength for the Kaneohe CU tests were considered quite large when compared to the CD strength data of the Kaneohe soil and both CU and CD strength data for the other soils. Even though all soil samples were consistently prepared following the procedures mentioned before, controlling the variability of the compacted soil still proved difficult.

Creep Tests on the Triaxial Apparatus

Creep, as defined for this study, is time-dependent deformation under a constant load. The purpose of this research was to determine the creep characteristics of some selected compacted Hawaiian soils under various loading conditions, and with different drainage conditions.

Creep tests were performed under drained and undrained conditions to simulate the possible responses. Consolidated-Undrained creep tests are applicable to saturated areas with poor drainage. An example of such areas is the windward side of Oahu. Also, valley areas and high grounds (i.e. mountain slopes) may be applicable to Consolidated-Undrained creep tests. Consolidated-Drained creep tests maybe applicable to areas where it is anticipated that sufficient drainage is available to dissipate the tendency for pore pressure build-up.

The triaxial apparatus was chosen to simulate these conditions, as several investigators have suggested this is appropriate for the investigation of time effects (Jamiolkowski et al., 1985; Feda, 1992). Axial load, axial deformation, and, depending on the type of drainage conditions, pore pressure or volume change measurements were recorded at time intervals. Referring to literature on creep, along with informative discussions with other researchers, it was decided that the loading on the compacted soils would be 100%, 90%, 80%, 70%, and 60% of the maximum load determined from static triaxial compression tests.

As this study was mainly concerned with the characteristics of engineered fill and to aid in providing comparable results, strength and creep tests were run using remolded saturated samples.

For both the Consolidated-Undrained and the Consolidated-Drained Creep tests, loading rates were intended to simulate, as accurately as possible, the type of drainage conditions and potential failure rates expected in the field. ASTM does not present any relevant equations for determining the time to failure, so other soil testing references were consulted. Equations were found for estimation of the time to failure by Al-Khafaji and Andersland (1992) and the CKC Triaxial Testing Manual (Chan, 1991). When comparing the two methods, it seems that the equations developed by Al-Khafaji and Andersland were a reasonable choice since there are parameters for radial flow through filter paper. However, the calculated time to failure for the Kaneohe and Waianae compacted soils would make the loading phase of the creep tests require at least three consecutive days. Hence, the equations from the CKC Triaxial Testing Manual were chosen, but not without serious considerations for practicality.

Using data from the consolidation tests, the time at 50% consolidation was determined. This value was used with the formulas found in the CKC Triaxial Testing Manual to determine the approximate time to failure. For the undrained conditions, the time to failure was taken as 4 times the elapsed time at 50% consolidation. For drained conditions, the time to failure was taken as 31 times the elapsed time at 50% consolidation. Time to failure for drained conditions is typically longer than for undrained conditions to allow pore pressures dissipation and to prevent pore pressure buildup. Simplified equations were found for time to failure for different drainage conditions, derived assuming the sample is single-drained and also draining radically due to filter paper on the sides.

Another practical question was how to apply the loads to the soil sample. While some investigators have chosen to apply the total test load at one time, this was deemed inappropriate for most cases since large pore pressures could adversely affect the tests and potentially fail the test specimens prematurely. Two alternative techniques were evaluated for use in this study. Loads can be applied in equal time increments, making the loading rate constant (e.g., an equal additional load every 10 minutes), or, referring to the loading schedule found in the CKC Triaxial Testing Manual (Chan, 1991), loads can be applied more rapidly in the earlier segments of the creep tests, but as the loading time gets close to the time to failure, the loading rate is decreased to accommodate the drainage conditions and to allow for more accurate control of the tests near the critical loads. (The Kaneohe Creep tests were started prior to consulting the CKC Triaxial Testing Manual so the incremental loading technique was only applied to the Waianae and Manoa soils). Both methods were dependent on the time to failure (t_f) and the laboratory loading capabilities. Regardless of technique, all design loads were applied to the samples in increments proportional with the expected time to failure.

Kaneohe CU Creep Tests

The loading technique used for the Kaneohe CU Creep tests was different than for the other soil samples. Based on the consolidation data for the Kaneohe test specimens, the time to failure was calculated to be less than five minutes. In addition, the Kaneohe soil had a moderately high undrained strength. These two factors together meant that larger loads needed to be applied quickly to the sample. Considering all of this, it was decided that for this case one large load increment would be applied to the sample, but not without limitations. The biggest concern with this type of loading is accuracy of the pore pressure measurements. Loading a saturated sample too fast causes unequal pore pressures to develop in the soil sample.

Another concern was the wide range of strengths obtained from the Kaneohe CU Strength Tests. With such a range of strengths, there was a possibility that the calculated 90% load may actually be the 100% load. However, after applying the 90% load, the soil sample did not fail thus giving confidence that the 90% load was indeed correct.

In support of the 100% max load chosen, it was demonstrated that the specimen failed immediately after application of the final increment of the 100% load.

Furthermore, the 90% load did not cause failure until after a duration of 3 weeks. Observations of the sample and test data show that the creep failure of the Kaneohe soil was both brittle and viscous. The 90% loaded sample displayed a failure plane and barreling effects. This type of creep characteristics are similar to those found in rocks and compacted clays. The 80% and 60% loads were still deforming when the creep tests were terminated after a period of over 4.5 months (Figure 3).

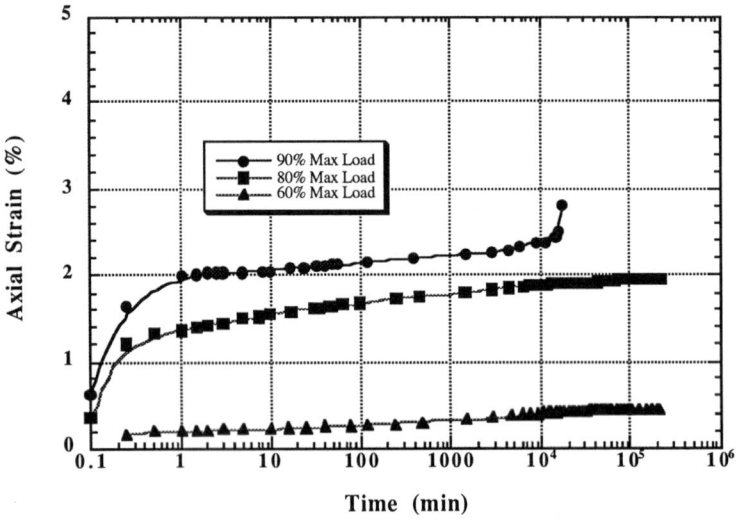

Figure 3 - CU Creep Test Results for Kaneohe Soil

The 90% load failed at an axial strain slightly less than 2.5%. Since the 80% load was still deforming at a cumulative strain of < 2%, this poses a question. Would the sample with 80% load also suffer from creep failure at an axial strain somewhat greater than 2%, or would the creep attenuate prior to reaching such an axial strain? It appeared from the data taken during the time frame of the creep tests performed, that the latter condition was more plausible since the strain rate appeared to be slowing. Much more time would be required in order to determine whether creep rupture or additional strength loss would eventually result for the 80% and 60% loaded samples.

Confining stresses tend to give a soil mass supporting strength below the subsurface. However, the action of positive pore pressures developing within a soil mass reduces the effective stress around the soil which can cause a destabilizing effect. Conversely, if pore pressures are reduced within a stable soil mass, the effective confining stress will increase thus giving the soil mass added strength. The effective confining stress of both the 100% and the 90% loads decreased over time, causing a less stable condition, and eventually led to failure. The 80% and the 60% loaded specimens each displayed a smaller increase in the effective confining stress after an initial decrease, but then later stabilized.

Kaneohe CD Creep Tests

Before beginning the CD Creep tests, several additional references on soil creep were consulted. As mentioned earlier, the loads were applied either in one increment in the beginning, or after short time intervals. In discussions with other investigators, it was determined that a rapid single load would be inappropriate for controlling the drainage conditions. Drained tests need to be run slowly so pore pressures do not build within the soil sample. For the Kaneohe CD Creep tests, the constant loading rate technique was used with 44.5 N weights applied to the soil samples in equal time increments until reaching the specified load.

The plot of creep test results for CD tests of the Kaneohe soil showed that the 100% load failed immediately after the final load increment was applied. The 90% load failed less than 2 hours after the final load increment. There is a possibility the calculated 90% load for the Kaneohe soil could have actually been higher than the calculated 90% load for that particular Kaneohe soil sample. The 80% and 70% loads appeared to be going into tertiary creep when the tests were terminated.

Waianae CU Creep Test

For the Waianae CU Creep tests, the CKC loading schedule was used, since it was assumed that this loading technique would adequately control the desired drainage conditions compared to the constant loading technique and would provide more detail and control near the critical stress levels. The specimen tested with the 100% load failed immediately upon application of the final load increment, thus verifying the 100% load derived from the triaxial test.

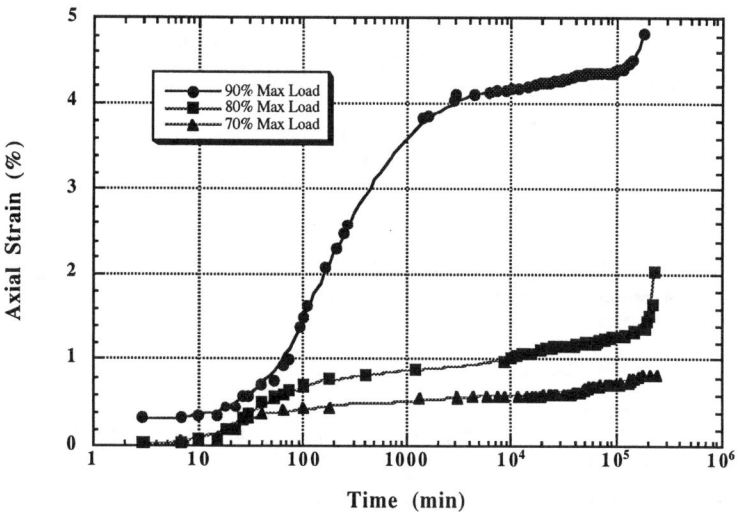

Figure 4 - CU Creep Test Results for Waianae Soil

For the 90%, 80%, and the 70% loads, all three plots appeared to be moving into tertiary creep by the time testing was terminated. Referring to the plot of the 90% load, the Waianae soil sample had an axial strain greater than 4%. Though the sample had not yet suffered from creep failure, some practicing engineers would have considered this axial strain as failure, as this amount of deformation may trigger the onset of structural damage. From visual observations of the creep tests, the 90% loaded Waianae soil appeared to going through barreling effects as it deformed. Although the Waianae soil was compacted, this deformation effect is common in soft soils.

Waianae CD Creep Test

For the Waianae CD creep tests, the CKC loading schedule was again used. Only the 100% and 90% load samples were tested. High strains were achieved in both samples.

The sample tested with the assumed 100% load failed immediately after the final load increment, verifying the maximum peak strength obtained from the triaxial stress controlled tests. The 90% load, though suffering from an axial strain of about 6%, did not suffer from a district creep rupture failure. As mentioned before, some practicing engineers may consider this axial strain as failure. Visual observations of the 90% loaded sample showed the sample suffered from barreling effects as it deformed as was seen in the CU tests of Waianae specimens.

Manoa CU Creep Tests

Since the previous load controlled testing performed on the Kaneohe and Waianae soil specimens had validated the ability to accurately derive peak 100% strength values, the 100% strength value of the Manoa specimens were derived directly from the load controlled tests. Specimens of the Manoa Clay were prepared at approximately 85% of the maximum dry density achieved by the Modified Proctor test, at a moisture within 2% of the optimum moisture content. While the tested specimens were remolded samples, in-situ samples of this soil taken from the same location were found to be at about this same density.

Undrained creep tests were loaded according to the same schedule as for the Waianae soil to 70%, 80%, and 90% of the 100% peak measured value and allowed to deform with time. Figure 5 shows the plots of axial deformation vs. time for these specimens. It can be seen from this figure that each of the loadings were still significantly deforming when the tests were terminated, and that the specimen loaded to 80% was well advanced into tertiary creep with failure imminent. Both the 90% and 70% specimens were terminated at earlier elapsed times due to time constraints and technical problems. Neither showed signs of attenuating, but rather appeared to be entering into tertiary creep modes.

Changes in Effective Stresses

An interesting observation was the generation of pore pressures for the compacted clay specimens during creep and ultimately at the onset of creep rupture. As an example the effective stress data for the undrained creep tests of the Manoa soil are shown in Figure 6 below. It can be seen that the effective stress decreases nearly uniformly for all specimens with a steep decline (generation of positive pore pressure) corresponding to the primary creep mode, a somewhat gentler slope during the secondary "attenuating" mode, and finally for the 90% and 80% loaded specimens an abrupt reversal of pore pressure generation as each of these specimens enter tertiary creep and rupture. Similar trends of pore pressure and changes in effective stress were noted for all of the soil samples tested.

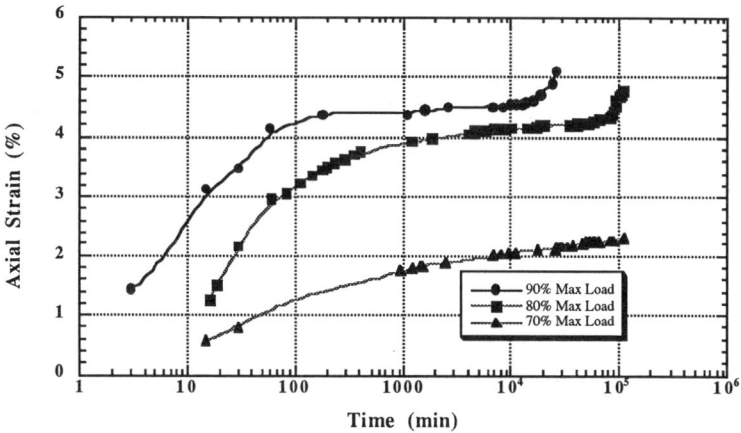

Figure 5 - CU Creep Test Results for Manoa Soil

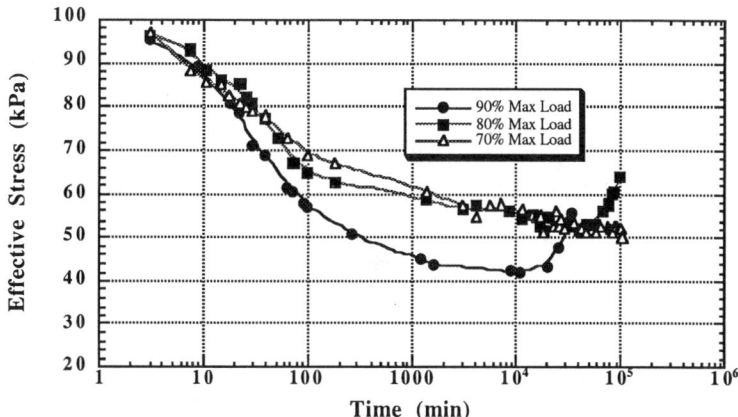

Figure 6 - Change in Effective stress for CU Creep Test of Manoa Soil

'm' Values

Previous investigators have suggested that the slope of the log strain rate vs. log time curve measured during the period of secondary creep (which is relatively linear) could be used as an indicator of a soil's creep potential. Singh and Mitchell (1968) have shown that soils with values of m < 1 display high creep potential. The values of "m" for the compacted Kaneohe and Waianae soils fall in the range of 0.95 to 0.97. An interesting observation is that the values of "m", for the different drainage conditions, are quite close in magnitude and approach m = 1, where soils typically show the same strength before and after creep. However, tropical soils differ greatly from the much tested and researched temperate zone soils used for the Singh and Mitchell study. Thus, tropical soils may show high creep potential for "m" values different than m <1.

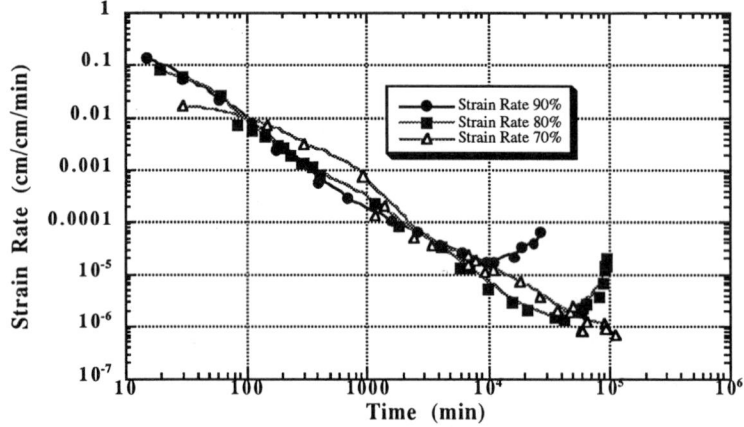

Figure 7 - 'm' Plot for CU Manoa Soil Creep Tests

Plots of the logarithmic strain rates vs. logarithmic time for the undrained Manoa creep tests (Figure 7) show "m" slope values approximately equal to 1.0 during a significant portion of the creep tests. Note that the 80% and 90% tests showed a clear decrease in "m" towards the later portions of the tests as the specimens move into tertiary creep. For some of the specimens tested, the slope of the strain rate actually steepened to m>1 for a significant portion of the test, but then reversed that trend to m<<1 as those specimens moved into tertiary creep mode signifying a physical change as the specimen approached rupture failure. This result raises the question as to the danger of assuming that if a specimen shows an "m" value of greater than 1 over a significant period of a test, that creep deformations will attenuate with time.

Additional Creep Testing

During a short period in the weeks before termination of this project, some additional testing was performed in order to evaluate the effects of different degrees of compaction and "as compacted water contents" on the creep characteristics of some of the soils.

Some specimens of the Kaneohe soil were compacted at 90% of the maximum density (same as earlier tests), but at a higher moisture content intended to represent the highest moisture at which the compaction curve intersected 90% $\gamma_{d,max}$. This condition was
more likely the condition in which this soil would be compacted in the field. Figure 8 shows the plot of axial strain vs. time for the creep test with a load equal to 90% of the maximum short term strength. Although this test was run for only a short duration, it can be seen from the plot that the specimen was deforming at an increasing rate at the time of termination.

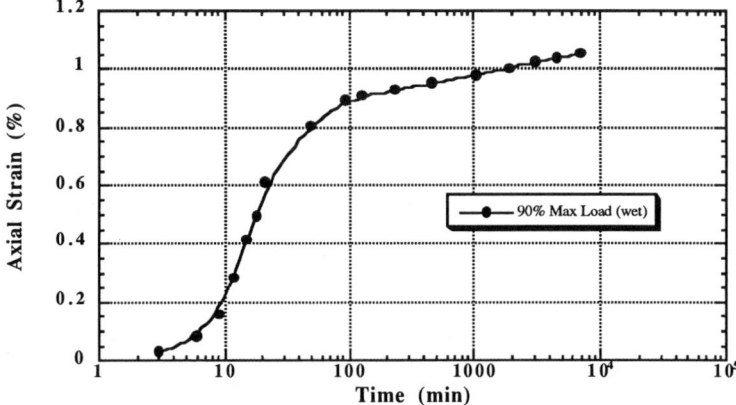

Figure 8 - Creep Test for Kaneohe Soil Wet of Optimum

In contrast, tests were performed on two specimens of the Manoa Clay compacted at lower densities approximately equal to 70% $\gamma_{d,max}$, at water content slightly above optimum. One of the specimens was tested undrained at a load of 70% of the maximum value obtained for the 85% $\gamma_{d,max}$ specimens. While it is difficult to accurately quantify what percentage of the actual maximum load this was for the 70% $\gamma_{d,max}$ specimen, the test was allowed to run for several days displaying a continuously high strain rate. An additional test was erroneously performed as a drained test. It was interesting to note that the creep strain rate for this test was just as great if not higher than for the undrained specimen. This result may indicate and support the premise that creep may be equally pronounced in soil samples even if full drainage is permitted. This is in contrast to speculations that undrained conditions generate more accelerated and greater tendency for soil creep.

Discussion and Conclusions

This report contains a detailed discussion of the work that has been done on testing and research into identifying the characteristics and mechanics of soil creep behavior. It is from these theories and prior work on which we can base our testing to identify soils which may be susceptible to creep, and to further our understanding of expected behavior based on laboratory testing.

Three different tropical soils which have been used for local earthwork construction and which have incurred incidental problems were tested in the laboratory during this investigation. Each of the soils showed significant creep potential (defined as continued deformation under a constant application of load), as well as potential for strength loss (defined as creep "rupture" failure under a load less than the short term tested peak strength). Each of the soils were of moderate to high plasticity and displayed time dependent stress-strain characteristics. The slope of the plots of the logarithmic strain rate versus logarithmic time, denoted "m", showed values of m slightly less than 1. Previous research has shown that soils which exhibit $m < 1$ typically display high creep potential and loss of strength over time.

While both drained and undrained specimens were tested, the majority of the tests were performed as undrained tests. This was done initially because it was believed that pore pressures generated during undrained shearing and deformations would tend to weaken the soils during testing and provide the greatest amount of creep. It was also considered that since the more troublesome soil problems, which may be related to creep phenomenon, occur in areas where the soils are at high saturation levels and where drainage is poor, so that undrained conditions would provide a better, albeit conservative, representation of actual field conditions. In retrospect, it may be more useful or practical to concentrate studies under drained conditions since: 1) a combination of testing results and theoretical mechanics indicate that creep phenomenon is not generally a function of pore pressure build-up but more likely rather a rearrangement/realignment of clay particles and structure, and 2) creep is a long-term process which is conventionally considered best represented by drained conditions.

The maximum elapsed time for each creep test performed as a part of this study was 2-1/2 to 3 months. This time limit was necessary due to the limited number of testing devices and to allow for the number of different specimens desired to test during the limited project period. This time span was relatively short considering the expected service life of 50 years or more. It is very likely that some of the specimens would have achieved considerably greater creep deformations and perhaps even eventual creep rupture failure if time had been allowed for testing to continue. This estimation is based on the fact that several of the specimens tested were still undergoing deformations at the time of their termination. What was informative from the shorter term tests performed was a reasonable indicator of the potential of the soils tested under the controlled conditions of saturation, confining pressure, drainage, and compaction conditions, to undergo extended creep behaviors.

The test results show that all of the soils tested display varying degrees of creep potential, and that under some conditions strengths are reduced. This can have practical significance to engineering design concerning critical strengths and deformations, and should be considered for those soils whose properties are indicative of possible creep potential.

It is recommended that future creep testing of these and other types of tropical soils should include further testing under drained conditions, an evaluation of soils prepared under various compaction conditions, and observation of behavior under much longer time durations (i.e. several months or years).

References

Al-Khafaji, A. W. and Andersland, O. B., (1992). *Geotechnical engineering and soil testing*, "Experiment 28 - consolidated-drained triaxial test", Harcourt, Brace and Jovanovich, 612-617.

Bishop, A.W., and Henkel, D.J., (1962). *The measurement of soil properties in the triaxial test,* Edward Arnold Ltd., London, 2nd Ed.

Chan, C.K., (1991). *CKC triaxial testing manual*, Soil Engineering Equipment Co., San Francisco, CA, 116 p.

Feda, J. (1992). *Creep of soils and related phenomena*, Elsevier Science Publishing Company, Inc., New York.

Holtz, R. D. and Kovacs, W. D. (1981). *Introduction to geotechnical engineering*, Prentice-Hall Inc., Englewood, 733 p.

Jamiolkowski, M., Ladd, C. C., Germaine, G. T., and Lancellotta, R. (1985). "New developments in field and laboratory testing of soils," *Proc., 11th Int. Conf. on Soil Mech. and Found. Engrg*, San Francisco, 1, 57-153.

Kuhn, M. K. and Mitchell, J. K. (1993). "New perspectives on soil creep", *J. Geotech. Engrg.*, ASCE, Vol. 119(3), 507-523.

Lum, W.B. (1982). "Engineering problems in tropical residual soils in Hawaii", *Proc., Engnrg. and Const. in Tropical and Residual Soils*, ASCE, 1-12.

MacDonald, G. (1982). *Volcanoes in the sea; the geology of Hawaii*, University of Hawaii Press, Honolulu, 544 p.

Mitchell, J. K. (1993). *Fundamentals of Soil Behavior*, 2nd Ed. John Wiley and Sons, New York, 437 p.

Mitchell, J. K., Campanella, R., and Singh, A. (1968). "Soil creep as a rate process", *J. Soil Mech. and Found. Div.*, ASCE, Vol. 94(1), 231-253.

Peck, R. B., and Wilson, S. (1968). "The Hind Iuka landslide and similar movements, Honolulu, Hawaii," *Report for City and County of Honolulu*, Honolulu, Hawaii.

Sholtz, C. H., and Engelder, J. T. (1976). "The role of asperity indentation and ploughing in rock friction - I, Asperity creep and stick-slip," *Int. J. Rock Mech. Mining Sci.*, 13(5), 149-154.

Singh, A., and Mitchell, J. K. (1968). "General stress-strain-time functions for soils", *J. Soil Mech. and Found. Engrg. Div.*, ASCE, Vol. 94(1), 21-46.

Tavenas, F., Leroueil, S., La Rochelle, P., and Roy, M. (1978). "Creep behavior of a lightly overconsolidated clay," Can. Geotech. Jour., 15 (3), 402-423.

Uehara, G. (1982). "Soil Science for the Tropics," *Proceedings, Engineering and Construction in Tropical and Residual Soils*, ASCE, 13-26.

Vyalov, S. (1986). *Rheological fundamentals of soil mechanics*, Elsevier Science Publishers B. V., The Netherlands, 564 p.

Strain Rate Effects on Stress-Strain Behaviour of Clay
as Observed in Monotonic and Cyclic Triaxial Tests

Satoru Shibuya[1], Toshiyuki Mitachi[2], Akihiko Hosomi[3] and Seong Chun Hwang[3]

Abstract

Undrained behaviour of reconstituted clay (I_p=27) was examined for a wide strain range between 0.0001% and 10%. In a series of undrained triaxial tests performed, the clay specimens were subjected to shearing in an axial-strain controlled manner. In monotonic loading tests, the rate of axial straining, fixed in each test, varied between 0.01% and 1% per minute. Similarly, in cyclic loading tests, the frequency of axial straining was altered over a range between 0.00025 and 0.1 Hz. Interpretation of the test results focused on the strain-rate dependent properties of the clay that were manifested in the stiffness and damping at strains less than 0.5% in particular. Applicability of Masing's rules by considering the strain-rate effects on the small-strain stiffness is in depth discussed. A hysteretic stress-strain model which accommodates the elastic-threshold strain dependent on the strain rate is newly proposed.

Introduction

There are some experimental data showing that the stiffness of geomaterials at strains less than 0.001% is independent of the rate of shearing, the number of cycles, N, including the specific case with N=0 for monotonic loading (ML), etc. (e.g., Teachavorasinskun et al., 1991, Tatsuoka and Shibuya, 1992, Shibuya et al., 1992, 1995, Lo Presti et al., 1993). Over a range of shearing speed seen in geotechnical

----

[1]Associate Professor, Hokkaido University, Sapporo, Japan
[2]Professor, ditto.
[3]Graduate Student, ditto.

engineering problems, the pseudo-elastic Young's modulus, E_{max}, involved with very small hysteretic damping can be characterized in this small-strain region, even in cyclic loading (CL) test on soft clays.

Yet, interrelationship of the soil stiffness in ML and CL tests has not been properly understood for intermediate strain range from 0.001% to 1%. As correctly pointed out by Isenhower and Stokoe (1981), this may be attributed to the current situation in soil testing practice; i.e., the soil response in ML test is examined under conditions of strain-controlled, whereas the cyclic behaviour for this small strain region is usually not so. Performance of strain-controlled CL test with the known results of comparative ML test like one reported by Dobry and Vucetic(1987) for the behaviour at a fixed strain rate is needed when examining this interrelationship at small strains over a wide range of strain-rate. Results of comprehensive tests as such will provide insight into modelling of hysteretic stress-strain relationship of cohesive soil on the basis of the strain-rate dependent backbone curve from the ML test with the corresponding strain rate.

Undrained triaxial tests performed

Reconstituted clay called NSF-clay (LL=56%, I_P=27 and ρ_s=2.78 g/cm³) was used in the current study. Sample preparation method used, together with the particle size distribution curve has been described by Shibuya et al. (1995).

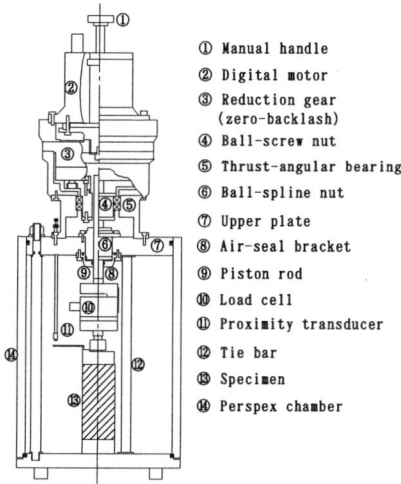

① Manual handle
② Digital motor
③ Reduction gear (zero-backlash)
④ Ball-screw nut
⑤ Thrust-angular bearing
⑥ Ball-spline nut
⑦ Upper plate
⑧ Air-seal bracket
⑨ Piston rod
⑩ Load cell
⑪ Proximity transducer
⑫ Tie bar
⑬ Specimen
⑭ Perspex chamber

Fig.1 -- Triaxial apparatus recently developed at Hokkaido Univ.(Toki et al., 1994).

Table 1. -- Undrained triaxial tests performed on NSF-clay.

Test[1]	e_c[2]	K[3]	$d\varepsilon_a/dt$ (%/min)	f (Hz)	E_{max}[4] (MPa)	q_{max}[5] (kPa)
ML	1.160	1.0	0.01	—	239	197
ML	1.165	1.0	0.13	—	239	197
ML	1.164	1.0	1.40	—	239	216
ML	1.163	0.8	0.13	—	258	223
ML	1.148	0.6	0.13	—	234	228
CL	1.148	1.0	—	0.01	239	—
CL	1.089	1.0	—	0.1	228	—
CL	1.171~1.182	1.0	0.02	—	—	—

1) ML : monotonic loading test , CL : cyclic loading test.
2) void ratio of specimen prior to shearing.
3) Stress ratio during consolidation , σ_h'/σ_v'.
4) Pseudo-elastic Young's modulus.
5) Maximum deviatoric stress in compression.

The triaxial apparatus used has recently been developed at Hokkaido University(Fig.1, Toki et al., 1994). Equipped with a digital servo-motor, it is capable of performing both ML and CL tests in an axial strain controlled mode. The minimum control of axial displacement is 1.5×10^{-7} mm, which in turn gives a resolution of axial-strain control of 1.5×10^{-9} for the specimen with 100 mm high and 50 mm in diameter. In addition, the direction of axial strain increment can be sharply reversed without any time-lag since the control system is virtually free from any backlash. This feature is crucially important for rigorous measurement of soil damping at a specified strain-rate.

The conditions for tests, together with some results, are shown in Table 1. All the specimens were consolidated to a common value of effective mean principal stress, $p'_c = (\sigma'_v + 2\sigma'_h)/3$ of 300 kPa. In the ML-tests, the soil response to failure was examined at the rate of axial straining, $d\varepsilon_a/dt$, of 0.01, 0.13 and 1.40 %/min. Two types of CL test performed were;

i) <u>multi-stage CL test</u>; the frequency, f, was fixed at 0.01 and 0.1 Hz using two specimens. In each test, single amplitude of cyclic axial strain, $(\varepsilon_a)_{SA}$, was increased in steps over a range between 0.005% and 1%, implying that $d\varepsilon_a/dt$ increased in a manner of $d\varepsilon_a/dt = 240 \cdot f \cdot (\varepsilon_a)_{SA}$ (%/min) (refer Shibuya et al., 1995), and

ii) <u>single-stage CL test</u>; a total of four specimens was subjected to cyclic loading using an axial strain rate of 0.02 %/min in common. The values of $(\varepsilon_a)_{SA}$, were 0.002%, 0.004%, 0.02% and 0.3%, resulting in a wide frequency range of 0.00025-0.05 Hz.

STRESS-STRAIN BEHAVIOUR OF CLAY 217

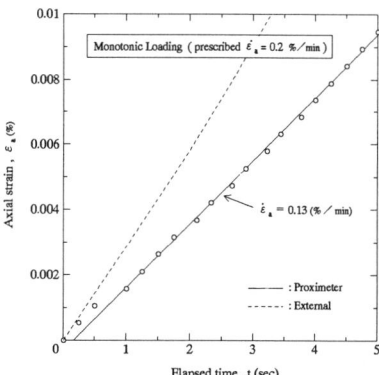

Fig. 2 -- Imposed axial strain with time in a ML test.

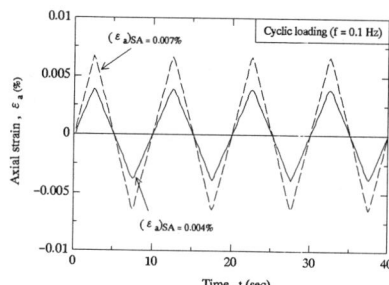

Fig. 3 -- Time history of axial strain in a multi-stage CL test.

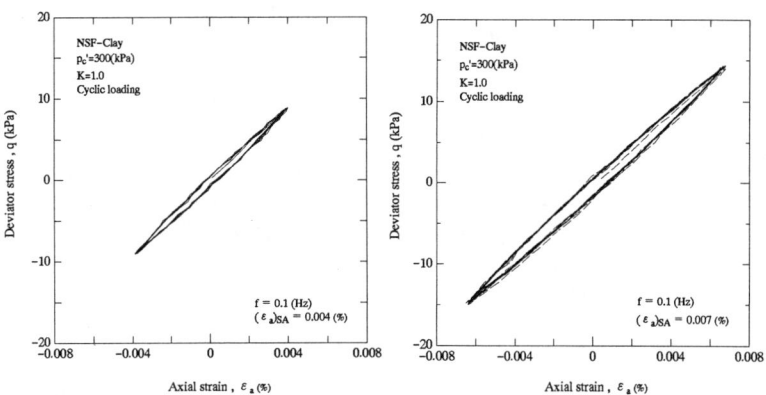

Fig. 4 -- Hysteresis loop using single amplitude cyclic strain of; a) 0.004% and b) 0.007%.

Presentation of Test Results

Referring to Fig.5 in Shibuya et al. (1995), definitions of the undrained stiffnesses and hysteretic damping used for the interpretation of the test results are;

Equivalent Young's modulus in the CL test; $E_{eq} = \Delta q_{SA}/(\varepsilon_a)_{SA}$ (1)
Secant Young's modulus in the ML test; $E_{sec} = \Delta q/\varepsilon_a$ (2)
Hysteretic damping ratio in the CL test; $h = \Delta W/(2\pi W)$ (3)

where Δq denotes deviatoric stress defined from the stress state at the start of shearing, and the subscript 'SA' stands for single amplitude. The symbol W denotes the elastic energy equal to $q_{SA} \cdot (\varepsilon_a)_{SA}$ imposed to the specimen during a cycle. Note that the values of E_{eq} and h are calculated directly from the digital data for the Δq-ε_a relation. Typically, 40 to 60 data points are recorded in each hysteresis loop.

Examples of time histories of the applied axial strain, together with the typical hysteresis loops in the CL test, are shown in Figs.2, 3 and 4. In the ML test, the value of $d\varepsilon_a/dt$ measured at right above the specimen top cap using a proximeter was slightly smaller than the externally imposed value of 0.2 %/min due to the compliance of load cell. However, it stayed at a constant value of 0.13 %/min in this small strain region (see Fig.2). In the multi-stage CL test, the prescribed frequency of f=0.1 Hz was successfully imposed, showing sharp peaks on reversal (see Fig.3). As a result of it, hysteresis loops also showed sharp peaks as examined for N from 1 to 11 in each stage (see Fig.4)

Fig.5 shows stress-strain relationship of three ML tests in compression using different values of $d\varepsilon_a/dt$. The effective stress paths are shown in Fig.6, in which the results of two tests with K=0.8 and 0.6 are also shown for comparison. The normally consolidated specimens reached to failure without exhibiting any distinct peak before the event. The strength is designated by the maximum value of deviator stress, q_{max}, at 15% strain. In tests with K=1.0, q_{max} was largest in the fastest test with $d\varepsilon_a/dt$ of 1.4 %/min. Despite the marked difference in $d\varepsilon_a/dt$, E_{max} of 239 MPa was observed in common with these tests. Conversely, the strain rate effects were obvious in respects that according to the decrease in the axial strain rate, elastic-threshold strain, $(\varepsilon_a)_{EL}$, beyond which the stress-strain relationship exhibited non-linearity, decreased, and the response became softer(see Fig.5b).

The variations of secant stiffnesses and damping with strain are shown in Fig.7, in which the E_{sec}-ε_a relation in the ML tests, together with the E_{eq}-$(\varepsilon_a)_{SA}$ relation in the CL tests are compared(see Fig.7a. Note that i) despite the marked difference in axial strain rate imposed to each specimen, E_{max} of 239 MPa was observed in common

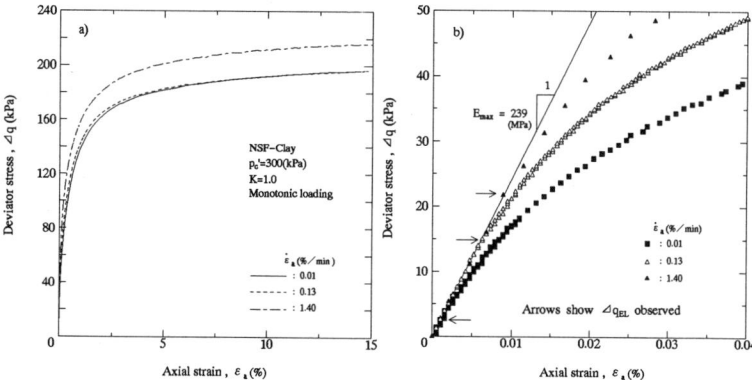

Fig. 5 -- Stress-strain relationship in ML tests ; a) a maximum strain of 15%, and b) a maximum strain of 0.04%.(low strain region of same tests)

with these ML and CL tests(see Table 1), ii) the axial strain at which each specimen commenced decay in stiffness was obviously larger in test with higher value of the axial strain rate, and iii) as a result, the damping values in the multi stage CL test with f=0.1 Hz were smaller compared to those in the other test with f=0.01Hz.

Accumulated excess pore water pressure, Δu, was examined with respect to axial strain (Fig.8). As already pointed out by Dobry et al.(1982) for sand and by Vucetic(1984) for clay, Δu was little in magnitude when the axial strain was smaller than about 0.02%. It should be reminded that the strain-rate effects were significant in this phase showing no dilatancy (see Figs.5b, 7a and b).

Discussions

a) Elastic threshold strain (ε_a)$_{EL}$

Fig.9 shows the relationship between axial strain rate and axial strain as observed in the ML and CL tests. In this double logarithmic plot, the contours of E_{sec} and E_{eq} are indicated using inclined solid lines for capital A=E_{sec}/E_{max} (or =E_{eq}/E_{max}) from 1.0 to 0.2 in every one-tenth. The hatched domain on the left means the elastic region where the stress-strain response was practically linear and recoverable, exhibiting very small hysteretic damping (see Fig.7a and b).

The elastic threshold strain, (ε_a)$_{EL}$, associated with the line of A=1.0 is therefore given by;

$$(\varepsilon_a)_{EL} = 0.0001 \times (d\varepsilon_a/dt)^{0.5} \qquad (4)$$

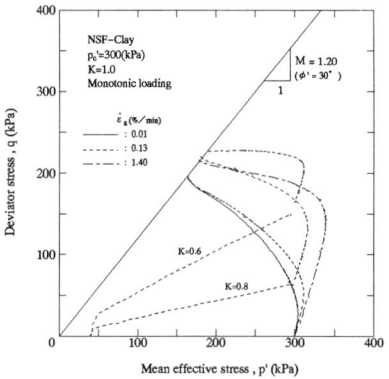

Fig. 6 -- Effective stress paths in ML tests.

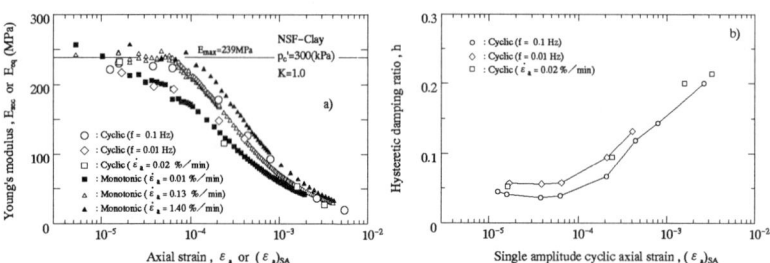

Fig. 7 -- Variations of stiffness and damping; a) secant stiffnesses with strain in ML and CL tests, and b) hysteretic damping with strain in CL tests.

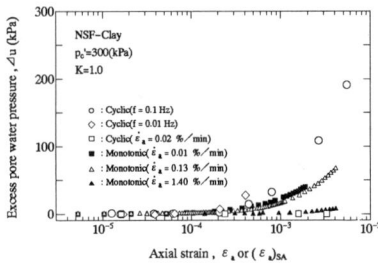

Fig. 8 -- Development of excess pore pressure with strain.

where $d\varepsilon_a/dt$ should be given in %/min. The line of A=1.0 was determined by means of a linear regression analysis applied to the $(\varepsilon_a)_{EL}$ values from a total of six ML and CL tests. The determination of $(\varepsilon_a)_{EL}$ in the ML tests can be seen in Fig.5b.

The elastic threshold strain was not influenced by anisotropic consolidation. Fig.10 shows determination of E_{max} in three ML tests with K equal to 1.0 (isotropic), 0.8 and 0.6. As a result of linear regression analysis applied to each data set, the $(\varepsilon_a)_{EL}$ values were approximately 0.006 % in these specimens as sheared with the axial strain rate of 0.13 %/min in common. Moreover, E_{max} was not significantly influenced by the K-value. This matches results of similar tests on kaolin clay reported by Tatsuoka et al. (1994).

Similar results on intact clay in Norway are shown in Fig.11. In this test, Drammen clay sample retrieved by displacement method was anisotropically consolidated with K=0.5 to a stress level four times of in-situ σ'_v. It was first subjected to undrained CL with $(\varepsilon_a)_{SA}$ of 0.002%, then it was sheared by increasing q monotonically using $d\varepsilon_a/dt$ of 0.10 %/min. The E_{eq} value in the cyclic stage coincided with E_{sec} in the subsequent stage of ML, implying that these are both pseudo-elastic Young's modulus. In addition, the elastic threshold strain was about 0.002%, which was close to the value of the reconstituted NSF-clay at the corresponding axial strain rate of 0.1%/min (refer Fig.9 and Eq.4). In quantitative sense, Eq.4 may be applicable also to the small-strain behaviour of intact clays.

b) Hysteretic stress-strain relation dependent on strain-rate

The stress-strain relationship in ML test may be used as the backbone curve to depict hysteretic stress-strain relation in CL test performed at the corresponding axial strain rate. The results shown in Fig.7a suggest it true at the intermediate strain of 0.001 and 0.1%; i.e., the E_{sec} - ε_a relation in the ML test with the axial strain rate of 0.01%/min coincided roughly with the E_{eq}-$(\varepsilon_a)_{SA}$ relation in a series of single stage CL tests performed at similar axial strain rate of 0.02%/min.

A similar comparison is made between the ML and multi-stage CL tests (Fig.12). The stress-strain curve of the ML test may be regarded as the backbone curve for a hysteresis loop of the CL test performed at similar strain-rate (i.e., the ML data denoted using open triangles as compared to hysteresis loop denoted using dashed line or the solid triangles vs. the solid line).

Applicability of Masing's rules is examined in Fig.13. In this figure, the variation of E_{sec} with q is examined for the results of a pair of comparative ML and CL tests performed at the axial strain of about 0.1%/min. The relation denoted using open triangles refers to the measured data from point 'a' to point 'b' in the ML test. Similarly, a set of data denoted by solid squares refers to the relation along the path from 'b' to 'c'

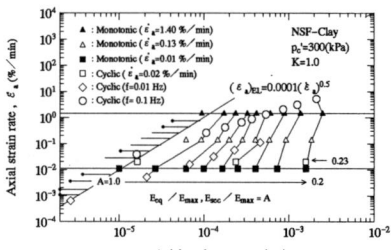

Fig. 9 -- Contours of secant stiffnesses.

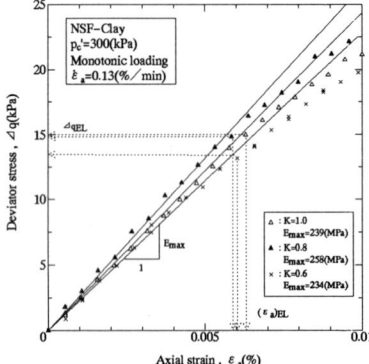

Fig. 10 -- Effects of anisotropic consolidation at small strains.

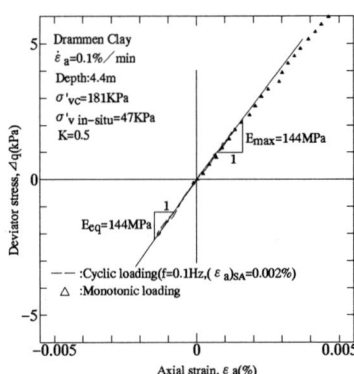

Fig. 11 -- Stress-strain relations at small strains in a test on intact Drammen clay.

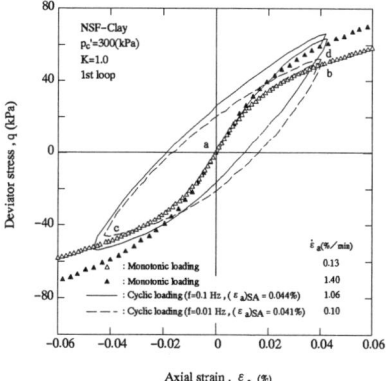

Fig. 12 -- Examination of backbone curves.

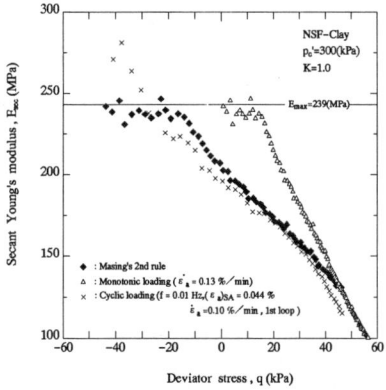

Fig. 13 -- Applicability of Masing's rules examined.

predicted by applying Masing's second rule to the measured data. It was produced by scaling up both of the measured q and ε_a by a factor of two. This can be compared to the actual relation measured in the CL test, which is denoted using a crossed mark. Predicted and measured relations over the entire strain range examined coincided well with each other, implying the validity of both Masing's first and second rules. It should be mentioned that the first rule assuming E_{max} commonly seen at points 'a', 'b' and 'c' in the event of reversal of shear direction has previously been validated in similar tests reported by Dobry and Vucetic(1987), Teachavorasinskun et al. (1991) and Shibuya et al. (1992). Independence of E_{max} on K-value also supports this in an implicit way (see Fig.10).

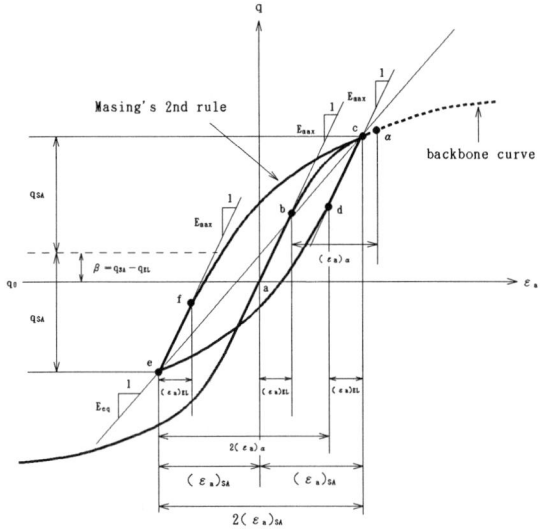

Fig. 14 – Strain-rate dependent (SRD) model newly proposed.

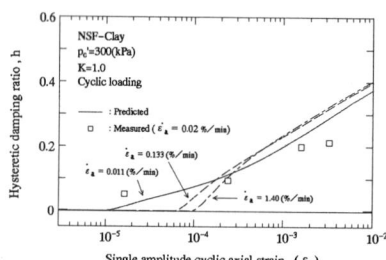

Fig. 15 – Prediction of hysteretic damping ratio by SRD model in single-stage CL tests.

c) A newly proposed hysteretic stress-strain model

A hysteretic stress-strain model is proposed on the basis of results of tests shown in Figs. 10, 12 and 13. This model termed as strain-rate dependent (SRD) model is shown in Fig.14. The important features of the SRD model are described in the following:

- it incorporates the elastic-threshold strain, $(\varepsilon_a)_{EL}$, at the start of shearing at point 'a' as well as at instants when the direction of axial strain increment is reversed at points 'c' and 'e',

- the (ε_a)EL value as well as the stiffness is common in magnitude for the linear parts of ab, cd and ef along the loop (i. e., the latter means incorporation of Masing's first rule),

- the (ε_a)EL value is dependent on the axial strain rate as expressed in Eq. 4, and

- the stress-strain relationship of ML test with relevant axial strain rate can be the backbone curve, and

- Masing's second rule can be applicable only to the non-linear parts; i. e., for example, the shape of the curve from d to e, also one from f to c, can be determined by scaling up the curve from b to c by a factor of two.

Therefore, the properties predicted using the SRD model will show the following characteristics ;

- hysteretic damping, h, is equal to zero when (ε_a)EL is smaller than a half of (ε_a)EL at a given strain rate, and

- the center of cyclic amplitude of deviatoric stress will not coincide with the ε_a axis.

Data in Fig.15 shows the variation of h with (ε_a)SA as measured in the single-stage CL tests performed with $d\varepsilon_a/dt$ of 0. 02%/min. Predicted relationships on the basis of the results of ML tests with $d\varepsilon_a/dt$ of 1.40, 0. 133 and 0.011%/min, coupled with the SRD model, are also shown for comparison. It is demonstrated that the SRD model is capable of providing a strain-rate dependent characteristic of h seen in particular at small strains. More importantly, the prediction for $d\varepsilon_a/dt$ of 0. 011%/min closely matches the measured data that were obtained at a similar strain rate of 0. 02 %/min.

Conclusions

Strain rate effects of the reconstituted clay subjected to monotonic and cyclic shear were examined in undrained triaxial test over a relatively narrow range of the axial strain rate between 0.01 and 1 % per minute.

Pseudo-elastic Young's modulus, E_{max}, was observed in a small-strain region where the maximum stiffness was scarcely influenced by the shearing rate imposed. In addition, the secant stiffness in the monotonic loading test was coincident with that in the comparative cyclic loading test.

Conversely, elastic threshold strain, beyond which the stress-strain relation exhibited non-linear and irrecoverable behaviour, was significantly influenced by the strain rate in a manner that it increased in magnitude in a exponential form with the strain rate. Quantification of this is given in Eq.4, which may be valid to not only the reconstituted clay but also the intact Drammen clay. On the other hand, the initial effective stress state was not influence to govern elastic threshold strain as examined in tests on the reconstituted clay for a range of $K = \sigma'_h / \sigma'_v$ between 0.6 and 1.0.

The stress-strain relationship in monotonic loading test may be used as the backbone curve to depict hysteretic stress-strain relation as the clay is subjected to cyclic loading at the corresponding strain rate. This can be achieved through the application of Masing's rules to the results of the monotonic loading test.

On the basis of these test results, a hysteretic stress-strain model named strain-rate dependent (SRD) model has been newly proposed. It is demonstrated that the SRD model is capable of reasonably predicting a strain-rate dependent characteristic of the clay damping seen in particular at small strains.

Acknowledgements

Drammen clay sample was supplied by Dr. H. Tanaka, Port and Harbour Research Institute in Yokoshuka, Japan. A part of the present research has been supported by a grant from TEPCO (Tokyo Electric Power Company) Research Foundation. The authors appreciate Mr. T. Kawaguchi, graduate student at Hokkaido Univ., for his help in preparing the manuscript.

References

Dobry, R. L., Ladd, R. S., Yokel, F. Y., Chung, R. M. and Powell, D. (1982): Prediction of pore pressure build up and liquefaction of sands during earthquakes by the cyclic strain method, National Bureau of Standards, Building Science Series 138, US Department of Commerce.

Dobry, R. L., and Vucetic, M.(1987): State-of-the-art report: Dynamic properties and seismic response of soft clay deposits, Proc. of the International Symposium on Geotechnical Engineering of Soft Soils, Mexico City, Vol.2, pp.51-87.

Isenhower, W. M. and Stokoe, K. H.(1981): Strain-rate dependent shear modulus of San Francisco Bay mud, Proceedings of International Conference on Recent Advances in Geotechnical Earthquake Engineering and Soil Dynamics, St. Louis, Vol.2, pp. 597-602.

Lo Presti, D. C. F., Pallara, O., Lancellotta, R., Armandi, M. and Maniscalco, R. (1993): Monotonic and cyclic loading behavior of two sands at small strains, Geotechnical Testing Journal, Vol. 16, No. 4, pp. 409-424.

Shibuya, S., Tatsuoka, F., Teachavorasinskun, S., Kong, X. J., Abe, F., Kim Y-S. and Park, C-S. (1992): Elastic deformation properties of geomaterials, Soils and Foundations, Vol.32, No.3, pp.26-46.

Shibuya, S., Mitachi, T., Fukuda, F. and Degoshi, T. (1995): Strain rate effects on shear modulus and damping of normally consolidated clay, Geotechnical Testing Journal, Vol.18, No.3, pp.365-375.

Tatsuoka, F. and Shibuya, S. (1992): Deformation characteristics of soils and rocks from field and laboratory tests, Keynote Lecture, Proc. of 9th Asian Regional Conf. on SMFE, Vol.2, pp.101-170.

Tatsuoka, F., Sato, T., Park, C. S., Kim, Y. S., Mukabi, J. N. and Kohata, Y.(1994): Measurements of elastic properties of geomaterials in laboratory compression tests, Geotechnical Testing Journal, Vol.17, No.1, pp.80-94.

Teachavorasinskun, S., Shibuya, S. and Tatsuoka, F. (1991): Stiffness of sands in monotonic and cyclic loading in simple shear, ASCE Geotechnical Engineering Congress, Boulder, Geotechnical Special Publication No.27, Vol.1, pp.863-878.

Toki, S., Shibuya, S. and Yamashita, S. (1994): Standardization of laboratory test methods to determine the cyclic deformation properties of geomaterials in Japan, Pre-failure Deformation of Geomaterials (Shibuya, S., Mitachi, T. and Miura, S. eds.), Balkema, Vol.2, pp.741-784.

Vucetic, M. (1994): Cyclic threshold shear strains in soils, ASCE Journal of Geotechnical Engineering, Vol. 120, No. 12, pp. 2208-2228.

Drained Creep Behavior of Marine Clays

Armand J. Silva[1] and Horst G. Brandes[2]

Abstract

Long-term, drained triaxial creep tests were conducted on undisturbed and reconstituted specimens of a North Central Pacific clay. The undisturbed specimens were obtained from cores taken in water depth of about 5800 m and the reconstituted samples were prepared from remixed and reconsolidated tanks of sediments. A power law, with m_a being the principal creep parameter, is a good predictor of the axial strain behavior with a constant value of m_a for up to 6-9 months, but decreasing values thereafter. The results from the reconstituted set of samples confirms that m_a is time dependent, and to a lesser degree, stress dependent. High temperature results in higher overall strain rates and lower m_a values at equivalent times, compared to a one-year undisturbed illite test. The results indicate that the strain rates would have reached the tertiary creep stage, resulting in rupture if given enough time, particularly at the higher stress level of 60% of the failure stress. The tests on both undisturbed and reconstituted samples showed that there was a significant increase in strength during long-term creep with essentially no volume change.

Introduction

Drained creep behavior of fine-grained ocean sediments has been studied by the Marine Geomechanics Laboratory (MGL) group at the University of Rhode Island for almost two decades and an extensive data set exists for a variety of sediment types and conditions. The drained creep laboratory was initially developed for the study of long-term displacements of nuclear waste canisters buried in deep sea sediments (Hollister et al, 1981) of the North Central Pacific. The study included high-temperature tests to simulate the effects of heat caused by the radioactive waste. Similar instrumentation was used in research of creep processes on continental slope sediments off the northeast coast of North America (Silva and Booth, 1986; Booth et al, 1984). The research was then expanded, with an increase in the modeling component, to include sediment samples from the Gulf of Mexico and more generic studies of time-dependent deformations of fine-grained ocean sediments. This recent

---

[1] Professor of Ocean and Civil Engineering, Director of Marine Geomechanics Laboratory, University of Rhode Island, Narragnasett, RI 02882, USA

[2] Assistant Professor of Civil Engineering, Department of Civil Engineering, University of Hawaii, Honolulu, HI 96822, USA

stage of the research was initiated as a result of interest in a particular deep water site where possible creep processes were being investigated.

Downslope displacements and failure of large masses of seabed sediments are a common occurrence on continental slopes and rises (Embly and Jacobi, 1986). Recent evidence off the eastern margin of the US adjacent to Cape Hatteras indicates that at least 50% of the continental slope and about 30% of the upper rise have been affected by mass wasting in the Late Pleistocene (Pratson and Laine, 1989). Although the dominant, gravitationally-driven processes in these provinces appear to be debris flows, rotational slumps, and translational slides (Laine, et al., 1986), there is also considerable geological and geophysical evidence that downslope sediment deformations due to creep occur on or within many submarine slopes (Damuth, 1980). By creep we mean very slow deformations of surficial or buried sedimentary layers that occur over many years under the driving force of gravity. These creep deformations may indeed be precursors to some of the more catastrophic failures that are clearly evident. From a stress-strain-time viewpoint, creep is the long-term continuing deformation due to sustained deviatoric stress conditions that occurs after dissipation of consolidation excess pore pressures.

In this paper we present a summary of key findings from the experimental program. The main thrust of the laboratory program has been to determine the long-term, drained, stress-strain behavior of ocean sediments for a variety of environmental conditions. The overall approach has included: a) a geological and geophysical analysis of sampling sites, b) a study of material behavior with controlled laboratory tests, c) constitutive modeling of material behavior, d) development of a finite element code to predict long-term deformations, and e) a laboratory physical model experiment of a simplified slope. A companion paper (Brandes and Silva, 1996) presents numerical predictions with the GEO-CP code for a typical isotropically consolidated, drained creep test from the undisturbed illite series discussed below.

Program and Equipment

The laboratory testing program has focused on determining the long-term deformation behavior of fine-grained ocean sediments under drained conditions, with most tests lasting over three months and a few over one year. The overall program has included a range of sediment types obtained from three different oceanic regimes: a) illites and smectites from the North Central Pacific (NCP) about 900 km north of Hawaii in water depth of 5800 m, b) illites from the North West Pacific (NWP) in water depth of 6100 m, c) silt-clays from the continental slope in the North Western Atlantic (NWA) off the northeastern United States in water depths of 1000 - 2000 m and d) illitic clays from the Green Canyon area, Western Gulf of Mexico (WGM) in water depth of about 600 m. A summary of the overall program along with references that contain details of the experimental work is shown in Table 1. The equipment, instrumentation, and analysis procedures have changed somewhat over the history of the program (Moran, 1981; Brandes, 1984; and Tian, 1992). In addition to the triaxial mode of testing (Fig. 1), a special direct simple shear device was fabricated to better simulate plane strain conditions present on infinite slopes (Tian et al, 1994). Some features of the triaxial testing include: use of sea water cell fluid of the same salinity as the pore fluid, back pressuring of pore fluid to assure saturation, oil-water interfaces through a burette system (used for volume change measurements), temperature controlled environment for room temperature tests, heating jacket for high temperature tests and refrigerated room for low temperature tests, dead load system for application of deviatoric loads, and automatic data acquisition of axial deformations and pressures. In the initial phases of the program, the confining and back pressures were applied through statically-loaded piston pressure cells (Moran, 1981). This was later replaced

Fig. 1 Drained triaxial creep equipment

with a pneumatic pressure regulator system (Brandes, 1984).

The "undisturbed" subsamples were obtained from box cores or gravity cores with a special piston apparatus and thin-walled stainless steel tubes. Most of the samples were taken soon after retrieval on board the ship but some were taken from linered sections at the shore-based laboratory. In all cases, care was taken to minimize disturbance due to handling and transportation and all subsamples were properly sealed to prevent dewatering. As indicated by the sharp curvature of e-log σ' plots from consolidation tests (Silva and Jordan, 1984), the samples are considered to be of very good quality with little disturbance. The reconstituted samples were obtained from large tanks of sediment that had been thoroughly mixed to a thick slurry consistency and reconsolidated to predetermined void ratios (Darlington, 1995).

The overall testing program (Table 1) included six different types of fine-grained ocean sediments, a wide range of stress levels (9-95% of the failure strength determined from triaxial compression tests), and temperatures from 4°C to 150°C. Stress level (S.L.) is the deviatoric stress defined as a percentage of the failure stress. Failure conditions (without creep) were determined from a large number of triaxial strength tests. Both isotropic and anisotropic tests were conducted for the creep and strength tests. For the creep tests, the samples are consolidated incrementally in small steps, either isotropically or anisotropically. In the anisotropically consolidated samples, the ratio of lateral to vertical stress was equal to 0.5. Each consolidation step was maintained until the volumetric strain rate was less than 10^{-5} %/min. All stresses were applied in small increments to minimize transient excess pore pressures. Adjustments were generally made to the applied load throughout the duration of each test to maintain constant deviatoric stresses. However, because of the much higher volumetric strains expected during the high temperature tests, adjustments were also made to stresses to account for the increasing strength caused by consolidation (determined from separate strength tests).

In this paper we present representative results of both room temperature (20°C) and high temperature (75°C and 150°C) drained creep tests on undisturbed and reconstituted sediments from the North Central Pacific.

Sediments Properties

The illite samples all had very similar properties with a clay content (2 μm) of about 70% and plasticity index of about 45% (Table 2). The smectite samples, obtained from a deeper zone in the same area, had a similar clay content but very different consistency limits with a plasticity index of about 135% and high natural water contents (160% and higher). Tests indicate that the surface sediments at the study site are overconsolidated, with high overconsolidation ratios near the surface, decreasing exponentially to near 10 at 0.5 m depth, and becoming normally consolidated below 10 m depth. This effect is attributed to *apparent* overconsolidation, caused by high interparticle bonding and possible cementation (Silva and Jordan, 1984). It should be noted however, that all the undisturbed creep tests were consolidated well beyond the in-situ preconsolidation stress. Permeability values range between 10^{-5} and 10^{-7} cm/s and compression indices between 0.5 and 3.1, with the smectites being more compressible than the illites. Undrained shear strength, determined from miniature vane and triaxial tests, increases gradually with depth, reaching 8 kPa at 5 m, 16 kPa at 10 m, and 32 kPa at 30 m (Silva and Jordan, 1984).

Discussion of Data Sets

The results of five sets of data representing 19 samples are shown in five figures (Figs. 2-6). Because of the differences between some of the test procedures and the fact that the data analyses techniques varied somewhat, each data set is discussed separately. However, in each case creep strains are observed to decrease exponentially with time and therefore we consider power laws of the Singh-Mitchell (1968) type:

$$\dot{\varepsilon}_d = Ae^{\alpha D}\left(\frac{t_o}{t}\right)^m \qquad (1)$$

where $\dot{\varepsilon}_d$ is the deviatoric strain rate, A, α, and m are material creep parameters, D is the stress level, and t_o is a convenient reference time. Since we are interested in long-term applications, we focus our discussion on the parameter m, which is the negative slope of log-strain rate versus log-time. This equation was originally developed from relatively short-term (1000 minutes) undrained creep tests. Singh and Mitchell (1968) note that this general relationship was also expected to apply for drained conditions and we can write a similar expression for the axial strain rate, $\dot{\varepsilon}_a$, during drained creep:

$$\dot{\varepsilon}_a = G\left(\frac{t_o}{t}\right)^{m_a} \qquad (2)$$

where G is a stress-dependent parameter, and m_a is the absolute value of the slope of the log of axial strain rate versus log of time:

$$\log\left(\frac{\dot{\varepsilon}}{\dot{\varepsilon}_o}\right) = -m_a \log\left(\frac{t}{t_o}\right) \qquad (3)$$

A similar expression to (2) can also be written for the volumetric strain. For triaxial conditions where the specimen is assumed to remain cylindrical; the axial, deviatoric, and volumetric strains are related by:

$$\varepsilon_d = \varepsilon_a - \frac{1}{3}\varepsilon_v \qquad (4)$$

The time dependent creep characteristics of the five sets of data are compared in terms of the parameter m_a. Although Singh and Mitchell (1968) propose that the parameter m is dependent only on the material type, it will be shown that m_a is generally a function of time and stress level.

Undisturbed Illite, Long Term Test

This sample was obtained during the MARA-02 cruise, core GC-08, at 31° 3.0' N and 157° 49.4' W in water depth of 5800 m. The objective was to conduct a long-term test (1-year and 19 days) at a low stress level (25%) to examine the nature of triaxial creep deformations over the very long term. The sample was consolidated isotropically to 93 kPa, followed by drained deviatoric loading to 62 kPa. The 93 kPa corresponds to an in situ effective stress at about 18 m depth in the sediment column and was chosen to assure that all samples were in the normally consolidated stress state. This load was applied all at once, hence the initial rapid increase in axial strain to above 5%. A step-loading procedure was used for all the other tests. The axial strain plot (Fig. 2a) shows a smooth transition from primary consolidation to creep. Results

DRAINED CREEP BEHAVIOR OF CLAYS 233

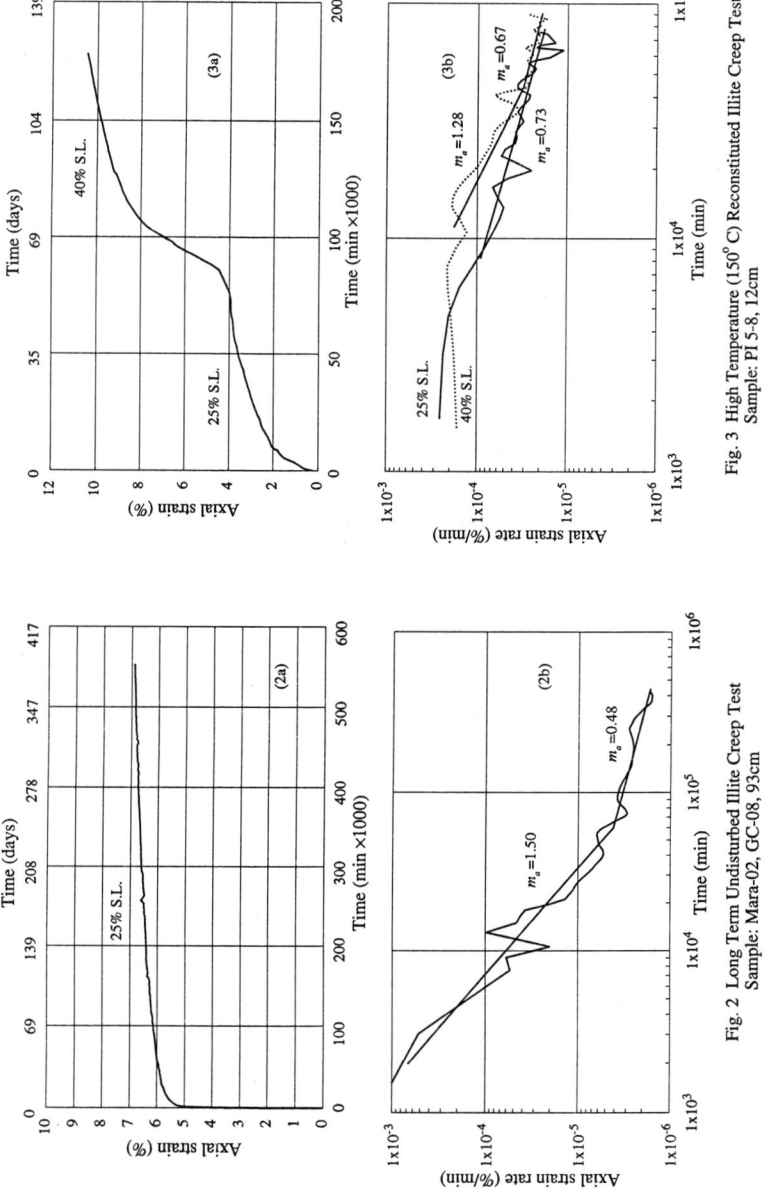

Fig. 2 Long Term Undisturbed Illite Creep Test
Sample: Mara-02, GC-08, 93cm

Fig. 3 High Temperature (150° C) Reconstituted Illite Creep Test
Sample: PI 5-8, 12cm

from the isotropic phase and numerical simulations (Brandes and Silva, 1996) indicate that most of the excess pore pressures had dissipated after 1000 minutes and therefore axial strains above 5% are due entirely to creep. Strain rates were calculated after averaging using a cubic spline fit procedure. Volumetric strain measurements were unreasonably large indicating that leakage had occurred. In general, measuring volume changes accurately over long periods of time is much more difficult, due to the possibility of leakage, than measuring axial strains.

The long-term behavior is best analyzed in terms of the parameter m_a (Fig. 2b). Contrary to the findings of Singh and Mitchell, in this test m_a was not a constant but instead varied with time. For the first 38 days $m_a = 1.50$, but from there on m_a decreases to 0.48. More accurately, m_a appears to vary parabolically with time, and the strain rate approaches a constant value of about 10^{-6} %/min. A non-linear parameter can significantly affect creep strain accumulations over engineering and geologic time frames. For example, by integrating Eq. (3), we find that after one year axial strain would amount to 6.3% for $m_a = 1.50$ and to 6.8% for $m_a = 0.48$. However, over the typical 50 year life span of marine engineering structures, axial strain increases to 6.4% for $m_a = 1.50$ and to 15% for $m_a = 0.48$. It is evident that over long time spans even small changes in m_a can have large effects on predicted axial strains. The time-dependency of m_a is an important issue that should be investigated further with additional long-term tests.

High Temperature (150°C), Reconstituted Illite Test

This test was conducted on an illite-rich fine-grained sediment that is found in the upper 5 m of the study site described earlier. It was recovered with a large-diameter gravity corer, reconstituted and reconsolidated. Two high temperature tests (75°C and 150°C) were conducted on reconstituted illite, but only the 150° C results are discussed here (the 75°C test showed similar trends). The sample was isotropically consolidated to 93 kPa, followed by five thermal consolidation steps until the temperature reached 150°C (Fig. 3). This temperature was maintained until the end of the test. Heat was applied through a heating jacket placed on the outside of the cell. Mechanical consolidation resulted in 14% volumetric strain and thermal consolidation produced an additional 10%. After consolidation, the sample was loaded deviatorically to 25% and 40% stress levels as shown in Fig. 3a. Volumetric strains were again unreasonably large, attributed to leakage caused by membrane break-down at elevated temperature. After 179,000 minutes (124 days), a 60% stress level was attempted but the sample failed abruptly.

The axial strain is seen to increase at a significant rate throughout the test compared to the room temperature test, and the strain rate never drops below 10^{-5} %/min (Fig. 3b). At 25% stress level axial strains are lower than in the room temperature test [both samples had similar water contents (Table 2)], which is attributed to the reconstitution process and the additional dewatering from the thermal consolidation. At 25% stress level the parameter m_a drops from 1.28 to 0.67 after 28 days, whereas at 40% stress level it remains more or less constant at 0.73. In general, m_a values below 1.0 were only observed after long periods of time and at elevated temperatures. All elevated temperature tests clearly indicate large strain rates compared to room temperature tests. This is as expected from rate process theory (Mitchell, 1976) and was also observed by Murayama (1969).

High Temperature (75°C), Reconstituted Smectite Test

Smectite-rich clay is found at the study site below 10 m. Although the concentration of smectite is only 5% (1% or less in the illitic clay), the clay has higher natural water content, Atterberg Limits, and specific gravity (Table 2) compared to the illite. A total of four tests were conducted at 75°, 100° (2 tests), and 150°C. Results for the 75°C test are shown in Fig. 4. The sample was isotropically consolidated to 93 kPa, followed by thermal consolidation to 75°C. Mechanical consolidation resulted in 12% volumetric strain and additional thermal consolidation of 11%.

The volumetric strains are considered reliable since the cumulative volume changes compare very closely to those calculated from the measured changes in water content during the test. Volumetric strain is about 1% higher than axial strain throughout most of the deviatoric phase, which is contrary to the findings in all the room temperature tests where axial strain is always larger than volumetric strain (Fig. 4). The level of axial strain at 20% and 40% stress is lower than for the 150°C illite test, but is similar to that of most of the room-temperature tests on illite. However, both axial and volumetric strains increase dramatically during the 60% stress level stage. The log of axial strain rate again varies non-linearly with the log of time (Fig. 4b), indicating that the m_a parameter is a function of time, whereby there is a gradual reorientation of the sediment fabric. After some 14 days m_a decreases to 0.54, 0.31 and 0.42 in each of the 20%, 40% and 60% stress level stages respectively. It should be expected that m_a decreases with increasing stress level. A closer examination of the 40% and 60% stress level tests (Fig. 4b) shows a clear trend toward essentially zero slope after 30,000 and 70,000 minutes respectively. Therefore, it is possible that if these two stress levels had been maintained longer, the samples would have experienced creep rupture. The m_a parameter varies somewhat with deviatoric stress level, but the trends are not clear.

Undisturbed Illite, Box Core Samples

A set of ten specimens were obtained from a box core (50 cm x 50 cm) sample of illitic clay from the NCP area at 30° 17' N, 157° 50' W. The specimen samples were obtained at the same depth (23 cm) with a special piston sampler, capped, sealed, and stored in sea water at 4°C until testing. This set of creep tests are considered to be on essentially identical samples with very little disturbance. The samples had the following properties: clay fraction (2 µm) = 68%, I_p = 46%, CAU strength parameters of ϕ' = 26.5° and c' = 3.75 kPa, C_c = 1.10, and overconsolidation ratio of about 10 (*apparent* overconsolidation). This apparent overconsolidation is typical of surficial (upper few meters) of fine-grained ocean sediments, which is attributed to interparticle bonding and sometimes cementation (Silva and Jordan, 1984).

Four samples were consolidated anisotropically (K_o = 0.5) to a normally consolidated condition and tested at various stress levels ranging from 25 to 95% in the triaxial creep apparatus under drained conditions. Three of the tests were loaded to two deviatoric stresses and one was a longer term test loaded to a single 95% stress level (Fig. 5). Actually, each deviatoric creep loading was applied very gradually over a period of 4 to 20 days to minimize excess pore pressures (Brandes and Silva, 1996).

The increments of axial strain at 10^4 days, beyond that achieved at completion of deviatoric loading, were 1.7, 2.2, 2.5, and 2.2% at 25, 40, 70, and 95% stress levels respectively. This suggests that time-dependent axial deformations beyond primary consolidation were relatively insensitive to the applied stress. Except for the 25% stress level, the volumetric strain behavior showed a similar pattern.

In this series of results, a power law fitting procedure was used to describe the time-dependent axial strain behavior with the following form:

$$\varepsilon_a = F + E\left(\frac{\bar{t}+t_c}{t_o}\right)^{1-m_a} \tag{5}$$

where t_c is a cutoff time taken as 2000 minutes for this series, which essentially removes the consolidation strains from consideration, E and F are creep parameters,

236 TIME DEPENDENT SOIL BEHAVIOR

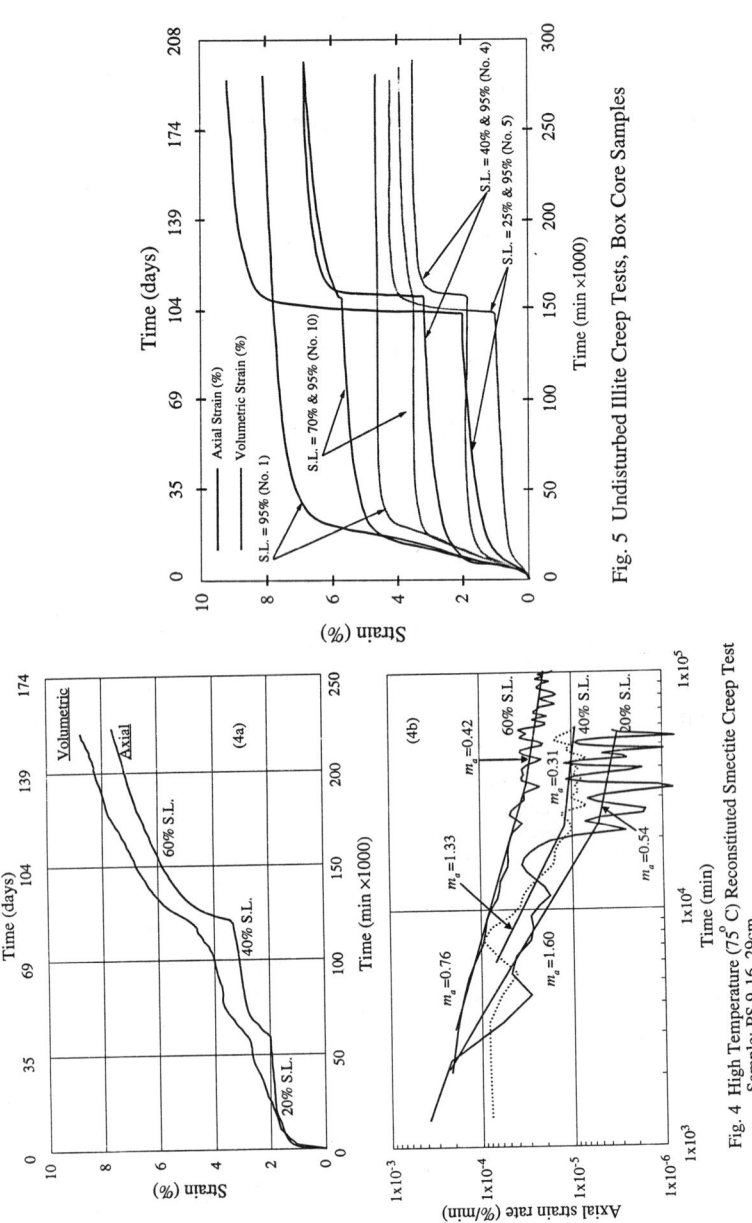

Fig. 5 Undisturbed Illite Creep Tests, Box Core Samples

Fig. 4 High Temperature (75° C) Reconstituted Smectite Creep Test
Sample: PS 9-16, 29cm

and $\bar{t}$ is the time after application of the final load. The first derivative of (5) with respect to time yields a relation similar to Eq. (2). The results shown in Fig. 5 represent best fit lines using Eq. (5) for axial strain and an equivalent equation for volumetric strain. In each case the fits are very good (Tian, 1992). The parameter m_a shows little variation for the four different stress levels (Table 3). However, the corresponding slope (m_v) for the volumetric data showed a significant trend of increasing values with increasing stress level, ranging from 0.78 for 25% S.L. to 1.63 for 95% S.L. The results suggest a tendency for the m and G parameters (Eq. 2) to be stress level dependent for volumetric strains. The trends for the first and second load steps were similar.

By fitting the data with relationship (5) we can not draw any conclusions as to the long term time dependency of m_a. However, the longer term test (sample No. 1, S.L. = 95%) appears to show a tendency for approaching a nearly constant strain rate toward the end of the test (195 days). At this time, the axial strain rate was about 10^{-6} %/min. This is less than that observed in the long term test of Fig. 2 for comparable times, but the strain rates near the end of the tests are similar at about 10^{-6} %/min.

Room Temperature, Reconstituted Illite Tests

Because of problems in obtaining replicate samples of undisturbed ocean sediments, two batches of reconstituted illitic sediments were prepared especially for this portion of the creep program (Darlington, 1995). One of the objects of this series of tests was to study the apparent increase in strength caused by creep which was observed by Tian (1992). A series of fourteen anisotropically consolidated, drained triaxial creep tests were conducted at three stress levels (18, 29, and 51%) for three different time intervals (approximately 5, 10, and 24 weeks) to examine both creep behavior and effects on strength due to different time durations (Fig. 6).

Each batch of sediment was consolidated to a vertical stress of 50 kPa. The average initial water contents of the two sets of samples was slightly different at 80 and 76% (corrected for 35 ppt salinity) and the corresponding void ratios were 2.27 and 2.07. All samples were back pressured to 415 kPa and consolidated anisotropically (K_o = 0.5) to a vertical stress of 83 kPa before applying deviatoric creep loadings. Therefore, all samples were brought into the normally consolidated stress state before proceeding with the creep tests.

As indicated in Figs. 6a and 6b, problems were experienced during some of the tests and therefore there are two gaps in results. The most obvious omission is the lack of a long term (24 week) test at the 51% S.L., caused by equipment failure and limited time to begin a new test. In addition, the volumetric data for the 24 wk (29% S.L.) test is omitted because of leakage problems. Following is a summary of the test program:

Stress Level (%)	Test durations (wks)
18	5*, 6.5, 9.5*, 10, 24
29	5*, 10*, 10, 24**
51	5, 8.5, 10, 10

Notes: * Curves overlap, Fig. 6a
** Volumetric data omitted, Fig. 6b

The best fit lines for axial and volumetric data (Figs. 6a and 6b) were determined with the creep power law of Eq. (5). For most tests, the data and best fit lines are virtually identical for times greater than t_c (Darlington, 1995).

The fitted creep data for each creep stress level were averaged to arrive at a representative axial and volumetric strain curve for each of the three stress levels and

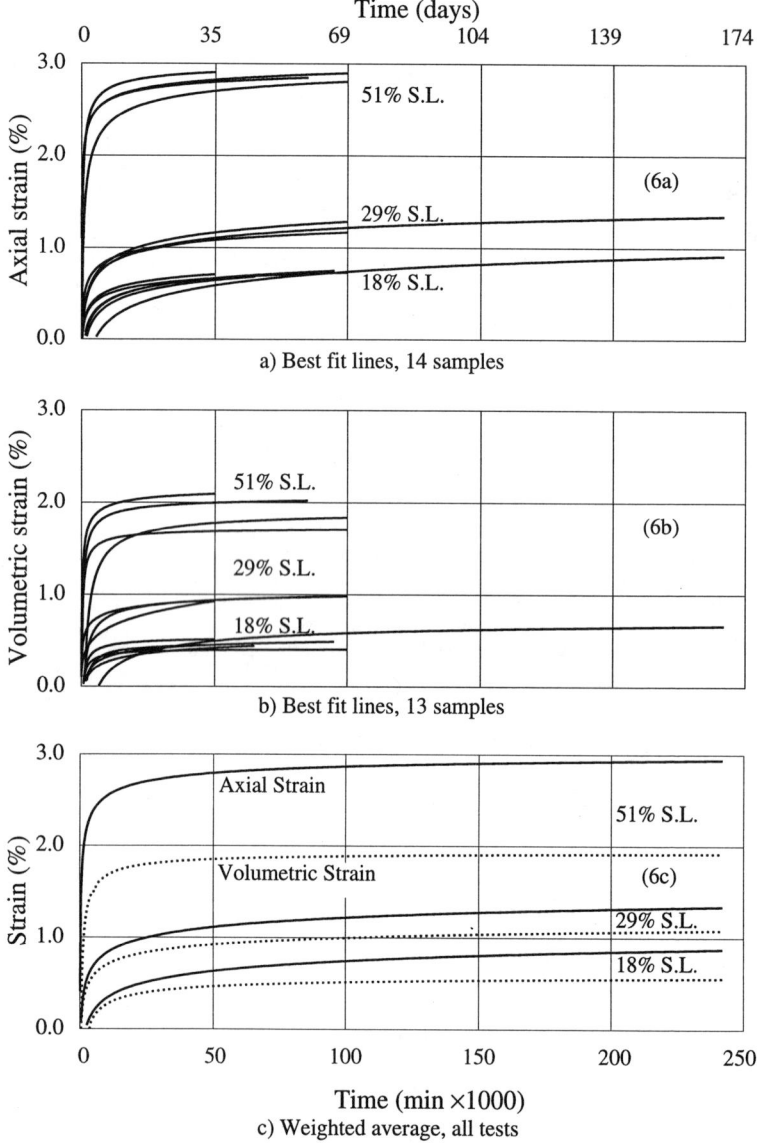

Figure 6. Reconstituted Illite Creep Tests

extrapolated to 24 weeks where necessary. A weighting factor was used in the averaging procedure in direct proportion to the length of the test (i.e. 5, 10, and 24 weeks), such that the longer tests were assigned greater importance. The resulting curves (Fig. 6c) are considered to represent drained creep behavior of these reconstituted NCP illitic sediments. The corresponding m_a parameters (Table 3) for the three stress levels range from 1.11 to 1.33 with an average of 1.19 (Table 3). These results compare very well with those of the undisturbed box core samples which had an m_a range of 1.01 to 1.19 and average of 1.10. The smaller total strains observed in the reconstituted samples are due primarily to the lower initial void ratios and higher consolidation stresses.

Preliminary CAU triaxial strength tests yielded an effective angle of internal friction (ϕ') of 29° and zero cohesion. Ten of the creep samples were tested in the triaxial compression mode after completion of the creep tests to examine any differences in stress-strain/strength behavior. All the creep samples showed significant increases in strength, beyond that predicted by changes in void ratio, with ϕ' angles of 35 to 37.9° (an increase of 6 to 8°). The average strength after creep was 36% greater than that determined from the preliminary CAU strength tests (Tian, 1992).

The results of this set of data on reconstituted illite did not compare well with the behavior observed by Singh and Mitchell (1968). Darlington (1995) observed that the slope of log-axial strain rate versus stress is not a constant but instead varies with time. He proposed the following relationship:

$$\ln \dot{\varepsilon}_a = n_t (S.L.) + k_t \qquad (6)$$

where n_t and k_t are functions of time.

Summary and Conclusions

There are both scientific and engineering interests in understanding and being able to predict long-term deformations of marine slope sediments. Geologically, creep processes have relevance for interpretation of seabed morphology and as a possible precursor to other mass-wasting processes. From an engineering point of view, slow downslope movements can have important implications for site selection and foundation design of offshore installations.

The experimental results presented here represent an extensive data set on the long-term, drained, stress-strain (creep) behavior of deep sea clays from the North Central Pacific. Following is a summary of key findings and conclusions.

- In the one-year illite and the two high-temperature tests, where strain rates were calculated directly from the data, the creep parameter m_a varies with time, and to a lesser degree, with stress level. Values of m_a decrease from above 1.0 during the first 35 days to values well below 1.0 for longer times. The strain rate approaches a constant value of about 10^{-6} %/min. For time spans of 50 years or more, this level of strain rate could result in large deformations, possibly leading to failure. It is noted that this parameter is material specific and the results presented are for this particular deep sea clay.
- The undisturbed box core and reconstituted tank test series on NCP illite indicate a small dependence of m_a on stress level. Values of m_a range between 1.01 and 1.33 for stress levels from 18 to 95% (Table 3). There does not appear to be much of a difference in terms of m_a between the undisturbed and the reconstituted samples, suggesting that the reconstitution process was successful in reproducing the in-situ creep characteristics of the material.

- High temperature results in higher overall strain rates and lower m_a values at equivalent times, compared to the one-year undisturbed illite test. It is quite possible that creep would have reached the tertiary stage resulting in rupture if given enough time, particularly at the higher stress level of 60%.
- The creep samples showed increases in strength of about 36% beyond that accounted for by changes in void ratio, indicating that the gradual changes in microstructure during creep result in strengthening.

Acknowledgments

Support for this research was provided mainly by the National Science Foundation and the Department of Energy. Some initial work was also sponsored by Shell Oil Company

References

Booth, J.S., Silva, A.J. and Jordan, S.A. (1984). Slope-stability analysis and creep susceptibility of quaternary sediments on the northeastern United States continental slope. In: *Seabed Mechanics*; B. Denness (Ed.), Proceedings of IUTAM and IUGG Symposium; London: Graham and Trotman, pp. 65-75..

Brandes, H.G. (1984). The drained triaxial creep behavior of deep sea clays. *M.S. Thesis*, Department of Ocean Engineering, University of Rhode Island.

Brandes, H.G and Silva, A.J. (1996). Simulation of pore pressures in triaxial creep tests. ASCE Washington Convention Session, *Measuring and Modeling Time-Dependent Soil Behavior*, this volume.

Damuth, J.E. (1980). Use of high-frequency (3.5-13 kHz) echograms in the study of near-bottom sediments in the deep sea: A Review. *Marine Geology*, 38:51-75.

Darlington, C. (1995). Creep behavior of reconstituted north central Pacific illitic sediments. *M.S. Thesis*, Department of Ocean Engineering, University of Rhode Island.

Embly, R.M. and Jacobi, R. (1986). Mass wasting in the western north Atlantic. *The Geological Society of America*, M:479-490.

Fagan, B. (1985). Effect of drainage conditions on creep behavior of deep sea clay. *M.S. Thesis*, Department of Ocean Engineering, University of Rhode Island.

Hollister, C.D., Anderson, D.R. and Heath, G.R. (1981). Subseabed disposal of nuclear waste? *Science*, 38: 1321-1326.

Laine, E.P., Damuth, J.E. and Jacobi, R. (1986). Surficial sedimentary processes revealed by echo- character mapping in the western north Atlantic ocean. *The Geological Society of America*, M:427-436.

Mitchell, J.K. (1976). *Fundamentals of Soil Behavior*. New York: John Wiley and Sons.

Moran, K. (1981). Drained creep of deep sea sediments. *M.S. Thesis*, Department of Ocean Engineering, University of Rhode Island.

Murayama, S. (1969). Effect of temperature on elasticity of clays. *Highway Research Board Special Report No. 103*, pp. 194-203.

Pratson, L.F. and Laine, E.P. (1989). The relative importance of gravity-induced versus current-controlled sedimentation during the quaternary along the mideast US outer continental margin revealed by 3.5 kHz echo character. *Marine Geology*, 89:87-126.

Silva, A.J., Akers, S. and Moran, K. (1983). Stress-strain-time behavior of deep sea clays. *Canadian Geotechnical Journal*, 20(3):517-531.

Silva, A.J. and Jordan, S.A. (1984). Consolidation properties and stress history of some deep sediments. In: *Seabed Mechanics*, B. Dennes (Ed.), Proceedings of

IUTAM and IUGG Symposium; London: Graham and Trotman, pp. 25-39.
Silva, A.J., Brandes, H.G. and Maswose, J.J. (1985). Geotechnical studies for subseabed disposal of high level radioactive wasters. *Progress Report No. 11*: FY 84 prepared or Sandia National Laboratories, New Mexico.
Silva, A.J. and Booth, J.S. (1986). Creep behavior of submarine sediments. *Geo-Marine Letters*, 4:215-219.
Singh, A. and Mitchell, J.K. (1968). General stress-strain-time function for Soils". *Journal of the Soil Mechanics and Foundations Division, ASCE*, SM1:21-46.
Tian, W-M. (1992). Long Term creep behavior of cohesive marine sediments under drained conditions. *Ph.D. Thesis*, Department of Ocean Engineering, University of Rhode Island.
Tian, W-M., Silva, A.J., Veyera, G.E. and Sadd, M.H. (1994). Drained creep of undisturbed cohesive marine sediments. *Canadian Geotechnical Journal*, 31:841-855.
Wildman, J.C. (1981). Effect of temperature on creep behavior of deep sea sediments. *M.S. Thesis*, Department of Ocean Engineering, University of Rhode Island.

Table 1. URI Creep Testing Program

Sediment Origin	Primary Clay Mineral	Type of Creep Tests	Number of Tests	Reference
North Central Pacific (5800 m)	Undist. Illite	CID	4	Moran (1981)
		CID-VT	4	Wildman (1981)
		CID	4	Brandes (1984)*
		CID	2	Silva et al. (1985)
		CID	4	Tian (1992)*
		CIU	4	Fagan (1985)
	Recon. Illite	CAD	14	Darlington (1995)*
		CID-HT	2	Brandes (1984)*
	Undist. Smectite	CID	3	Moran (1981)
	Recon. Smectite	CID	1	Moran (1981)
		CID-HT	2	Brandes (1984)*
		CID-HT	2	Silva et al. (1986)
NW Pacific (6100 m)	Undist. Illite	CID	1	Brandes (1984)
W Gulf of Mex. (600 m)	Undist. Illite	CAD	12	Tian (1992)*
NW Atlantic (1000-2000 m)	Undist. Illite	CID	3	Booth et al. (1984)

*Results presented in this paper

Table 2. North Central Pacific (NCP) Sediment Properties

Test Series	Water Content (%)	Liquid Limit	Plastic Limit	Silt/Clay (%)	Specific Gravity
Undisturbed Illite	93	86	40	33/67	2.76
Reconstituted Illite	78	81	43	30/70	2.77
High Temp Illite (Recon.)	85	90	38	30/70	2.82
High Temp Smectite (Recon.)	160	230	95	31/69	2.87

Table 3. Summary of Axial Strain Rate Creep Parameter m_a

Test Series		Stress Level (%)	Parameter m_a
Undisturbed Illite	Long-term test	25	1.50 - 0.48
Undisturbed Illite (Box Core) Tests	No. 5	25	1.01
	No. 4	40	1.19
	No. 10	70	1.15
	No. 1	95	1.03
Reconstituted Illite	6 tests	18	1.11
	4 tests	29	1.12
	4 tests	51	1.33
High Temperature Illite (Reconstituted)	150 Deg C.	25	0.73
		40	1.28 - 0.67
High Temperature Smectite (Reconstituted)	75 Deg C	20	1.60 - 0.54
		40	1.33 - 0.31
		60	0.76 - 0.42

Rate-dependent deformation of structured natural clays

Kenichi Soga[1], Assoc.M., ASCE and James K. Mitchell[2], Hon.M., ASCE

Abstract

Structured natural soils exhibit large time-dependency with respect to deformation as a result of softening or destruction of the metastable soil structure. Strain rate controlled triaxial compression tests and drained triaxial creep tests were performed on undisturbed Pancone clay samples from Pisa, Italy to investigate the time- (or rate) dependency of the clay. The test results were then used along with other published data to examine the rate-dependency of yield stresses for soils with different degrees of metastability. It was found that both the preconsolidation pressure and the undrained shear strength of a soil became more deformation rate-dependent as the degree of metastability increased.

Introduction

Soil fabric is the form in which soil particles associate and arrange. Natural soils have a certain soil fabric which developed over geologic time under the influences of both the depositional environment and post-depositional processes. The stability of soil fabric is sensitive to changes in applied forces and in the chemical environment. Both soil fabric and its stability control the mechanical behavior of natural soils. In this paper, the term soil structure is used to describe the combined effects of soil fabric and its stability.

At a given effective stress, soil can have different void ratios depending on its soil structure. The influence of soil structure on void ratio under a given effective stress is shown in Fig. 1 (Mitchell, 1993). At point A, the soil has a large void ratio, because the particles are arranged in edge-to-edge and edge-to-face associations. When the arrangement is disrupted by remolding and reworking, the void ratio will decrease towards point B in Fig. 1, which is the state of fully destructured soil. At point B, the soil is considered to be stable, and the void ratio at this state is determined by the

---

[1] Lecturer, University of Cambridge, Engineering Department, Trumpington St., Cambridge CB2 1PZ, U.K.
[2] University Distinguished Professor, Virginia Tech., Blacksburg, VA, 24061-0105, USA

densest possible packing of particles at the given effective stress. If the state of a soil is above the line of fully destructured soil, it will have some degree of metastability, and the soil can be further consolidated if disturbed and recompressed. Hence, a soil that has a large open fabric is likely to be metastable, and it is defined as a structured soil in this paper. A soil with particles arranged in an efficient way is stable and defined, and it is defined as a destructured soil.

Past studies show that natural structured clays exhibit significant time-dependent deformation behavior; e.g., Graham *et al.*, 1983, Leroueil *et al.*, 1985, and others. When soil is subjected to a constant stress it continues to deform with time. This phenomenon is termed creep. The inverse phenomenon, stress relaxation, involves a decrease in stress with time when a soil is maintained at a particular constant strain level. Creep and relaxation are two manifestations of the same phenomenon; i.e., the time-dependent change of clay structure. When a metastable soil structure breaks down under a stress and the accompanying deformation, it consolidates and becomes more stable if drainage is possible, whereas, under undrained conditions, there is an increase in pore water pressure, decrease in effective stress, and loss of strength.

For the study described in this paper, strain rate controlled triaxial compression tests and drained triaxial creep tests were performed on undisturbed Pancone clay samples from Pisa, Italy to investigate the time- (or rate-) dependency of the clay. The test results are then used along with other published data to examine the magnitude of rate-dependency on yield stresses for soils with different degree of metastability.

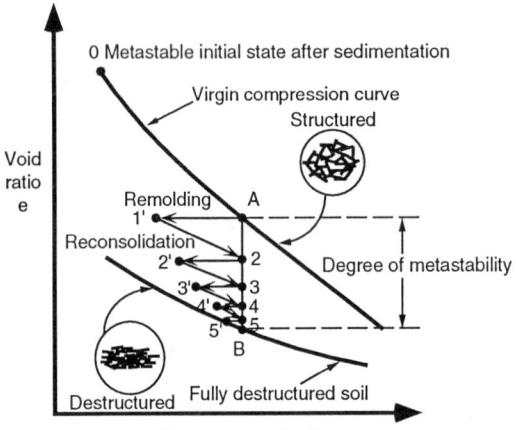

Figure 1 The influence of metastable fabric on void ratio under an effective consolidation pressure (after Mitchell, 1993)

Experimental investigation on undisturbed Pancone clay samples

Undisturbed Pancone clay samples were obtained from a location several tens of meters to the southwest of the Pisa Tower in Italy. The sampling was done using the Laval sampler (La Rochelle *et al.*, 1981). The sample depths and their physical properties are listed in Table 1 (Calabresi *et al.*, 1992).

Table 1 Pancone sample identification and the physical properties

Sample name	depth (m)	w0 (%)	LL (%)	PI
10m	10.69-10.89	59.5-66.1	92	48
14m	14.73-14.93	38.7-40.4	52	21
19m	19.22-19.42	61.0-63.5	96	55

Strain rate controlled undrained triaxial compression tests

Undrained triaxial compression tests were performed on the samples to examine the undrained stress-strain-time relationship. Axial load was measured under two different axial strain rates, 0.05 and 0.005%/min, after initial anisotropic consolidation to the estimated in-situ stresses. The approximate size of the cylindrical specimens used in the tests was 5 cm in diameter and 10 cm in height. A strip filter paper around the specimen allowed radial drainage during initial consolidation.

The measured stress-strain curves, stress paths, and pore pressure generation curves are shown in Fig. 2. All samples exhibited some rate effect on the stress required to cause deformation. The maximum deviator stress at failure increased about 8.6~10.9%, for a one order of magnitude increase in strain rate. This result is consistent with other data on the effect of strain rate (Kulhawy and Mayne (1990)).

The measured stress paths in Fig. 2 (e) show that the slope of the failure line M is independent of the applied rate of strain. Therefore, the undrained strength decrease with decreasing strain rate is due to creep-driven pore pressure development under a long period of loading, as was also found in past studies (Richardson and Whitman, 1963; Akai *et al.*, 1975; Vaid and Campanella, 1977; Vaid *et al.*, 1979; Nakase and Kamei, 1986).

For all samples except the 19 m sample, pore pressure generation was independent of strain rate and depended only on axial strain, as shown in Fig. 2 (b). Thus, pore pressure buildup is strain driven, as noted by Lo (1969), Akai *et al.* (1975) and others. This, coupled with the observation that M was independent of strain rate, leads to the additional interpretation that strain at failure should increase with decreasing strain rate. As it was not possible to determine the exact failure strain from the constant strain rate triaxial tests, further testing is needed to evaluate this hypothesis.

When the specimen from 19 m depth approached failure, a higher pore pressure was measured at a given axial strain for deformation at a slower rate, as shown in Fig. 2(d). Several sudden stress drops observed in the deformation curve, Fig. 2 (c), were due to failure along slickensides observed in the sample. When a distinct shear zone began to form, local variation of excess pore pressure developed within the specimen, and different pore pressures were measured at the bottom of the specimen

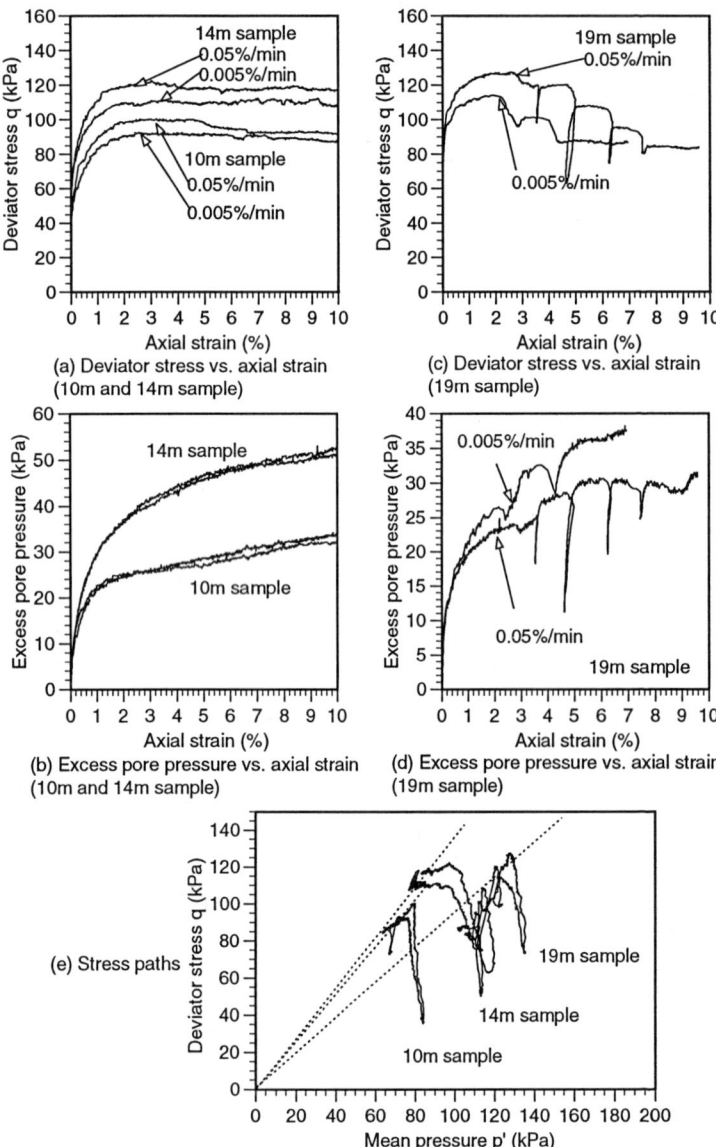

Fig. 2 Test results of undrained compression tests on Pancone clay samples

due to incomplete pore pressure equalization within the specimen. Similar observations were made by Lefebvre and LeBoeuf (1987) for a very high rate of shearing. Therefore, the large rate dependency observed after 1% axial strain was considered not to be a property of the intact soil.

Drained creep triaxial tests

Drained triaxial creep tests were performed on undisturbed Pancone clay to investigate the following three aspects of the drained creep behavior of the clay : (a) effect of plasticity, (2) effect of stress magnitude, and (3) effect of stress ratio *(q/p')*.

The approximate size of the cylindrical specimens was 3.4 cm in diameter by 8.9 cm in height. Drainage was allowed at the top and bottom of the specimen (one dimensional flow). Throughout the test, a back pressure of 98 kPa was applied to maintain saturation. Volume change was measured by the amount of water expelled from the specimen, determined using a volume gauge transducer. However, it was found that the volume change data was unreliable because of leakage of water through the membrane and generation of gas within the sample due to its high organic content of 2.2% .

Five creep tests were conducted on Pancone clay samples as shown in Table 2. For all tests, the magnitude of applied load was increased in increments, but the stress ratio (q/p') remained approximately the same for each load increment. In Tests 1~3, a stress ratio corresponding to the estimated K_0 condition was applied on the 10m, 14m and 19 m specimens, whereas, a stress ratio larger than K_0 was applied on the 14m sample in Tests 4 and 5. All the initially overconsolidated samples became normally consolidated after the application of the first load increment, and the samples exhibited large deformations in the normally consolidated region.

The loading duration of each step was 7 days. At each loading increment, the load was applied instantly. Thus, a sample experienced stages from undrained deformation at the beginning of the loading, partially drained consolidation, to fully drained conditions after the dissipation of excess pore pressure. Primary consolidation required not more than one day even for the highly plastic 10m and 19m samples.

Table 2 Summary of creep tests

	soil	Initial	Load step 1		Load step 2		Load step 3		Load step 4	
		(p',q)	(p',q)	q/p'	(p',q)	q/p'	(p',q)	q/p'	(p',q)	q/p'
1	10m	105,9	154,78	0.51	199,106	0.55	218,121	0.56	251,148	0.58
	εa/logt		1.16		1.19		1.16		1.31	
2	14m	125,18	166,78	0.47	206,108	0.52	215,116	0.54	223,126	0.56
	εa/logt		0.286		0.418		0.310		0.331	
3	19m	151,31	168,86	0.51	206,108	0.52	212,111	0.52	215,115	0.53
	εa/logt		0.66		2.22		1.47		1.14	
4	14m	135,49	163,106	0.65	190,135	0.71	199,147	0.74	206,157	0.76
	εa/logt		0.475		0.650		0.650		0.775	
5	14m	135,49	163,106	0.65	190,135	0.71	195,137	0.71	190,127	0.67
	εa/logt		0.366		0.590		0.321		0.067*	

* unloading resulted in small creep rate

The measured axial strain versus time curves for the 10m sample are shown in Fig. 3. The creep rate was determined from the slope of the axial strain versus logarithm of time relationship after the end of primary consolidation of approximately 1,000 minutes. The slope was determined by assuming a straight line in the time range of 5,000 minutes to 10,000 minutes after load application, which should be well into the stage of secondary compression.

The measured logarithmic axial strain rates are listed in Table 2. For the 14m and 10m samples, the logarithmic axial strain rates were approximately the same throughout the test, which reflects the fact that at each stage the q/p' was about the same. The logarithmic axial strain rate was independent of magnitude or increment of loads in the normally consolidated region. As a soil becomes more plastic, the rate increased at a given q/p'. The measured creep rate of the 10m sample was approximately 1.3% per logarithmic time cycle, 3 times larger than that of the 14 m sample. For the 19m sample, a large creep rate was observed after Step 2 loading. It was found from the undrained compression tests that the 19m sample was highly structured, which might have contributed to this irregular creep behavior.

The logarithmic axial strain rate increased with q/p' for the 14m samples as shown in Fig. 4. The average axial strain rate was 0.5% per logarithmic time cycle under q/p' of approximately 0.7, whereas the rate was approximately 0.3% per logarithmic time cycle for q/p' of 0.5. Similar results are reported by Ladd et al. (1977). Therefore, for normally consolidated soils, the magnitude of the creep rate is determined principally by the current stress ratio.

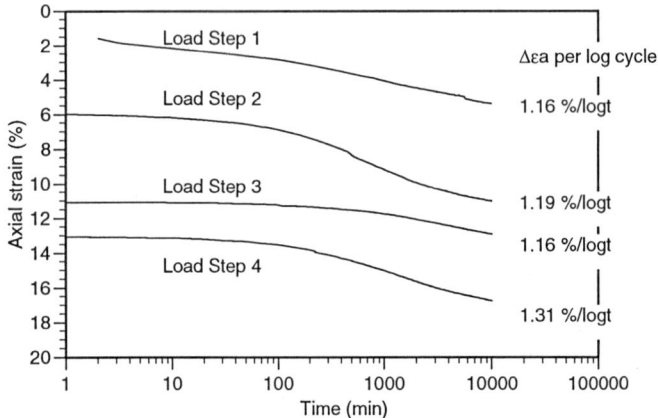

Fig. 3 Settlement curves from drained creep triaxial tests on 10m Pancone clay sample

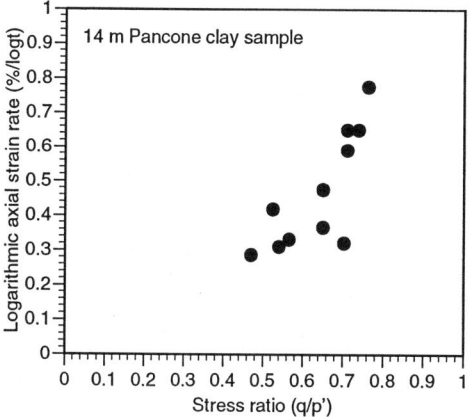

Fig. 4 Logarithmic axial strain rate vs. stress ratio (q/p') of 14m Pancone clay samples

Effect of soil structure on rate dependent deformation

Degree of metastability

A soil that has a large open network is considered to be metastable and defined as a structured soil in this study. Remolding turns a metastable structured soil into a more stable destructured soil, because soil structure breaks down by disturbance. Many structured natural soils, particularly marine soils, are metastable because of their initial flocculated condition and chemically rich environment.

Normalizing parameters, such as sensitivity and liquidity index, have been used in the past to compare that degree of metastability of a variety of structured soils. Houston and Mitchell (1969) showed that there was an unique normalized compression curve for fully destructured soil; i.e., sensitivity of 1.0, in terms of liquidity index (LI) vs. logarithmic effective stress ($\log\sigma$) as shown in Fig. 5(a). Burland (1990) found a similar relationship, which he called "intrinsic behavior".

In this paper, the degree of metastability is measured by comparing the existing condition to the equivalent state of the fully destructured soil. A schematic diagram of the compression behavior of a structured soil is shown in Fig. 5(b). The preconsolidation pressure of the structured soil is point P with the liquidity index of LI_p. The position of P compared to the state of fully destructured soil as given by Houston and Mitchell (1969) provides a measure of the degree of metastability of the soil. Therefore, the degree of metastability (MI) is quantified by the following index.

$$MI = LI_p - LI_0 \qquad \text{Eq. (1)}$$

where LI_p is the liquidity index of a given soil at the preconsolidation pressure and LI_0 is the liquidity index of fully destructured soil at the same effective pressure.

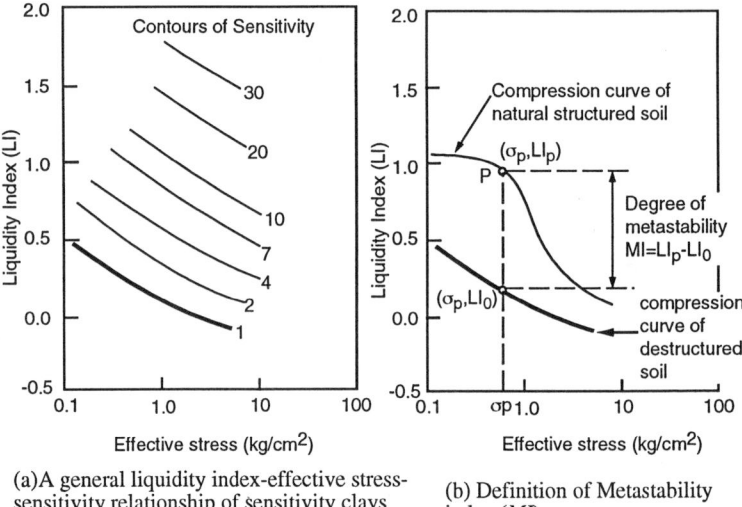

(a) A general liquidity index-effective stress-sensitivity relationship of sensitivity clays (after Houston&Mitchell (1968))

(b) Definition of Metastability index (MI)

Fig. 5 The degree of metastability of structured clay

This metastability index characterizes the soil structure which depends both on the present state of stress and openness of the fabric. The degree of metastability of a given soil can be obtained from the void ratio (or water content) and the preconsolidation pressure determined from the consolidation test. In some reported cases of rate effects on deformation, the void ratio at the preconsolidation pressure was not described in the literature. In this situation, the initial water content was used to compute the metastability index of a reported soil because the decrease in water content during recompression was considered to be small in the overconsolidated range.

Effect of soil structure on rate dependent preconsolidation pressure

If it is assumed that the relationship between strain and logarithmic time during secondary compression is linear over the time ranges of interest and the secondary compression index C_α is constant regardless of load, contours of equal axial strain rate in the ε_l-$\log\sigma_v$ plane can obtained as shown in Fig. 6. If the test were performed with a constant rate of loading, the state of the soil follows the compression line for that particular rate. Thus the location of both the virgin compression line and the value of the preconsolidation pressure σ_{vp} are strain rate dependent, and the values determined in the laboratory are influenced by the rate of loading during the one-dimensional consolidation test, as shown by Graham *et al.* (1983) and Leroueil *et al.* (1985).

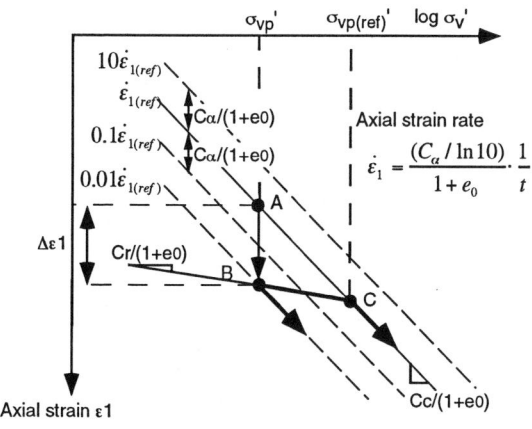

$$\Delta \varepsilon_1 = \frac{C_a}{1+e_0}\log\left(\frac{\dot{\varepsilon}_1}{\dot{\varepsilon}_{1(ref)}}\right) = \frac{C_c - C_r}{1+e_0}\log\left(\frac{\sigma_{vp}}{\sigma_{vp(ref)}}\right)$$

Fig. 6 Axial strain-stress-strain rate relationship of soils in one-dimensional consolidation test

From simple geometry shown in Fig. 6, the rate dependent preconsolidation pressure σ_{vp} can be related to the axial strain rate.

$$\frac{\sigma_{vp}}{\sigma_{vp(ref)}} = \left(\frac{\dot{\varepsilon}_1}{\dot{\varepsilon}_{1(ref)}}\right)^{\frac{C_a}{C_c - C_r}} = \left(\frac{\dot{\varepsilon}_1}{\dot{\varepsilon}_{1(ref)}}\right)^{\alpha} \qquad \text{Eq. (2)}$$

where C_c is the virgin compression index, C_r is the recompression index and $\sigma_{vp(ref)}$ is the preconsolidation pressure at a reference strain rate. A similar relation was proposed by Mesri and Choi (1984) and Silvestri et al. (1986), and the rate dependency of preconsolidation pressure is a function of creep rate C_α as well as the plastic deformation represented by C_c-C_r. In Eq. (2), the rate effect increases with the value of α (=$C_\alpha/(C_c-C_r)$).

Available published data were compiled to examine the effect of metastability on rate-dependent preconsolidation pressure (see Appendix A). The variation of preconsolidation pressure at different strain rate is plotted in Fig. 7 (a). The data define straight lines, and the slope of the lines give the parameter α.

The effect of the degree of metastability on rate dependent preconsolidation pressure is examined by plotting the α parameter against the metastability index (MI),

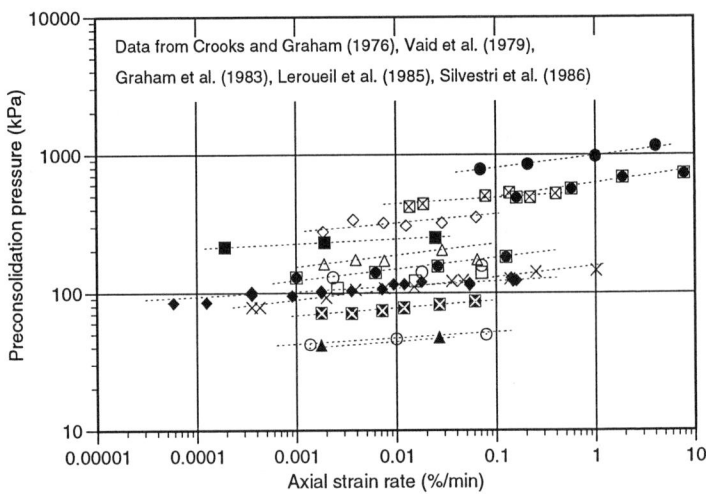

(a) Rate dependent preconsolidation pressure from 1-D constant strain rate test

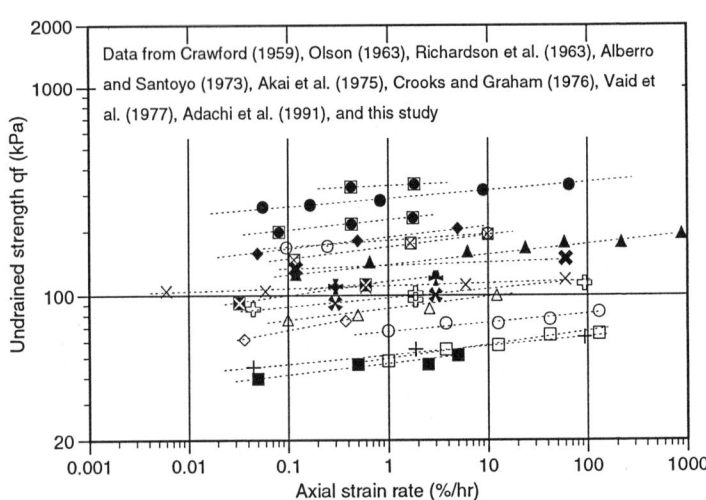

(b) Rate dependent undrained shear strength from constain strain rate CU test

Fig. 7 Rate dependency on preconsolidation pressure and undrained shear strength

as shown in Fig. 8. The more metastable the structure, as indicated by an increase in *MI* value, the greater the rate dependency, as indicated by increasing α value.

Effect of soil structure on rate dependent undrained shear strength

A structured soil will exhibit high rate dependency of undrained strength if it is initially consolidated close to the preconsolidation pressure where the soil structure is most likely to be in a metastable state. When structured soil is sheared rapidly, the structure offers more resistance to the applied load. But when it is sheared slowly, the structure breaks down as a result of more creep. Consequently, larger pore pressures develop at a given deviator stress, and undrained strength decreases as a result of decreased effective confining stress. Thus, a structured soil that is initially consolidated to a metastable condition will exhibit a greater strain rate effect on strength than does a destructured soil.

Herein, similar to the mathematical form used in Eq. (2), a rate parameter β is defined as the variation of the deviator stress at failure q_f at a particular strain rate with reference to q_f at a reference strain rate by the following equation.

$$\frac{q_f}{q_{f(ref)}} = \left(\frac{\dot{\varepsilon}_1}{\dot{\varepsilon}_{1(ref)}}\right)^\beta \qquad \text{Eq.(3)}$$

The test results reported in this paper and other published data were compiled to examine the effect of the degree of metastability on rate-dependent undrained strength (see Appendix B). The variation of undrained shear strength at different strain rates is shown in Fig. 7(b) to obtain the β rate parameter in Eq. (3). For a given soil, Eq.(3) seems to fit the data reasonably well and the slope of the fitted line in the figure becomes β. All the soils examined here are normally consolidated or slightly overconsolidated.

The variation of the β rate parameter is plotted as a function of *MI* in Fig. 8. Similarly to the α rate parameter, higher β values were found as a soil became more metastable.

Rate dependent yield envelope

It was shown in the previous sections that undrained strengths and preconsolidation pressures decrease with decreasing strain rate or increasing duration of testing. However, preconsolidation pressures obtained from one-dimensional consolidation tests and undrained shear strengths obtained from triaxial tests are just two points on a soil's yield envelope in stress space.

It is important to note here that the rate effects characterized by empirical equations Eqs. (2) and (3) are of the same mathematical form. The compiled data plotted in Fig. 8 suggest that the values of rate parameters α and β are approximately the same for soils with similar degree of metastability. Therefore, the rate effect on yield stresses is almost independent of stress ratio, and the data at least suggests that a single mechanism of time-dependent softening of soil structure controls the general time-deformation behavior of soils in the range of strain rate investigated in the tests. The variation of rate dependent yield stresses can be written in the following general mathematical form.

$$\frac{\sigma^y}{\sigma^y_{(ref)}} = \left(\frac{\dot{\varepsilon}}{\dot{\varepsilon}_{(ref)}}\right)^\gamma \quad \text{Eq. (4)}$$

where σ^y is the yield stress of σ at a strain rate of interest and $\sigma^y_{(ref)}$ is the yield stress at a reference strain rate, $\gamma\,(=\alpha=\beta)$ is the rate parameter which generally depends on the degree of metastability of a soil as shown in Fig. 8. Equation (4) is similar to the mathematical formulations of several general creep models proposed in the past (e.g. Singh and Mitchell, 1968; Sekiguchi, 1977; Adachi and Oka, 1982; Nova, 1982; Matsui and Abe, 1985; Kutter and Sathialingam, 1992 ; Soga, 1994). Therefore, the creep parameters used in these models should be able to correlate with the degree of metastability of natural structured clays.

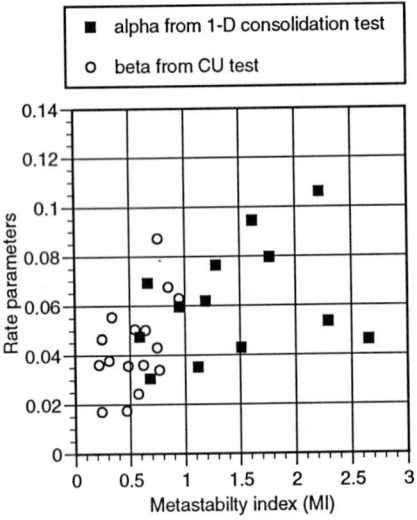

Fig. 8 Effect of the degree of metastability on rate parameters α and β of Eqs. (2) and (3)

Conclusions

Time-dependent deformation of soil is a complex process which results from the interactions among soil structure change, stress history, and drainage conditions. Investigation of the time-dependent behavior of undisturbed Pancone clay samples showed behavior similar to that observed in past studies. The slope of the failure line

M was independent of the applied rate of strain. The measured undrained strength decreases with decreasing strain rate resulted from creep-driven pore pressure development under sustained loading. Pore pressure generation was independent of strain rate and totally dependent on axial strain in undrained compression. The creep rate of normally consolidated soils was dependent on stress ratio (q/p'), but independent of the stress magnitude for a given stress ratio.

The rate dependency of preconsolidation pressure and undrained shear strength was characterized using empirical equations (2) and (3). It was found that the preconsolidation pressure and undrained strength of a soil became more rate-dependent as the degree of metastability increased. Thus, structured natural soils are expected to exhibit large time-dependency of deformation. Similar values of the rate parameters α and β for a given soil indicate that the degrees of rate effect are approximately the same even at different stress ratios and that a single mechanism of time-dependent softening of soil structure controls the general time-deformation behavior of soils in the range of strain rate investigated in the tests.

Acknowledgment
    The undisturbed Pancone clays were supplied by Professor G. Calabresi and Professor S. Rampello at the University of Rome. Their support is greatly acknowledged.

References
Adachi, T. and Oka, F. (1982) : "Constitutive equations for normally consolidated clay based on elasto-viscoplasticity," *Soils and Foundations*, Vol. 22, pp. 57-70
Adachi, T., Oka, F., Hirata, T., Hashimoto, T., Pradhan, T.B.S., Nagaya, J. and Mimura, M. (1991) : "Triaxial and torsional hollow cylinder tests of sensitive natural clay and an elasto-viscoplastic constitutive model," *Proc. European Conference on SMFE*, pp.3-6
Akai, K., Adachi, T. and Ando, N. (1975) : "Existence of a unique stress-strain-time relation of clays," *Soils and Foundations*, Vol. 15, No. 1, pp. 1-16.
Alberro, J. and Santoyo, E. (1973) : "Long term behavior of Mexico City clay," *Proc. 8th ICSMFE*, Vol. 1, pp. 1-9
Burland, J.B. (1990) : "On the compressibility and shear strength of natural clays," *Geotechnique*, Vol. 40, No. 3, pp. 329-378
Calabresi, G., Rampello, S. and Callisto, L. (1992) : "Geotechnical characterization of the Tower's subsoil within the framework of the critical state theory," Universsita' di Roma, Dipartmento di Ingegneria Strutturale
Crooks, J.H.A. and Graham, J. (1976) : "Geotechnical properties of the Belfast estuarine deposits," *Geotechnique*, Vol. 26, No. 2, pp. 293-315
Crawford, C.B. (1959) : "The influence of rate of strain on effective stresses in sensitive clays," *ASTM Special Publication*, STP254, pp. 36-48
Graham, J., Crooks, J.H.A. and Bell, A.L. (1983) : "Time effects on the stress-strain behavior of natural soft clays," *Geotechnique*, Vol. 33, No. 3, pp. 327-340
Houston, W.N. and Mitchell, J.K. (1969) : "Property interrelationships in sensitive clays," *Journal of the Soil Mechanics and Foundation Division*, Proc. of ASCE, Vol. 95, No. SM4, pp. 1037-1062
Kulhawy, F.H. and Mayne, P.W. (1990) : "Manual on estimating soil properties for foundation design," *EPRI report*
Kutter and Sathialingam (1992) : "Elastic-viscoplastic modelling of rate-dependent behavior of clays," *Geotechnique*, Vol. 42, No. 3, pp. 427-441

La Rochelle, P., Sarrailh, J., Tavenas, F., Roy, M. and Leroueil, S. (1981) : "Causes of sampling disturbance and design of a new sampler for sensitive soils," *Canadian Geotechnical Journal*, Vol. 18, No. 1, pp. 52-66

Ladd, C.C., Foott, R., Ishihara, K., Schlosser, F. and Poulos, H.G. (1977) : "Stress-deformation and strength characteristics," *Proc. 9th ICSMFE*, pp. 421-494

Lefebvre, G. and LeBoeuf, D. (1987) : "Rate effects and cyclic loading of sensitive clays," *Journal of Geotechnical Engineering*, Vol. 113, No. 5, pp. 467-489

Leroueil, S., Kabbaj, M., Tavenas, F. and Bouchard, R. (1985) : "Stress-strain-time relation for the compressibility of sensitive natural clays," *Geotechnique*, Vol. 35, No. 2, pp. 159-180

Lo, K.Y. (1969) : "The pore pressure-strain relationship of normally consolidated clays. Part II. Experimental investigation and practical applications," Canadian Geotechnical Journal, Vol. 7, pp. 395-412

Matsui, T. and Abe, N. (1985) : "Elastic/viscoplastic constitutive equation of normally consolidated clays based on flow surface theory, *Proc. 5th Int. Conf. Numeri. Methods Geomech.*, Vol. 1, pp. 407-413

Mesri, G. and Choi, Y.K. (1984) : "Discussion-Time effect on the stress-strain behavior of natural soft clays," *Geotechnique*, Vol. 34, pp. 439-442

Mitchell, J.K. (1993) : *Fundamentals of Soil Behavior*, Second Edition, John Wiley and Sons, 437 pp.

Nakase, A. and Kamei, T. (1986) : "Influence of strain rate on undrained shear characteristics of K0-consolidated cohesive soils," *Soils and Foundations*, Vol. 26, No. 1, pp.85-95

Nova, R. (1982) : "A viscoplastic constitutive model for normally consolidated clay, *Proc. IUTAM Conf. on Deformation and Failure of Granular Materials*, pp. 287-295

Olson, R.E. (1963) : "Shear strength properties of a sodium illite," *Journal of the Soil Mechanics and Foundations Division*, Proc. of ASCE, Vol. 89, No. SM1, pp. 183-208

Richardson, A.M. and Whitman, R.V. (1963) : "Effect of strain rate upon undrained shear resistance of a saturated remolded fat clay," *Geotechnique*, Vol. 13, No. 4, pp. 310-324

Sekiguchi, H. (1977) : "Rheological characteristics of clays," *Proc. 9th ICSMFE*, Vol. 1, pp. 289-292

Silvestri, V., Yong, R.N., Soulie, M. and Gabriel, F. (1986) : "Controlled-strain, controlled-gradient, and standard consolidation testing of sensitive clays," *Consolidation of Soils: Testing and Evaluation*, ASTM STP 892, Yong, R.N. and Townsend, F.C. (Eds.), pp. 433-450

Singh, A. and Mitchell, J. K. (1968) : "A General Stress-Strain-Time Function for Soils," *Journal of the Soil Mechanics and Foundations Division,* ASCE, Vol. 93, No. SM1, pp. 21-46.

Soga, K. (1994) : "Mechanical behavior and constitutive modelling of natural structured soils," Ph.D. thesis, University of California at Berkeley

Vaid, Y.P. and Campanella, R.G. (1977) : "Time-dependent behavior of undisturbed clay," *Journal of Geotechnical Engineering*, Vol. 13, No. GT7, pp. 693-709

Vaid, Y.P., Robertson, P.K. and Campanella, R.G. (1979) : "Strain rate behavior of Saint-Jean-Vianney clay," *Canadian Geotechnical Journal*, Vol. 16, pp. 34-42

APPENDIX A : Rate-dependency of preconsolidation pressure for several clays

Clay type	w0	PI	LI	MI	σvp at 1%/hr	α	Reference
Saint Jean Vianney	42	16	1.38	1.61	910	0.0945	Vaid et al.
Winnipeg clay	60	45	0.62	0.68	278	0.0306	Graham et al.
Holywood	37	22	0.77	0.59	56.3	0.0476	Crooks&Graham
Lousville	82	33	1.7	1.76	292	0.0796	Silvestri et al.
Backebol	102	65	1.04	0.94	101	0.0596	Graham et al.
Batiscan clay	79.6	21	2.7	2.65	136	0.0462	Leroueil et al.
St Cesaire	84.8	43	1.3	1.28	168	0.0764	Leroueil et al.
Joliette	65	19	2.3	2.29	177	0.0535	Leroueil et al.
Louisville	76.5	43	1.1	1.11	206	0.0352	Leroueil et al.
Mascouche	67.6	30	1.4	1.51	385	0.043	Leroueil et al.
Ottawa	65	37	1.02	1.19	561	0.0618	Graham et al.
Drammen	51	31	0.65	0.66	204	0.0694	Graham et al.
Ottawa	55	31	2.04	2.22	601	0.0106	Graham et al.

APPENDIX B : Rate-dependency of undrained shear strength for several clays

Clay type	wc	PI	LI	σv (kPa)	MI	qf at 1%/hr	β	Reference
Mississippi	28.6	38	0.12	413	0.24	139	0.0171	Richardson et al.
Holywood	37	22	0.77	43	0.54	11.6	0.0505	Crooks et al.
Haney clay	32	18	0.33	514.5	0.48	145	0.0357	Vaid et al.
Fukakusa clay	33.5	27	0.24	196	0.25	72	0.0466	Akai et al.
Eastern Osaka clay	69.7	49	0.89	78.4	0.76	40.8	0.0873	Adachi et al.
Eastern Osaka clay	71.3	49	0.92	117.6	0.85	57.9	0.0675	Adachi et al.
Ottawa	62.7	30	0.94	196	0.95	84.5	0.0629	Crawford
Ottawa	53.7	30	0.71	269.5	0.76	90.4	0.034	Crawford
Ottawa	46.7	30	0.53	392	0.64	113.4	0.0502	Crawford
Ottawa	41.8	30	0.4	588	0.57	167	0.0244	Crawford
Illite	57.6	40	0.45	207	0.47	59	0.0175	Olson
Mexico City	330	279	0.95	49	0.74	26.1	0.0431	Alberro et al.
Mexico City	265	279	0.72	98	0.62	48.2	0.0361	Alberro et al.
Pancone clay 19m	60	55	0.34	185	0.34	58	0.0555	this study
Pancone clay 14m	38	20	0.34	154	0.31	57.6	0.0378	this study
Pancone clay 10m	58	48	0.3	109	0.22	48	0.0362	this study

Creep Deformation and Stress Relaxation
in Preloaded/Prestressed Geosynthetic-Reinforced Soil
Retaining Walls

Fumio Tatsuoka[1], Taro Uchimura[2]
Masaru Tateyama[3] and Katsumi Muramoto[3]

Abstract

A new construction method which aims at increasing substantially the stiffness of a geosynthetic-reinforced soil retaining wall by vertical preloading and prestressing is described. The behaviour of field full-scale models with gravel and clay backfill during preloading and under prestress for more than a half year are presented. Rheological properties of the backfill soil, which plays an essential role for this method, were determined from triaxial tests.

Introduction

Over about the last ten years, geosynthetic-reinforced soil (GRS) retaining walls have been constructed for a total length of more than 20 km in Japan mainly to support railways (Tatsuoka et al., 1992, 1996b). The GRS retaining walls have a rigid full-height facing which is cast-in-place directly on the wall face of the GRS wall after the wall and is constructed with help of gravel-filled gabions placed on the shoulder of each soil layer (Tatsuoka, 1993). By taking advantage of the contribution of a rigid facing to the stability of GRS retaining wall, thirteen GRS bridge abutments have been constructed (Tatsuoka et al., 1996b). The longest bridge girder supported is, however, still only 13.2 m long. If GRS bridge abutments are to be used for longer and heavier bridge girders, they should be made more rigid. It is known that an increase in the amount of horizontal reinforcement is not sufficient to significantly increase the vertical stiffness of reinforced soil (Huang and Tatsuoka., 1990).

On the other hand, in one of the most seriously damaged areas during the 1995 Great Hanshin-Awaji earthquake, a GRS retaining wall having a full-height rigid facing performed very well (Tatsuoka et al., 1996a). Although the wall did not collapse at all, it showed some shear deformation with slight outward displacement and tilting. To prevent the above, it would be effective to increase the shear rigidity of the backfill by increasing the confining pressure.

---
1 and 2; Professor and Graduate Student, Department of Civil Engineering, University of Tokyo, 7-3-1, Bunkyo-ku, Hongo, Tokyo 113 Japan, 3; Research Engineer, Railway Technical Research Institute

From the two points discussed above, a new construction method named "preloaded and prestressed (PLPS) reinforced soil method", was developed (**Fig. 1**)(Tatsuoka et al., 1996c, Uchimura et al., 1996). This aims at constructing a reinforced soil structure, such as a GRS bridge abutment, that is very stiff under both vertical loading and lateral shear loading. Described herein are the mechanisms of this construction method, the behaviour of ful-scale model walls during preloading and their long-term behaviour under prestress. Also presented are the creep and relaxation properties of two different backfill soils, gravel and clay, that were determined from triaxial tests.

Mechanisms of PLPS reinforced soil

In order to construct a reinforced soil mass that is nearly elastic under working loads, sufficiently large preload is first applied by introducing tension into metallic tie rods which penetrate the reinforced soil and are connected to a bottom reaction block (Fig.1). As the backfill soil is reinforced, it can sustain large preload without failure. The tensile force T in the tie rods and the corresponding compressive load C in the backfill soil function as the prestress. This prestress keeps the confining pressure high, which in turn achieves high soil stiffness. This consideration is based on the fact that Young's modulus for vertical elastic strain increments of a granular material is a unique function of vertical normal stress (Tatsuoka and Kohata, 1995 and Tatsuoka et al., 1996c).

When either compressive load P_C or tensile load P_T is applied to the top block (Fig. 1), the load is supported by both the tie rods in tension and the soil mass in compression. That is, upon the application of P_C, the tie rod tension decreases to $T + \Delta T$ ($\Delta T < 0$) associated with compression of the backfill, while the increase in the vertical load applied on the backfill is reduced to $P_C + \Delta T$ from the value P_C in the case without prestressing. Similarly, upon the application of P_T, the compressive force in the backfill decreases to $C + \Delta C$ ($\Delta C < 0$) associated with elongation of the tie rods, while the increase in the tie rods tension is reduced to $P_T + \Delta C$ from the value P_T in the case without prestressing. Preloading also introduces prestress in the reinforcement, which contributes to maintaining the integrity of the backfill soil. Under high confining pressure by prestressing, the shear rigidity of the backfill soil is also increased. A more detailed discussion and the results of numerical analyse are reported by Tatsuoka et al. (1996c).

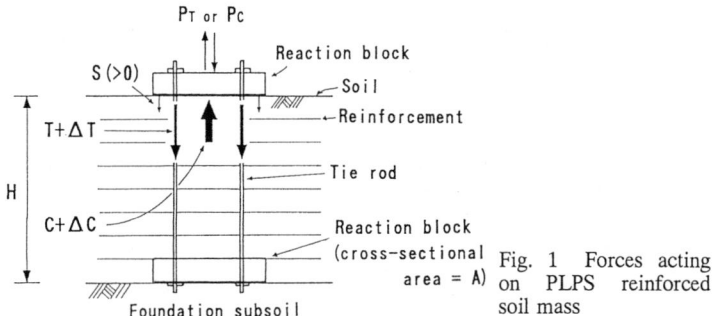

Fig. 1 Forces acting on PLPS reinforced soil mass

Full-scale field model tests

To validate the above concepts, a full-scale test wall facility having four test setions was constructed in 1995 at IIS Chiba Experiment Station, University of Tokyo. Each test section is 5 m-high and 4 m-wide, and is separated from respective adjacent sections by 60 cm-thick RC walls with a fixed distance between their tops. Two test sections having either clay or gravel backfill were constructed for studying the effects of PLPS (**Fig. 2**). In total four pairs of top and bottom RC reaction blocks were constructed (**Fig. 3**). The north segment 2N of the clay section and the south and north segments 3S and 3N of the gravel section have the same basic configuration having a wrapped-around wall face constructed with gravel-filled gabions placed on the shoulder of each soil layer. A full-height rigid facing had not been constructed at the time of this investigation. The central test segment 3M of the gravel section is separated from segments 3S and 3N by a 0.2 m-wide unreinforced gravel zone.

For the clay section, on-site volcanic ash clay, called Kanto loam, was used, which has a natural water content = 123 %, a maximum particle size = 0.84 mm, a sand, silt and clay contents = 8.0 %, 62.5 % and 29.5 %, and liquid and plastic limits = 168 % and 115 %. The clay was compacted to an average water content = 80 to 90 % and an average dry unit weight = 7.84 kN/m^3 (0.80 gf/cm^3). **Fig. 4** shows the result of a one-dimensional

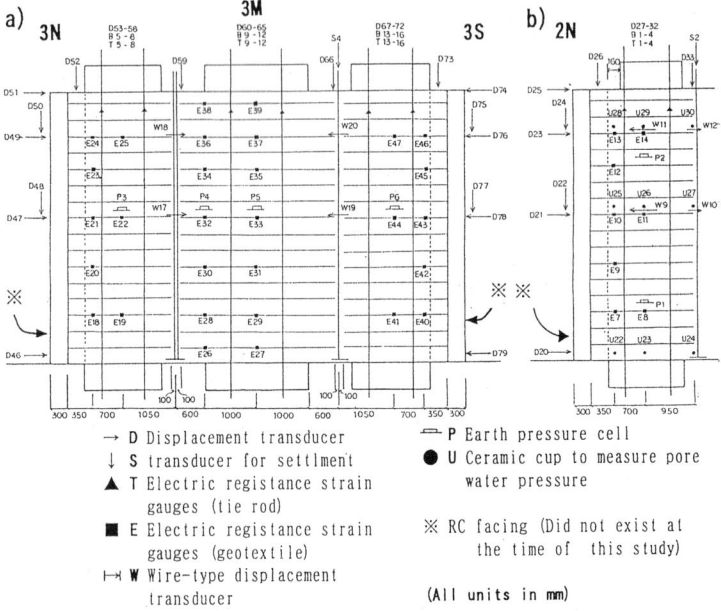

Fig. 2 Cross-sections of test sections with a) gravel backfill and b) clay backfill

compression test performed at an axial strain rate of about 0.1 %/min on an undisturbed sample having a diameter of 30 cm and an initial height of 30 cm obtained by block sampling from the compacted soil layer. The soil exhibited an yield pressure of about 29.4 kPa (0.3 kgf/cm^2).

The clay fill was reinforced with a composite consisting of a woven geotextile sheet sandwiched between two sheets of spun-bonded 100 % polypropylene non-woven geotextile at a vertical spacing of 30 cm. The nominal rupture strength of each woven geotextile sheet is 64.7 kN/m (6.6 tonf/m) at an elongation strain of 6.7 %, while the rupture strength of the two non-woven geotextile sheets is 4.37 kN/m (446 kgf/m) at an elongation strain of 123 %. In practice, the strength and stiffness of the non-waven part may be ignored for design, while the hydraulic conductivity of the waven part may be ignored.

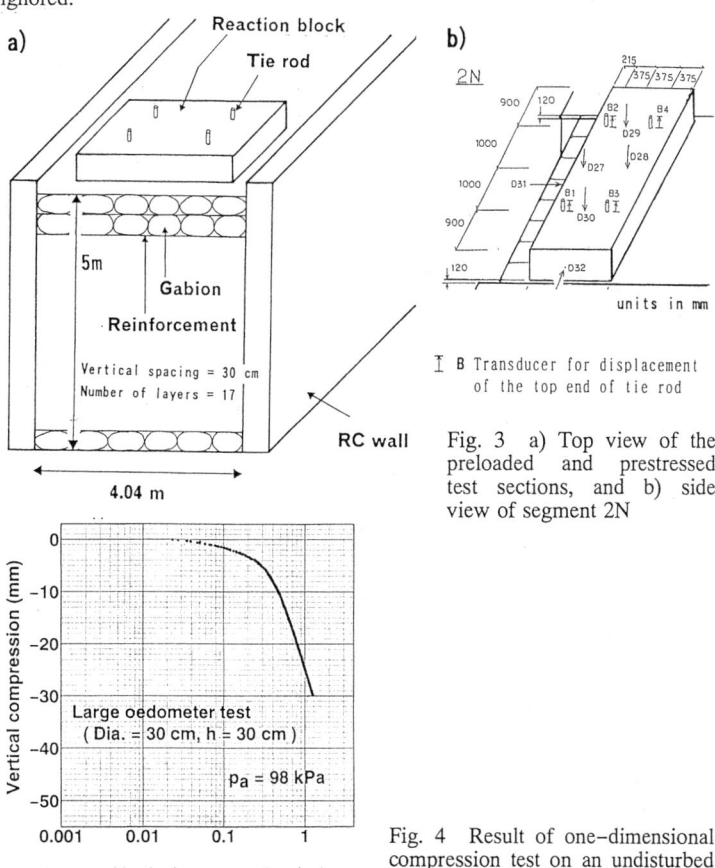

Fig. 3 a) Top view of the preloaded and prestressed test sections, and b) side view of segment 2N

Fig. 4 Result of one-dimensional compression test on an undisturbed clay specimen

For the gravel section, a well-graded crushed sandstone (D_{max} = 37.5 mm, D_{50} = 4 mm, C_u = 6.4 mm/0.09 mm = 71 and a fines content = 8.0%) was compacted to a dry density of 18.4 kN/m^3 (1.88 g/cm^3) with an average water content of 7 %. The grid was made of polyvinyl alcohol with the trade mark 'Vinylon' having a nominal tensile rupture strength of 73.5 kN/m (7.5 tonf/m). This type of grid has been used to construct a number of prototype GRS retaining walls and bridge abutments in Japan.

Fig. 5 shows a summary of the time histories of the average contact pressure at the bottom of the top reaction block from the end of August 1995, when preloading was started. The reaction blocks have an average contact pressure due to the self weight P_0 equal to (the own weight P_0)/(contact area A) = 11.8 kPa (0.12 kgf/cm^2). Segment 3S with gravel backfill was first preloaded to $(P_{PL} + P_0)/A$ = 704 kN/5.7 m^2 = 123.5 kPa (1.26 kgf/cm^2) by introducing a tension force P_{PL} in the four tie rods by means of four center-hole hydraulic jacks. The preload P_{PL} was maintained only for about ten minutes. Subsequently, the top ends of the tie rods were fixed to the top block, while releasing the pressure in the hydraulic jacks. During this operation, the tie rod tension T dropped about 25% of P_{PL}, which was also observed for the other two segments of the gravel section. Next, the relaxation of the tie rod tension T with time was monitored. Similar test procedures with different periods for creep deformation at the preload stage were used for segments 3N and 3M. The influence of typhoon No.12 is due to wetting of the backfill by heavy raining and associated deformation of the backfill. The details of these processes are reported in Tatsuoka et al.,1996c and Uchimura et al.,1996.

From 16 October 1995, segment 2N of the clay section was preloaded. The average pressure at the top reaction block was increased step by step to an average contact pressure of 60.8 kPa (0.62 kgf/cm^2) with an increment of about 0.98 kPa (0.01 kgf/cm^2) per day (**Fig. 6**). This relatively slow rate of loading was chosen as relatively large deformations of the wall were observed

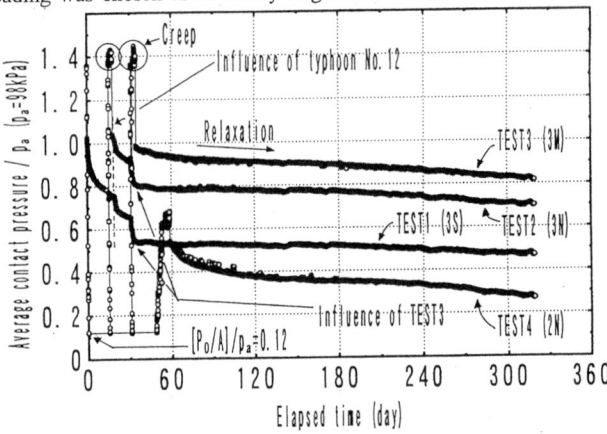

Fig. 5 Summary of average contact pressure $(T + P_0)/A$ at the top RC block of segments 3S, 3N, 3M and 2N versus time

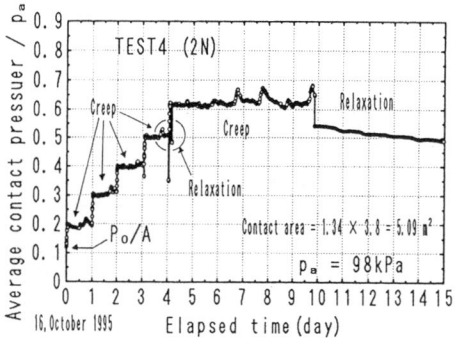

Fig. 6 Detailed time histories of average contact pressure $(T + P_0)/A$ at the top RC block of segment 2N with clay backfill

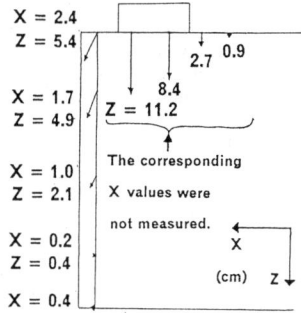

Fig. 7 Deformation during preloading of segment 2N with clay backfill

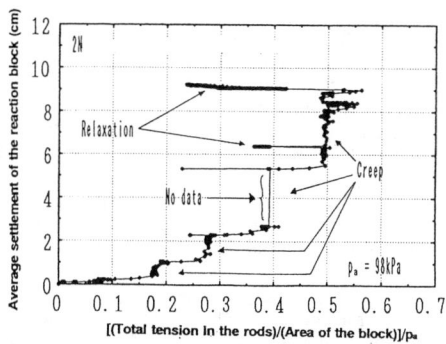

Fig. 8 Relationship between applied pressure and average settlement of the top RC block, segment 2N

shown in **Figs. 7** and **8**. It may be seen from Fig. 7 that the settlement of the top reaction block was larger near the wall face, which is reflected in the outward deformation of the wall face near the top. It may also be seen from Fig. 8 that the yield pressure is about 30 kPa (about 0.3 kgf/cm^2), which is consistent with the value from the one-dimensional consolidation test (Fig. 4).

The relaxation rate of the tie rod tension immediately after preloading and partial unloading was very large, about 1.57 kPa/hour (0.016 kgf/cm^2/hour). This rate is much larger than that observed in the gravel segments, likely due to much larger viscous effects on the deformation of the clay. After a relaxation period of two hours, preload was applied again at a contact pressure of about 60.8 kPa (0.62 kgf/cm^2). Creep deformation of the clay backfill was allowed to occur for about six days. During this period, the top reaction block settled about 4 cm (Fig. 8). Subsequently, the second stage of relaxation testing was started.

Fig. 9 shows the relationships between earth pressure increment measured at P1 and P2 indicated in Fig. 2b and applied pressure T/A at the top reaction block in segment 2N. It may be seen that the pressure in the backfill is about half of the contact pressure applied at the top and bottom reaction blocks. This may be mainly due to the following two factors: a) as the gravel-filled gabions placed at the wall face are much stiffer than the clay backfill, load shedding to the gabions can take place, and b) part of the applied pressure may be distributed to the fill beyond the contact area.

The following trends of behaviour may be noted;
1) In segment 2N, by allowing some amount of creep deformation to occur during the second preloading stage, the rate of stress relaxation immediately after the start of the second relaxation became much lower (Fig. 6). Similar behaviour was observed for the gravel segments as typically seen from the comparison of a very low relaxation rate immediately after the start of the second relaxation in segment 3N and 3M with a relatively high relaxation rate in segment 3S (Fig. 5).
2) In clay segment 2N, the over-all relaxation rate for a long duration was much larger than that seen in the gravel segments (Fig.5).
3) During the second relaxation stage in segment 2N, the relaxation rate decreased continuously. After about four months of relaxation, the relaxation rate became very low. The decrease rate is not sufficiently low when considering the long life time of a civil engineering structure. It is considered, however, that the relaxation rate can be made much smaller by applying a larger preload with larger creep deformation, as discussed later.
4) The vertical stiffness of segment 2N became much larger when reloaded than when initially loaded (Fig. 8). This is due to effects of preloading. Similar behaviour was observed in the gravel segments (Tatsuoka et al., 1996c). This result indicates that the backfill under prestressed conditions will behave as a nearly elastic material against a certain range of external load applied at the top reaction block.

Fig. 10 shows the relationships between tensile strain increment in the reinforcement at E13 and E14 and applied pressure T/A in segment 2N. It may be seen that the tensile strain after unloading of the pressure at the top reaction block is noticeably larger than that at the same contact pressure

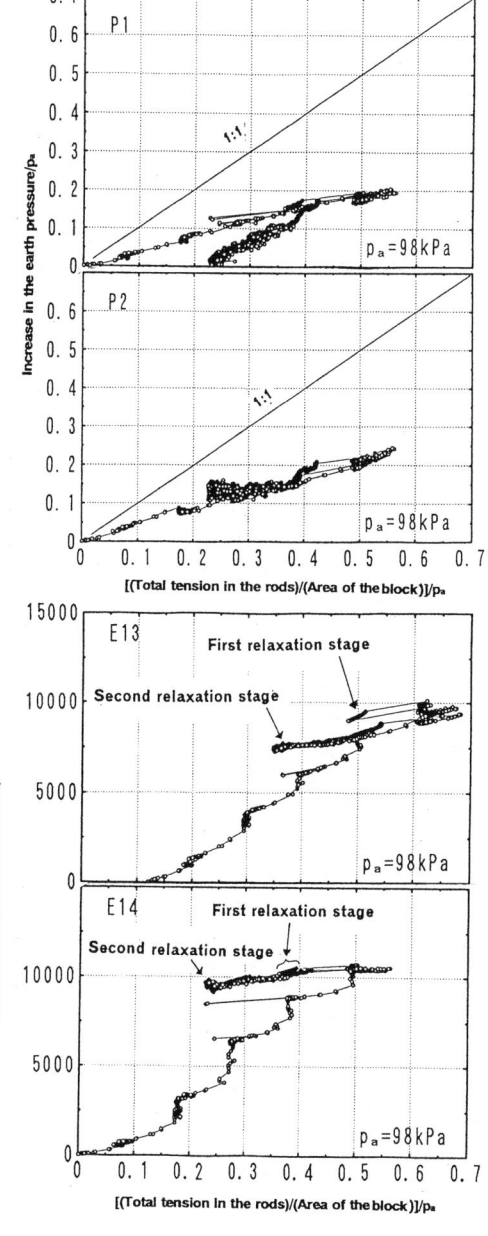

Fig. 9 Relationship between earth pressure increment and applied pressure T/A; P1 and P2 in segment 2N with clay backfill

Fig. 10 Relationship between tensile strain increment in the reinforcement and applied pressure T/A; E13 and E14 in segment 2N with clay backfill

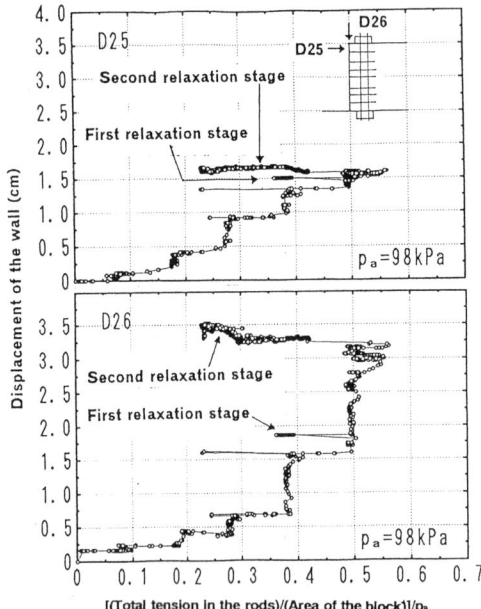

Fig. 11 Relationship between outward lateral wall displacement and applied pressure T/A; D25 and D26 in segment 2N with clay backfill

during the primary loading. This result shows that a tension prestress was also introduced in the reinforcement by preloading. It may also be noted that tensile strains in the reinforcement decreased slightly during the second relaxation stage. The behaviour may be due to the compressive creep deformation in the horizontal direction of the clay backfill caused by the prestress in the reinforcement. At the same time, the wall face did not show any outward displacement during this relatively long relaxation stage (**Fig. 11**).

In the design of GRS retaining walls, outward creep deformation at the wall face has been considered to be one of the potential problems for GRS walls constructed with clay backfill. The results shown above suggest, however, that creep can be minimized by properly preloading and prestressing the reinforced soil.

Creep and relaxation tests on the backfill soil

It is essential to maintain high tension prestress in the tie rods during a long relaxation stage. To understand the relaxation behaviour of the full-scale models described above and to know whether relaxation in prestress can be made practically zero for a very long time, the effect of preloading and associated creep deformation on the creep and relaxation characteristics of the backfill soils as observed at subsequent unloaded stages were investigated by means of triaxial tests.

For the gravel used in the field test, a rectangular prism-shaped specimen (60 cm in height and 23 cm x 23 cm in cross-section; **Fig. 12**) with a water content of 6.9% and a dry unit weight of 17.6 kN/m^3 (1.8g/cm^3) was prepared by vibrational compaction. Axial strain ε_1 and lateral strain ε_3 were measured locally by means of a set of local deformation transducers (LDTs) contacting the surfaces of the specimen to removethe effects of bedding error at the specimen boundary. The specimen was consolidated isotropically to 24.5 kPa (0.25 kgf/cm^2), and then a deviator stress q was applied at a constant stress rate dq/dt = 49 kPa/min (0.5 kgf/cm^2/min) (**Figs. 13a** and **b**). The results of a similar test on a specimen consolidated to 49 kPa (0.5 kgf/cm^2) are reported by Uchimura et al. (1996).

The time histories of q, ε_1 and ε_3 for the gravel specimen are shown in **Figs. 14a, b** and **c**. At several stages during primary loading and after unloading, either q was allowed to relax at a constant axial strain (relaxation stages R1 – R4) or creep deformation was allowed to occur at a constant q (creep stages C1 – C8). The confining pressure was maintained at 24.5 kPa (0.25 kgf/cm^2) throughout the test. The q values at stages C1, C2 and C4 and their initial values at stages R1, R2, R3 and R4 were the same so as to examine solely the effects of preloading and associated creep deformation. Elastic deformation characteristics evaluated during small unload/reload cycles are beyond the scope of this paper.

The following trends of behaviour can be noted:
1) The rate of stress relaxation is relatively large in the first and second relaxation stages R1 and R2, and the rate of creep deformation is also high in the first creep stage C1.
2) The rate of creep and relaxation decreased significantly at stages R3 and C2 after the first preloading and became nearly zero at stages C4 and R4 after the second preloading, at which some creep deformation was allowed to occur. Although the behaviour is subtle, the compressive axial strain decreased while the absolute value of lateral strain also decreased with time at stage C4, while the axial stress increased with time at stage R4. Such "creep recovery" was observed more clearly at stages C5, C6, C7 and C8 at lower stress levels. This

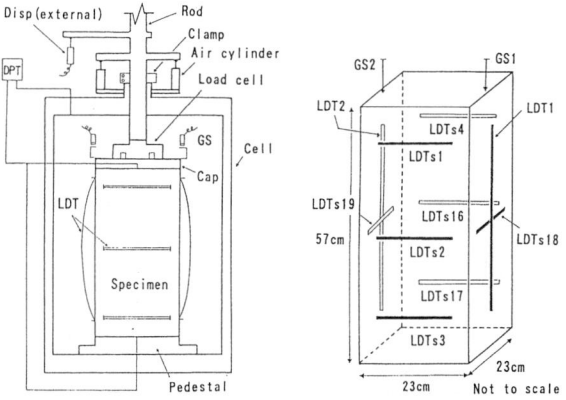

Fig. 12 Triaxial testing method for gravel using local deformation transducer, LDTs (Goto et al., 1991)

creep recovery was also observed in other similar tests (Uchimura et al., 1996). This behaviour can be modelled by a three-component model consisting of an elasto-plastic component connected to a pair of another elasto-plastic component and a dashpot in parallel (**Fig. 15**) (Di Benedetto and Tatsuoka, 1996). When EP and EP1 are linear elastic materials having stiffness E and E_1, "creep recovery" and "relaxation recovery" take place when the current stress state is located below or to the right of the line having a slope $E_m = E \cdot E_1 / (E + E_1)$; the curve E_m represents the ultimate stress state of the soil at time equal to infinity (**Fig. 16**). The rheological behaviour of soil is much more complicated than that shown in Fig. 16, but the fundamental aspects are captured using this simple model.

The triaxial test results that the relaxation rate decreases significantly due to preloading is consistent with that observed in the field gravel segments. The laboratory test results also suggest that for field full-scale PLPS reinforced soil with gravel backfill, prestress may even increase with time when a sufficient amount of creep deformation is allowed to occur during preloading.

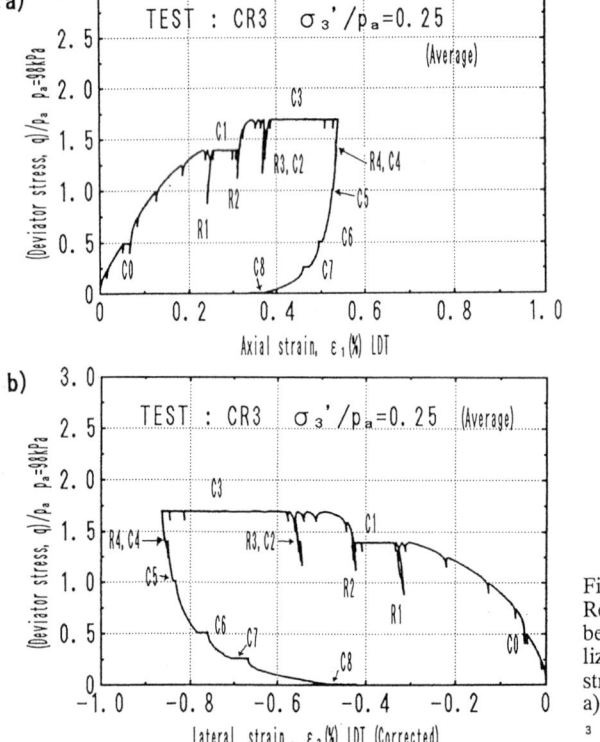

Fig. 13 Relationship between normalized deviator stress q/p_a and a) ε_1 and b) ε_3 from a triaxial test on gravel

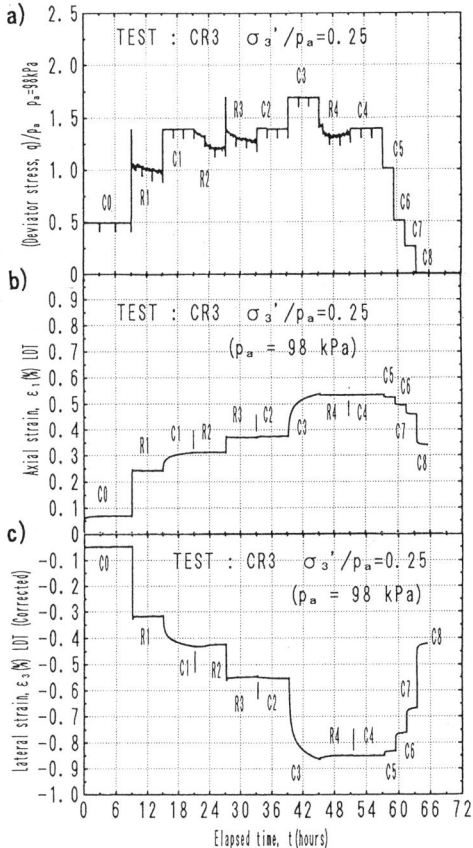

Fig. 14 Time histories of a) q/p_a, b) ε_1, and c) ε_3 from a triaxial test on gravel

Creep triaxial compression tests were performed on reconstituted clay samples, 5 cm in diameter and 10 cm in height, which were prepared by compacting in five layers clay obtained from the test wall. The initial water content was 84 %, the initial dry unit weight was about 7.84 kN/m³ (0.8 gf/cm³) and the initial total unit weight was about 14.4 kN/m³ (1.472 gf/cm³), which were similar to the field values. After the samples were water-saturated, they were anisotropically consolidated at a constant stress ratio K = σ'_3 / σ'_1 equal to either 0.3 or 0.6 from an initial isotropic stress state at 9.8 kPa (0.1 kgf/cm²). These K values were selected as the typical field values. Loading was performed incrementally and creep deformation was allowed to occur at each constant stress state (**Figs. 17 and 18**). From the peak axial stress state at σ'_1 = 78.4 kPa (0.8 kgf/cm^2), the sample was unloaded using the same K value as primary loading.

By comparing the creep value of ε_1 at values of σ'_1 = 19.6 or 39.2 kPa (0.2 or 0.4 kgf/cm^2) before and after preloading (Figs. 18a and 18b), it may be seen that by preloading, the creep rate decreased substantially, while noticeable "elastic recovery" took place immediately after the start of creep testing. Other similar tests showed similar results. The results shown above are qualitatively similar to the behaviour of the full-scale clay model wall and suggest that for clay backfills, the rate of relaxation can be reduced substantially by proper preloading where sufficiently large creep deformation is allowed to occur. It was found, however, that creep axial deformation re-initiated after a some period at a constant value of σ'_1 (**Figs. 19a and 19b**). The Burger model, which has another dashpot connected to the EP component of the three-component model shown in Fig. 15, may be appropriate to describe this behaviour. Further research is required to investigate whether such a re-initiation of creep can be suppressed by applying larger preload.

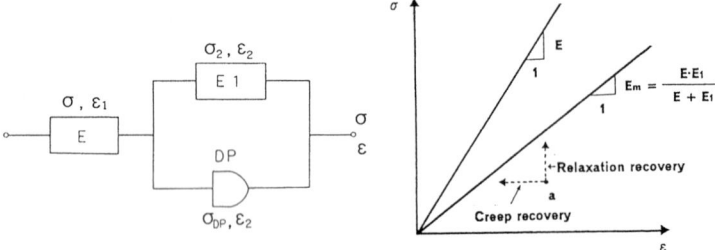

Fig. 15 Three-component rheology model (Di Benedetto and Tatsuoka, 1996)

Fig. 16 Creep recovery and relaxation recovery in a linear three-component rheological model

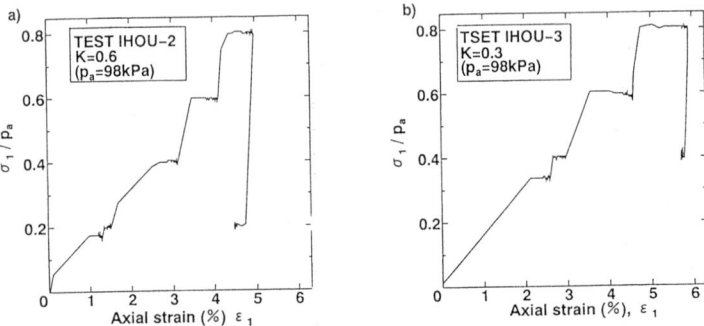

Fig. 17 Relationship between σ_1/p_a and ε_1 from triaxial tests on clay; a) K = 0.6 and b) K = 0.3

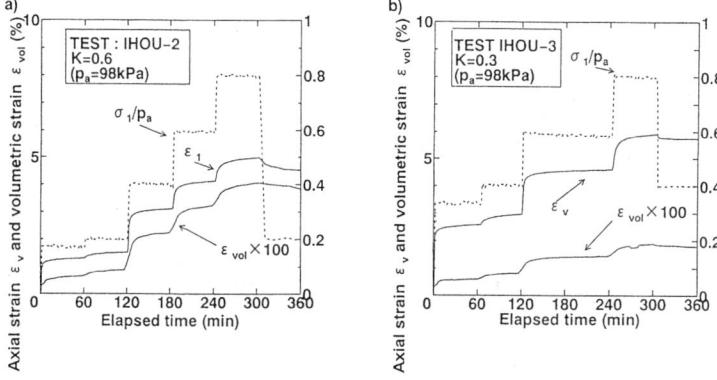

Fig. 18 Time histories of σ_1/p_a, ε_1 and ε_{vol} at the same stress state before and after preloading from triaxial tests on clay; a) K = 0.6 and b) K = 0.3

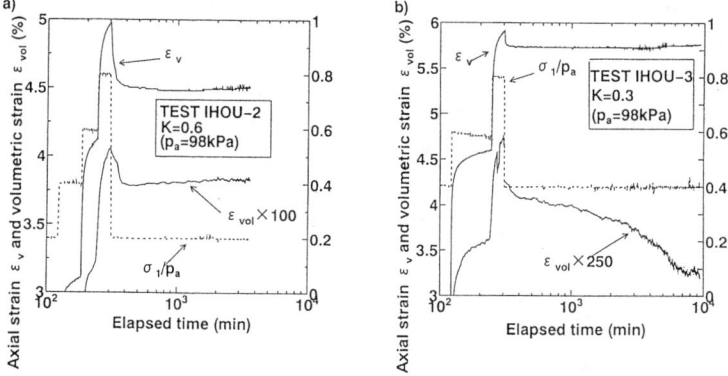

Fig. 19 Time histories of σ_1/p_a, ε_1 and ε_{vol} at after unloading from triaxial tests on clay; a) K = 0.6 and b) K = 0.3

Conclusions

The rigidity of reinforced soil can be increased substantially if; 1) plastic deformation of soil upon loading is reduced substantially by preloading, 2) high stiffness of the backfill soil is maintained by prestressing, and 3) high stiffness of the backfill soil is realized by prestressing both the soil mass and the tie rods. The behaviour of both field full-scale models having compacted well-graded gravel backfill reinforced with geogrid reinforcement and volcanic clay backfill reinforced with a non-woven and woven geosynthetic composite supports this conclusions. The field test results showed that the relaxation rate of prestress can be minimized by allowing a sufficient amount of creep deformation to occur during preloading. This result was supported by the results of triaxial creep and relaxation tests on gravel and clay specimens made by using the backfill soils. These results indicate that the application of this new construction method using well compacted gravel backfill to actual construction projects is quite feasible. The first prototype PLPS geogrid-reinforced soil bridge pier using gravel backfill was constructed in 1996 (Tatsuoka et al., 1996d). Further field and laboratory tests will be required, however, to substantiate this method for clay backfill.

Acknowledgements

The authors are deeply indebted to their colleagues in helping us to perform this study, particularly, Dr. J.Koseki, Dr. T.Kodaka, Mr. T.Sato, Mr. R.Motohiro, and Mr. H.Shibata. Valuable suggestions given by Prof. H. Di Benedetto, DGCB ENTPE, France, are gratefully acknowledged.

References

Di Benedetto,H. and Tatsuoka,F. (1996): Small strain behavior of geomaterials: modelling of the strain rate effects, Soils and Foundations (to appear)

Goto,S., Tatsuoka,F., Shibuya,S., Kim,Y.-S. and Sato,T. (1991): A simple gauge for local small strain measurements in the laboratory, Soils and Foundations, 31-1, pp.169-180

Huang,C.-C. and Tatsuoka,F. (1990): Bearing capacity of reinforced horizontal sandy ground, Geotextiles and Geomembranes 9, pp.51-82.

Tatsuoka,F., Murata,O. and Tateyama,M. (1992): Permanent geosynthetic-reinforced soil retaining walls used for railway embankment in Japan, Geosynthetic-Reinforced Retaining Walls. (Wu ed.), Balkema, pp.101-130

Tatsuoka F. (1993): Roles of facing rigidity in soil reinforcing, Keynote Lecture, Proc. Int. Sym. on Earth Reinforcement Practice, IS Kyushu '92, Balkema, Vol. 2, pp.831-870.

Tatsuoka and Kohata (1995): Stiffness of hard soils and soft rocks in engineering applications, Proc. Int. Symp. on Pre-Failure Deformation Characteristics of Geomaterials, IS Hokkaido '94, Balkema, Vol.2, pp.947-1063

Tatsuoka,F., Tateyama,M., and Koseki,J. (1996a): Performance of soil retaining walls for railway embankments, Special Issue for the 1995 Great Hanshin Awaji earthquake, Soils and Foundations, (to appear)

Tatsuoka,F., Tateyama,M., Koseki,J. and Uchimura,T. (1996b): Geotextile-reinforced soil retaining walls and their seismic behaviour, Special Lecture, Proc. of 10th Asian Regional Conf. on SMFE, 1995, Beijing, Vol.II (to appear)

Tatsuoka,F., Uchimura,T. and Tateyama,M. (1996c): Preloaded and prestressed reinforced soil, Soils and Foundations (to appear)

Tatsuoka,F., Tateyama,M., Uchimura,T. and Koseki,J. (1996d): Geosynthetic-reinforced soil retaining walls as important permanent structures, the 1996-1997 Mercer lecture, Proc. of the EuroGeo 1 Conf., Maastricht (to appear)

Uchimura,T., Tatsuoka,F., Tateyama,M., Sato,T. and Tamura,Y. (1996): Performance of preloaded and prestressed geosynthetic-reinforced soil, Proc. of IS-Kyushu '96

SUBJECT INDEX

Page number refers to the first page of paper

Axisymmetry, 181

Clay soils, 61
Clays, 96, 122, 137, 151, 166, 214, 243
Compacted soils, 151
Compression tests, 151
Constitutive models, 61
Creep, 96, 109, 166, 181, 195, 228
Cyclic loads, 214

Deformation, 243

Earth structures, 195

Failures, 195
Finegrained soils, 96
Finite element method, 96
Full-scale tests, 258

Geosynthetics, 258
Glacial till, 122

Ice sheets, 122

Laboratory tests, 195
Load tests, 214
Loading, 166

Marine clays, 181, 228
Mississippi River, 109
Modeling, 61

Oedometers, 137

Plane strain, 151
Pore pressure, 96
Preloading, 258
Prestressing, 258

Reinforcement, 258
Retaining walls, 258
Rheology, 122

Safety factors, 109
Slope stability, 195
Slope stabilization, 109
Soil compressibility, 1
Soil nailing, 109
Soil structure, 137, 243
Soil tests, 1
Soils, 258
Stability analysis, 109
Stiffness, 166
Strain rate, 1, 137, 166, 214, 228, 243
Stress, 151
Stress strain relations, soils, 214

Temperature effects, 1
Time dependence, 61, 122, 137, 181, 228, 243
Triaxial tests, 96, 181, 228
Tropical soils, 195

Viscoplasticity, 122
Viscosity, 1, 61

AUTHOR INDEX

Page number refers to the first page of paper

Adachi, Toshihisa, 61

Bessho, Kaoru, 181
Brandes, Horst G., 96, 228

Cali, Peter R., 109
Cavallaro, Antonio, 166
Clark, P. U., 122

Fujii, Clint F., 195

Ho, C. L., 122
Hosomi, Akihiko, 214
Hwang, Seong Chun, 214

Jamiolkowski, Michele, 166
Jenson, J. W., 122

Leroueil, Serge, 1, 137
Ling, Hoe I., 151
Lo Presti, Diego C. F., 166
Locat, Jacques, 137

Mimura, Mamoru, 61
Mitachi, Toshiyuki, 214
Mitchell, James K., 243

Murakami, Satoshi, 181
Muramoto, Katsumi, 258

Nicholson, Peter G., 195

Oka, Fusao, 61

Pallara, Oronzo, 166
Perret, Didier, 137

Russell, Philip W., 195

Shibuya, Satoru, 214
Silva, Armand J., 96, 228
Soares Marques, Maria Esther, 1
Soga, Kenichi, 243

Tateyama, Masaru, 258
Tatsuoka, Fumio, 151, 258

Uchimura, Taro, 258

Vela, J. C., 122

Yasuhara, Kazuya, 181

AVAILABLE BOOKS OF THE 1996 ASCE ANNUAL CONVENTION & EXPOSITION

WASHINGTON, DC ★ NOVEMBER 10 - 14, 1996

Proceedings of sessions held in conjunction with the
1996 ASCE Convention in Washington, DC

Analysis & Design of Retaining Structures against Earthquakes
Shamsher Prakash, Editor
Geotechnical Special Publication No. 60
ISBN 0-7844-0206-X

Case Histories of Geophysics Applied to Civil Engineering and Public Policy
Paul Michaels and Richard Woods, Editors
Geotechnical Special Publication No. 62
ISBN 0-7844-0208-6

Civil Engineering History: Engineers Make History
Jerry R. Rogers, Donald Kennon, Robert T. Jaske, and Francis E. Griggs, Jr., Editors
First National Symposium on Civil Engineering History sponsored by the ASCE & the US Capitol Historical Society
ISBN 0-7844-0209-4

Civil Engineers Influencing Public Policy
Maureen K. Cotton, Editor
Sponsored by the Construction Division
ISBN 0-7844-0204-3

Design with Residual Materials: Geotechnical and Construction Considerations
Gordon Matheson, Editor
Geotechnical Special Publication No. 63
ISBN 0-7844-0207

Engineered Contaminated Soils and Interaction of Soil Geomembranes
Jay N. Meegoda, Luis E. Vallejo, and L. N. Reddi, Editors
Geotechnical Special Publication No. 59
ISBN 0-7844-0213-2

Materials for the New Millennium
Ken P. Chong, Editor
Proceedings of the Fourth Materials Engineering Conference sponsored by the Materials Engineering Division
ISBN 0-7844-0210-8

Measuring and Modeling Time Dependent Soil Behavior
Thomas C. Sheahan and Victor N. Kaliakin, Editors
Geotechnical Special Publication No. 61
ISBN 0-7844-0205-1

Non-Aqueous Phase Liquids (NAPLs) in Subsurface Environment: Assessment and Remediation
Lakshmi N. Reddi, Editor
Proceedings of the Specialty Conference sponsored by the Environmental Engineering Division
ISBN 0-7844-0203-5

For pricing and availability, contact: **AMERICAN SOCIETY OF CIVIL ENGINEERS INTERNATIONAL HEADQUARTERS** 1801 Alexander Bell Drive, Reston, VA 20191-4400 ★ Phone: 800.548.2723; 703.295.6029 (international) ★ Internet email: marketing@asce.org